Quantum Dialogue

Science and Its Conceptual Foundations
A series edited by David L. Hull

Quantum Dialogue

The Making of a Revolution

Mara Beller

THE UNIVERSITY OF CHICAGO PRESS
Chicago & London

Mara Beller is the Barbara Druss Dibner Professor in History and Philosophy of Science at the Hebrew University of Jerusalem.

The University of Chicago Press, Chicago 60637
The University of Chicago Press, Ltd., London
© 1999 by The University of Chicago
All rights reserved. Published 1999
Printed in the United States of America
08 07 06 05 04 03 02 01 00 99   1 2 3 4 5
ISBN: 0-226-04181-6 (cloth)

Photo credits: All courtesy American Institute of Physics, Emilio Segrè Visual Archives. *Heisenberg and Bohr,* photo by Paul Ehrenfest Jr. *Bohr and Einstein,* photo by Paul Ehrenfest Jr. *Stern, Pauli, and Heisenberg,* photo by Paul Ehrenfest Jr. *Dirac and Heisenberg,* Physics Today Collection. *Bohr, Franck, and Hansen,* Margrethe Bohr Collection. *Schrödinger, king of Sweden, and Heisenberg,* Max-Planck-Institut für Physik. *Bohr and Pauli,* CERN. *Einstein and Pauli,* photo by Paul Ehrenfest Jr. *Zernike,* W. F. Meggers Collection.

Library of Congress Cataloging-in-Publication Data

Beller, Mara.
    Quantum dialogue : the making of a revolution / Mara Beller.
      p.   cm.—(Science and its conceptual foundations)
    Includes bibliographical references and index.
    ISBN 0-226-04181-6 (cloth : alk. paper)
    1. Quantum theory.   2. Communication in physics.   3. Physics—
Philosophy.   I. Title.   II. Series.
    QC174.13.B45   1999
    530.12—dc21
                                                        99-35499
                                                            CIP

In memory of my mother
Ida Baruch

Thinking and discourse are the same thing. . . . What we call thinking is precisely the inward dialogue.

Plato

The roads by which men arrive at their insights . . . seem to me almost as worthy of wonder as these matters in themselves.

Johannes Kepler

Revolutions are true as movements and false as regimes.

Maurice Merleau-Ponty

CONTENTS

〜

ILLUSTRATIONS

〜

Diagrams

Photographs (following page 190)

PREFACE AND
ACKNOWLEDGMENTS
~~

This book is about scientific dialogues that made the quantum revolution. The description of the intricate flux of these dialogues reveals the dynamic and intensely personal nature of scientific theorizing. During the work on this book I came to the rewarding realization that an ongoing responsiveness to the concerns of others, in addition to being a basic human value, is a precondition of scientific creativity. This book deals with the anatomy of scientific discovery (part 1) and with the strategies of consolidation of the orthodox quantum philosophy (part 2).

My approach is based on both admiration for the creation of scientific novelty and criticism toward the establishment of scientific dogmas. This book offers a thorough historical reevaluation of the emergence and consolidation of the Copenhagen interpretation of quantum physics—the dominant interpretation since its launching in 1927 by Niels Bohr, Werner Heisenberg, and Max Born.

It is not a historian's task to offer a specific alternative to the orthodox Copenhagen interpretation. I take no stand on the existing alternatives to the orthodox philosophy. This book does not deal with the extensive and lively contemporary research on the philosophical problems of quantum physics. Rather, my historical, philosophical, and sociological analysis of the Copenhagen philosophy demonstrates the possibility and the need of a viable alternative to the orthodox interpretation.

As my work on this book progressed, my revision of the history of the quantum revolution gradually evolved into a general critique of the revolutionary narratives for the description of the scientific change. Focusing on the quantum revolution, this book provides an analysis of how revolutionary stories in history of science are constructed, how division between "winners" and "losers" is fabricated, how the opposition is misrepresented and delegitimized, and how the illusion of the existence of a paradigmatic consensus among participants is achieved.

The bulk of my critical analysis of the Copenhagen philosophy of complementarity and indeterminism is found in part 2, which is more accessible to the general reader and can be read separately from the rest of the book. Nevertheless, parts 1 and 2 are mutually illuminating and conceptually interconnected. The introductory chapter 1 offers an overview of the major themes of this book, while the concluding chapter 15 is an argument for the dialogical approach for studying the advance of scientific knowledge.

This book was long in the making, and I have accumulated a long list of debts. There are a few scholars to whom my debt and gratitude are especially deep. Many ideas in this book were developed while I imagined addressing Arthur Fine. Arthur's writings and personal encouragement were invaluable. I owe a great debt to Jim Cushing, for his unbelievably meticulous and caring reading of the draft of this book, for his continuous support, and for his important work on related issues. Yemima Ben-Menahem, a colleague and friend, provided valuable criticism with a rare combination of philosophical wisdom and graceful encouragement. Sam Schweber and John Stachel, who took an early interest in my work, were always very generous with their intellectual and personal support. Tom Ryckman attentively read the draft of this book and provided valuable comments. I am grateful to Allan Franklin for many stimulating and pleasant conversations we had during my stay in Boulder.

I am fortunate to have a friendly and stimulating environment, with Yehuda Elkana, a teacher and a founder of an extensive academic activity in history and philosophy of science in Israel, and with colleagues at The Hebrew University of Jerusalem such as Yaron Ezrahi, Rafi Falk, Michael Head, Itamar Pitowsky, Mark Steiner, and Isachar Unna.

My list of debts is a long one, from Stephen Brush, my Ph.D. adviser in the early days of my academic life, to Shelly Goldstein and Anna Sfard, with whom I recently engaged in fruitful dialogues. I am also grateful to the following people for their stimulation, support, or assistance at some stage of working on this project: Pnina Abir-Am, Gideon Akavia, Diana Barkan, the late Yosef Ben-David, Michel Bitbol, Jed Buchwald, Catherine Chevalley, Robert Cohen, Olivier Darrigol, Alon Drory, Detleff Dürr, Gideon Freudenthal, Michael Friedman, James Fuchs, Ruthi Glasner, Galina Granek-Tiroshi, Gerald Holton, Don Howard, Roger Hurwitz, Tanya Karachentzeva, Arnon Keren, Alexei Kojevnikov, Edward MacKinnon, Avishai Margalit, Edna Ullman-Margalit, Jürgen Renn, Esther Rosenfeld, Cristoff Schmidt, Zur Shalev, Roger Stuewer, and Linda Wessels.

I gratefully acknowledge the support of the National Science Foundation (grant numbers 9011053 and 9123124) and the support of the

National Endowment for Humanities (grant number FA 31327-92) during my work on parts of this book. I am also grateful to David and Frances Dibner and the Dibner Foundation for the endowment of the Barbara Druss Dibner Chair in History and Philosophy of Science at the Hebrew University of Jerusalem, which I have the honor to hold.

Part of chapter 7 is based on my joint paper with Arthur Fine (Beller and Fine 1994). Portions of chapters 2, 4, 6, 8, 9, 13, and 14 are based in part on my previously published articles (Beller 1983, 1985, 1990, 1992a, 1996a, 1997a, and 1997b).

I am grateful to John Sanders for permission to quote materials from his collection (1987) of Bohr's published papers and unpublished manuscripts. The originals of Bohr's unpublished manuscripts are deposited in the Niels Bohr Archive in Copenhagen. They are available in microfilm in the Archive for the History of Quantum Physics (AHQP), copies of which are deposited in several universities throughout the world.

I am grateful to Felicity Pors from the Niels Bohr Archive in Copenhagen for her assistance. I also want to express my gratitude to the librarians of the Van Leer Institute, Edelstein Library, and the Niels Bohr Library in the American Institute of Physics for their help. I also gratefully acknowledge the courtesy of the Emilio Segré Visual Archives in the American Institute of Physics for the photographs appearing in this book, and I thank Jack Scott, the photo administrator, for his helpful assistance.

Very special thanks are due to my friend Shuli Barzilai, who returned two years ago from her trip to Russia with a unique souvenir—the photo she took of the statue of Einstein and Bohr in a park in Moscow. The photo decorates the jacket of this book.

I am one of those lucky authors who has had the pleasure and the privilege of working with Susan Abrams, executive editor at the University of Chicago Press, and to benefit from her sharp mind and big heart. I am grateful to the wonderful staff at the University of Chicago Press: to Rodney Powell and Charles Clifton, editorial associates for the sciences; to Leslie Keros, who worked on the proofs; to Martin Hertzel, who designed the book; to Joan Davies, who supervised the production; and to David Aftandilian, who handled the promotion. I am also grateful to Diana Gillooly for her meticulous and dedicated work on the manuscript and to James Farned for the index.

Last but not least, I want to thank my family for years of patience and support. My husband, Aaron, has always taken a caring interest in this work, providing uncompromising criticism and technical assistance. My gratitude to Aaron is beyond words.

CHAPTER 1

༄

Novelty and Dogma

I believe that to solve any problem that has never been solved before, you have to leave the door to the unknown ajar. You have to permit the possibility that you do not have it exactly right.

Richard Feynman 1998, 26–27

A dialogue can be among any number of people, not just two. Even one person can have a sense of dialogue within himself, if the spirit of the dialogue is present. The picture or image that this derivation suggests is of a stream of meaning flowing among and through us and between us.

David Bohm 1996, 6

Dialogical Creativity

"Science is rooted in conversations." These words were written by Werner Heisenberg, a great physicist of the twentieth century and a founder of the quantum revolution (1971, vii). How exactly is science rooted in conversations? And how did an extended conversation among scientists result in the quantum revolution? These are the major issues of this book.

Science is also rooted in doubt and uncertainty. "And it is of paramount importance, in order to make progress, that we recognize this ignorance and this doubt. . . . what we call scientific knowledge today is a body of statements of varying degrees of certainty. Some of them are most unsure; some of them are nearly sure; but none is absolutely certain. Scientists are used to this." These words belong to another great scientist of the twentieth century, the quantum physicist Richard Feynman (1998, 3, 27). "We know that it is consistent to be able to live and not know," continued Feynman: "I always live without knowing. That is easy. How you get to know is what I want to know" (1998, 27–28).

How is the presence of uncertainty and doubt built into scientific theorizing at its most basic level? How do scientists live without knowing? And how do they get to know? I argue throughout this book that the question of how scientists live in uncertainty, the question of how they create their knowledge, and the question of how science is rooted in conversations are, in fact, one and the same. I elaborate an answer to this question for the case of the quantum revolution.

This book examines the fluid, open-ended, and often ambiguous process of scientific creativity, treating it as being rooted in, and perhaps indistinguishable from, an ongoing scientific conversation about theories, experiments, and instruments. I address the central issue of how theoretical knowledge is achieved, articulated, and legitimated. I also deal with another major issue: the conceptual and emotional turmoil created by attempts to interpret the potent quantum formalism.

I describe and analyze the intricate flux of dialogues among quantum physicists—dialogues that resulted in scientific breakthroughs of unprecedented scope and in a radical quantum philosophy. These dialogues underlay both the open-minded foundational research and the erection of the orthodox interpretation of quantum physics: the Copenhagen interpretation. Tracing the web of dialogues reveals a story about the workings of free scientific imagination and about the consolidation of scientific dogma.

One of the major puzzles in the history of quantum physics is the existence of numerous contradictions in the Copenhagen interpretation. What is the source of these contradictions? And why are they impotent to detract from the spectacular power of quantum physics? A large part of this book constitutes an answer to these questions.

The analysis and the narrative in this work are permeated with the notion of communicability. I argue that dialogue underlies scientific creativity and that the emergence of scientific novelty cannot be understood without scrutinizing the ways scientists respond to and address each other. This book analyzes the complex, multidirectional dialogical nature of scientific theorizing (part 1) and the strategies by which this dialogical flux is flattened into a monological narrative (part 2).

This book is based on a close study of primary sources—correspondence between the participants, notebooks, original papers. Historians of quantum physics are fortunate to have access to the Archive for the History of Quantum Physics (AHQP), where correspondence and original manuscripts are collected. Thus I had the opportunity to study in detail the intricate paths along which ideas emerged as the founders of quantum physics addressed each other in their letters. I gradually realized that dialogical addressivity permeates not only scientific correspondence but also published scientific papers, and more generally, I came to the realization that scientific creativity is fundamentally dialogical in the sense elaborated in this book.

The dialogical approach to the history of science is "bottom up"—it searches for the most basic details in order to conceptualize the process through which knowledge grows. A dialogical analysis, by closely following ideas as they gradually form in numerous dialogues between scientists, deals primarily with the cognitive content of science. It requires painstaking attention to every nuance of the primary sources. In fact, it demands closer attention to the minutia of scientific reasoning than the older, "internal" (evolutionary) history of science and than "rational reconstruction" accounts.

My exposition differs from the usual accounts by describing the flux of ideas without presupposing underlying conceptual frameworks, schemes, or paradigms. In fact, these notions, whether on a global or a more restricted scale, are not easily compatible with the dynamics of ceaseless scientific change. Living in doubt and uncertainty is not compatible with the accepted historiographical notions of "beliefs" and "commitments." Nor are Kuhnian and post-Kuhnian "agreement" and "consensus" suitable to describe the dynamics of living without knowing. Doubt and uncertainty should be incorporated into the basic terms used to describe the growth of knowledge. From the dialogical perspective, it is "creative disagreement"—with oneself (doubt) or with others (lack of consensus)—that plays the crucial role in the advance of knowledge. The privilege to be unsure, to theorize freely, to explore different options at the same time, is incorporated into the notion of creative dialogical flux. I will elaborate on the philosophical, historiographical, and sociological advantages of such a dialogical approach in the concluding chapter of the book.

This book simultaneously revises the story of the quantum revolution and outlines a tentative program for a dialogical historiography of science. My work began with a revision of the history of matrix mechanics (Beller 1983) and progressed to revisions of other major episodes in the history of quantum physics, such as the emergence of Born's probabilistic interpretation (Beller 1990) and the birth of Bohr's complementarity (Beller 1992a). It gradually became clear to me that the need for ongoing revision has a fundamental historiographical cause. This cause is intimately connected with the complex dialogical nature of thought and with the strategies used to flatten it into linear monological narratives.

I begin my description of the quantum revolution with an analysis of matrix theory in flux (chapter 2), arguing against the received story of the existence of two totally distinct theoretical frameworks—the matrix and the wave theoretical. In chapters 2 and 3 I describe how a strong distinction between the matrix and wave approaches crystallized as the end result of a conceptually fascinating and emotionally intense confrontation among quantum physicists. In the fruitful ambiguity of the newly created knowledge, there was no place for strong "beliefs" in indeterminism or "commitments" to positivism. Nor is it correct to see the matrix theoreticians as committed to a particle ontology, as opposed to Schrödinger's wave approach. We will see that Heisenberg's, Born's, and Pauli's pronouncements on foundational and interpretive issues were all fluid and uncommitted. Similarly, I will argue, Born's probabilistic interpretation did not stem from his "belief" in particles and "commitment" to indeterminism, as the received history of quantum physics implies (chapter 2).

The flux of ideas in the emergence of matrix theory and in the formation of Born's probabilistic interpretation demonstrates the primacy of mathematical tools over fundamental interpretive ideas. These tools can be borrowed, developed, and successfully applied without a clear-cut stand on basic interpretive issues. It was on the efficiency of the mathematical tools, and not on metaphysical "paradigmatic" issues, that there was agreement in the community of quantum physicists. And on this point the orthodox and the opposition were united; agreement on the potency of these tools prevented scientific practice from disintegrating, be the philosophical disagreements as large as they may.

Theoretical tools (equations, methods of solution, and approximations) have their own momentum, while philosophical ideas are adapted a posteriori (chapter 3). The fact that theoretical tools have some autonomy allows scientists to theorize without taking an interpretive stand. The English physicist Charles Darwin wrote to Niels Bohr: "It is a part of my doctrine that the details of a physicist's philosophy do not matter much" (Darwin to Bohr, December 1926, AHQP). This belief in the primacy of mathematical tools was especially strong in Göttingen, inspiring Born's and Heisenberg's search for a new quantum theory. "Mathematics knows better than our intuition" was Born's motto (interview with Born, AHQP).

The neutrality of a mathematical formalism with respect to possible interpretations has another far-reaching consequence: scientists may give all authority in interpretive matters to a few leaders, whose philosophy they are willing to accept. Such humble resignation from philosophical exploration is often nothing but a convenient choice not to deal with confusing and perhaps irrelevant matters.[1] It is this attitude that creates room for an authoritative and privileged interpretation, such as the Copenhagen orthodoxy.

---

1. This position was taken by Darwin: "I'm quite ready in advance to believe that your criticisms are quite right, but I feel that perhaps this does not matter much. Because the best sort of contribution that people like me can make to the subject is working out of problems, leaving the questions of principles to you. In fact even if the ideas on which the work was done are wrong from the beginning to end, it is hardly possible that the work itself is wrong in that it can easily be taken over by any revised fundamental ideas that you may make" (Darwin to Bohr, December 1926, AHQP).

The dispensability of paradigmatic interpretive agreement also explains what Steven Weinberg, not without a touch of disdain, called "the unreasonable ineffectiveness of philosophy" in scientific practice (Weinberg 1992, 169). The application of theoretical tools involves flexibility and conceptual opportunism—no system of philosophical preconceptions can survive in such a fluid environment (chapter 3). Philosophical "influence" indeed cannot determine a scientific knowledge claim. Yet philosophy can be suggestive in a limited way, as inspiration along some path in the dialogical web of creativity. In chapter 4 I argue that the ideas of the German philosopher Fichte were important for Heisenberg's treatment of measurement in the uncertainty paper. This treatment, or "reduction of wave packets," became the source of the major conceptual hurdle of quantum physics—the notorious measurement problem.

Scientific creativity as a dialogical flux is exemplified by the emergence of Heisenberg's uncertainty principle, which I describe in chapter 4. We can see how Heisenberg theorized without a clearly delineated conceptual framework, without "beliefs" and "commitments." Instead, we observe the indispensability of open-ended disagreement and ongoing doubt for achieving a theoretical breakthrough. The existing historiography presents one-dimensional pictures: Jammer described the uncertainty formula as emerging naturally from the imperative to adapt the new mathematical formalism to the possibilities of measurement (Jammer 1966).[2] The perspective offered by social historians similarly focuses only on a single aspect: a predilection for acausality among quantum physicists determined their efforts to interpret the new quantum mechanics along probabilistic lines (Forman 1971; Feuer 1963).

I describe Heisenberg's discovery of the uncertainty relations as a multidirectional process, which took place in a communicative network with many interlocutors, including such prominent names as Einstein, Schrödinger, Pauli, Dirac, Bohr, Born, and Jordan. Less known, yet no less important for Heisenberg's discovery, are the names Campbell, Duane, Zernike, and Sentfleben. In chapter 4 we follow the process of discovery and observe how fragments of insight gradually emerge, how ideas clash, change, disappear, or survive. The gradual forming of a preference, of choosing one intellectual option over another, of defining what the options are—all occur in a coalescence of insights, arrived at in different dialogues and at different times.

Heisenberg's discovery of the uncertainty principle was a complex

2. Jammer described the controversy between Bohr and Heisenberg with respect to the uncertainty paper as a tool for elucidating Bohr's and Heisenberg's respective positions and as anecdotal history. From Jammer's analysis it is not clear how the disagreements contributed to the creation of uncertainty and complementarity.

process of disagreements, qualifications, elaborations, supplementa-
tions, and borrowings. He had no foundational commitments, even on
such basic issues as discontinuity and indeterminism; rather his preference for discontinuity and acausality took shape gradually in many
fruitful dialogues. Though immersed in interpretive efforts, Heisenberg
was uncertain even about what the word "interpretation" means (chapter 5). Disagreements with interlocutors—a militant one with Schrödinger, a subdued yet painful one with Bohr, a restrained one with Jordan and Born—were triggers for Heisenberg's reasoning. Agreement
too played a part, in the form of Heisenberg's partial, often only temporary, acceptance of the ideas of others, especially those of Dirac,
Campbell, and Duane. Heisenberg's case demonstrates how genuine
novelty emerges through dialogical creativity. Dialogical creativity is
not an instantaneous "eureka" experience; it is rather a patiently sustained process of responsiveness and addressivity to the ideas of others,
both actual and imagined.

One might expect that in a published scientific paper, all previous
cognitive tensions would be resolved and a coherent unequivocal message expressed. Yet an analysis of Heisenberg's uncertainty paper finds
clear traces of past struggles, conflicting voices on the same issue, and
unresolved tensions (chapter 5). The polyphony of the creative act echoes in the paper itself. Similarly, at least two conflicting, in fact incompatible, voices can be heard in Bohr's response to Einstein-Podolsky-
Rosen's challenge to the Copenhagen interpretation (chapter 7).

One might object, then, that such unresolved tensions perhaps characterize scientific papers written during revolutionary upheavals, but
when things settle down and the revolution is over, a new paradigm
triumphs and the foundational debate is closed. There is, the argument might continue, at the present time only one correct, agreed-upon
meaning of the uncertainty principle and of wave-particle complementarity. But is there? One can find a "correct" meaning in textbooks, or in
some philosophical writings on the quantum theory—in short, in the
graveyards of science. On the research frontier nothing is immune to
reappraisal—be it uncertainty, complementarity, or even the determinism of classical physics (chaos theories) and the indeterminism of quantum physics (Bohm's theory and Bohmian alternatives, such as Dürr,
Goldstein, and Zanghi 1992a, 1992b, 1996). Numerous meanings of the
uncertainty formulas are proposed in current research papers (Home
and Whitaker 1992; Valentini 1996). The same is true of Born's probabilistic interpretation and his solution of the collision problem (Daumer
1996). Similarly, there is no agreement about the meaning or even the
validity of wave-particle complementarity (chapter 11). Even such basic
formulas as Schrödinger's equation are open to modification (Ghirardi,

Rimini, and Weber 1986). The flux of creative research cannot be forced into an unequivocal, final conceptual scheme.

The description of Heisenberg's creative theorizing calls for a re-evaluation of the role of "lesser" scientists in the growth of scientific knowledge (chapter 4). We will see that some of the most important insights pertaining to Heisenberg's formulation of the uncertainty principle belonged to scientists whose names do not appear in the received history of quantum mechanics—Sentfleben and Campbell. The issue is not one of priority, even though the distribution of credit is often unfair. Neither Campbell nor Sentfleben, from their positions in the communicative web, could have accomplished exactly what Heisenberg did. Yet neither could Heisenberg have developed his ideas had he not been responding creatively to Campbell's and Sentfleben's insights. Heisenberg's discovery is organically linked to the ideas of the "lesser" scientists. From the epistemological point of view, the notion of a scientific collective is intrinsic to the dialogical approach.

We can reevaluate the prevalent idea of a lonely creative individual, and of solitude as a precondition of creativity. Conventional opinion holds that "the spark of creativity burns most brightly in a mind working in solitude" (Storr 1988). When Heisenberg "fabricated" the new quantum mechanics (his expression, van der Waerden 1967, 15), he was isolated on the island of Helgoland. After their stormy debates on the interpretation of quantum mechanics in the fateful year of 1926, Heisenberg and Bohr needed to get away from each other. Separated, Heisenberg wrote the essentials of his uncertainty paper and Bohr elaborated his complementarity. Yet solitude does not imply cognitive isolation. If Heisenberg needed time away from Bohr, it was in order to strike a proper, uncoerced balance in his own communicative network of cognitive responses (chapters 4 and 6).

The dialogical perspective provides a new way to read published scientific texts. Concealed doubt becomes visible, and a paper becomes a fascinating scientific and human document, resounding with conflicting inner voices, populated by many "virtual" interlocutors (chapters 5 and 6). We will see in chapter 5 that Heisenberg's uncertainty paper is permeated with doubts and indecision on such central issues as indeterminism, realism, visualizability, and the status of classical concepts in the quantum domain. A comparison of the (seemingly) confident published paper with the almost identical draft (in a letter to Pauli) filled with doubt regarding all the basic interpretive issues reveals how misleading it is to ascribe any "beliefs" or "commitments" to Heisenberg. My analysis applies to the issues of acausality and positivism that according to the accepted history of quantum physics, are the two central pillars of Heisenberg's uncertainty paper.

My examination of Heisenberg's uncertainty paper reveals the argumentative strategies by which interpretive freedom is concealed, and
the illusion created that the orthodox interpretation is "inevitable"—
the issue to which the bulk of part 2 of this book is devoted. Tension
between the conceptual freedom experienced by the Copenhagen physicists and their desire to advocate one privileged interpretation is one
of the major sources, I argue, of the numerous contradictions and inconsistencies in the Copenhagen interpretation of quantum physics.

Yet the dialogical analysis of a published paper does more than reveal
the inconclusive nature of scientific theorizing. Such an analysis can
also substantially modify our understanding of the content of a published paper. A dialogical reading is a potent tool for deciphering especially obscure and opaque texts. Chapter 6 is devoted to a dialogical
analysis of Bohr's Como lecture, in which Bohr announced his principle
of complementarity.

The Como lecture is considered one of the most incomprehensible
texts in twentieth-century physics. My dialogical reading constitutes a
basic revision of the accepted reading of this text, by presenting the
Como lecture, not as the unfolding of a single argumentative structure, but as the juxtaposition of several simultaneously coexisting arguments, addressed to different quantum theorists about different issues.
A dialogical analysis reveals that the central message of Bohr's paper
was not the resolution of wave-particle duality by the complementarity
principle, as usually assumed, but rather Bohr's extensive defense of his
concepts of stationary states and discontinuous energy changes (quantum jumps) against Schrödinger's competitive endeavors.

A dialogical reading allows the reevaluation of some central events
in the history of quantum physics, such as the famous clash between
Heisenberg and Bohr over the uncertainty paper. This reevaluation
merges "motives" and "reasons," connecting "conceptual" and "anecdotal" history into one meaningful, comprehensible story. This story
throws new light on Einstein's and Schrödinger's initial reservations
about the early interpretive attempts of Bohr and Heisenberg—reservations due to difficulties and contradictions in the emerging Copenhagen interpretation rather than to Einstein's and Schrödinger's "conservatism."

My analysis of the initial efforts of Bohr and Heisenberg to unveil
the physical meaning of the quantum formalism demonstrates the vast
freedom of interpretive endeavors. Yet this freedom is not arbitrary. We
will see the great extent to which the formulations of Born's probabilistic interpretation, Heisenberg's uncertainty, and Bohr's complementarity were woven around detailed analyses of experimental situations
(a fact not sufficiently apparent in the existing histories of quantum

mechanics). These first attempts at interpretation revolved around pivotal experiments by Franck and Hertz (1913), Bothe and Geiger (1925a, 1925b), and Compton (1923), as well as around experimental work by the lesser known Moll and Burger (1925). There is a crucial difference between evidence-based interpretive efforts and closed dogmatic systems, although the first can degenerate into the second, as with Bohr's, Heisenberg's, and Pauli's later philosophical writings, which came close to preaching a rigid ideology (part 2).

Chapter 7 (the last chapter of part 1) analyzes the process by which sincere and open-minded, though interest-laden, interpretive attempts hardened into an ideological stand intended to protect quantum theory from challenge and criticism. This chapter, partly based on my paper with Arthur Fine (Beller and Fine 1994), is devoted to an analysis of Bohr's reply to the Einstein-Podolsky-Rosen (EPR) challenge. This chapter is located, so to speak, on a "cut" between parts 1 and 2 and can be moved from the end of part 1 to the beginning of part 2 without distorting the argument of my book. This means that dialogical emergence and rhetorical consolidation are not completely distinct processes. Consolidation was already present in the initial interpretive attempts of the Göttingen-Copenhagen camp, and limited dialogical responsiveness accompanied later elaborations of the Copenhagen interpretation. Still, chapter 7 reveals a vast difference between Bohr's reasoning in his Como lecture and his reply to the EPR challenge. My analysis uncovers a transition from legitimate, though often confused, arguments for the consistency of quantum theory, to argumentative strategies promoting the inevitability of the orthodox stand. This transition naturally contains both old insights and new conquests, and it is not surprising that two different, often incompatible voices permeate Bohr's response to EPR. The voices correspond to Bohr's two different answers to EPR: one relying on the concept of disturbance, the other dispensing with it. In contrast to the received story, which, following the orthodox narrative, affirms Bohr's "victory" in this confrontation, the analysis in chapter 7 reveals that none of Bohr's answers can be considered satisfactory.

I further analyze Bohr's about-face on the central interpretive issues—the problems of reality, acausality, and measurement—raised by the EPR challenge. I extend this analysis in chapter 8, arguing that the Copenhagen interpretation is in fact a compilation of various philosophical strands, given a public presentation that often hid shifting disagreements between its main architects. The Copenhagen interpretation of quantum physics did not originate from a disinterested search for philosophical foundations—from the very beginning it was constructed in the heat of a fierce confrontation. As the nature of the opposition's challenge changed, so did the local responses of the orthodox. It is not surprising therefore that what is called the Copenhagen interpretation is so riddled with vacillations, about-faces, and inconsistencies (chapter 8). The orthodox aimed to present a united front to the opposition, concealing the substantial differences of approach among its members. I analyze some of the strategies by which such differences were suppressed by relying on a distinction between what scientists "must not" and what scientists "need not" do (chapter 8).

Rhetorical Strategies

How does one construct from among these numerous contradictory arguments a narrative that seems to irrevocably imply the pillars of the Copenhagen dogma? How does one reconstruct history so that the central tenets of the Copenhagen interpretation, such as indeterminism and the impossibility of an objective, observer-independent description, seem not merely highly persuasive but outright inevitable?

In part 2 of the book I contend that all the Copenhagen arguments of "inevitability" are in fact fallacious—they rely either on circular reasoning or on highly appealing but misleading metaphorical imagery (chapters 9 and 12). They are strongly supported by falsified history, which renders certain developments as dictated by the inner logic of the development of ideas (chapters 10 and 11). Discrediting the opposition and caricaturing the opposition's criticism of the Copenhagen stand is yet another potent rhetorical device to strengthen the orthodoxy (chapter 13). These chapters reveal how fruitfully ambiguous and wisely uncommitted interpretive efforts are concealed by rigid reconstructed stories. Complex, many-voiced, multidirectional theorizing is thus conflated into an orthodox, one-dimensional narrative.

"History is written by winners"—this cliché finds powerful confirmation in the case of the quantum revolution. We have numerous reminiscences by the winners—Bohr, Heisenberg, Born, Jordan, and others. There are hardly any reminiscences by Einstein and Schrödinger about the same events—we do not hear the opposition's side of the story. In the quantum revolution, the orthodox constructed the narrative, eliminating dissident voices and largely suppressing the crucial contributions of the opposition and of lesser scientists. In part 2 I describe the strategies by which the past is manipulated in order to make the winners look naturally right. By such a reconstruction of the past, the cornerstones of the Copenhagen interpretation—quantum jumps, the impossibility of causal space-time models, indeterminism, and wave-particle complementarity—were even more firmly entrenched (chapters 10 and 11).

I describe how the opposition's stand is delegitimated and trivialized. In their fabricated narratives, the winners construct the profile of the opposition and the description of past science concurrently. The ideas of the opposition are projected as most characteristic of the overthrown past, and thus the opposition naturally appears reactionary—disposing of the old and discrediting the opposition are, in fact, one and the same process. As a result, not only is the opposition caricatured but past science is trivialized. Hero worship of the winners further delegitimates the opposition and prevents criticism of the orthodox stand (chapter 13).

Historians of science rarely question the narrative of the winners: Jammer (1966) and Mehra and Rechenberg (1982), for example, closely follow the orthodox line. Jammer, in the preface to his classic book, quotes Einstein's penetrating warning not to rely on the recollections of the participants: "To the discoverer in this field the products of his imagination appear so necessary and natural that he regards them . . . not as creations of thought but as given realities." Yet Jammer "felt entitled to ignore this warning," and he "discussed the subject with quite a number of prominent physicists who contributed decisively to the development of the theory" (1966, viii). Pais's recent book (1991) is written exclusively from Bohr's perspective. This is not to say that these accounts do not use primary sources; in fact, Jammer, Mehra and Rechenberg, and Pais use the sources extensively. Yet what they see in those sources, and more important, what they ignore therein, is dictated by the overwhelming authority of the winner's perspective. A notable exception in this respect is Cushing's (1994b) important book about Bohmian mechanics as a viable alternative to the Copenhagen hegemony.

The "naturalness" and even "finality" of the orthodox point of view is advanced through powerful strategies of persuasion, which I refer to as the "rhetoric of inevitability" (chapter 9). The ingenious technique was to disguise arguments of consistency as those of inevitability. What is taken as objective quantum philosophy (and "inevitable" at that) turns out to be ideology—where by ideology I mean a system of assertions that imply, from within, their own justice, truth, and self-evidence. With respect to the entrenchment of the Copenhagen dogma, epistemology and sociology often merge—considerations of epistemic warrant and of social legitimacy are, at times, indistinguishable. The foundations of the Copenhagen paradigm were chosen and elaborated in direct contrast to the opposition's stand. The construction of the winner's narrative and philosophical arguments of inevitability serve the same end.

In chapter 11, I contrast the dramatic narrative of the "inevitability" of wave-particle complementarity with the freedom and plurality of theoretical approaches to the wave-particle issue. The "logical" arguments for the inevitability of wave-particle complementarity are built on Bohr's peculiar doctrine of the "indispensability" of classical concepts—a doctrine that few theorists, including Bohr's closest colla-borators, found persuasive (chapter 8). Those mathematical physicists who gave up this rigid doctrine suggested a rich variety of solutions to the wave-particle dilemma (chapter 11). The discussion in chapter 11 demonstrates that there is a fundamental ambiguity, and therefore a lack of "paradigmatic" agreement, concerning even such basic physical terms as "wave" and "particle" (scientists hold different opinions, each from his own theoretical viewpoint, of what are to be considered nec-essary or sufficient attributes of those terms). Despite this lack of clarity, or perhaps because of it, conversation on this issue continued, and new theoretical breakthroughs occurred.

The same freedom that created room for open-ended creative theo-rizing allowed the construction of a variety of ad hoc strategies for the legitimation of the Copenhagen stand. The proliferation of such ad hoc moves is yet another source of the inconsistencies that still plague the Copenhagen interpretation today. Both Heisenberg and Pauli sup-ported Bohr's philosophy of wave-particle complementarity in public, while often expressing, behind closed doors, views that were contrary to Bohr's. Heisenberg, Born, and Pauli, as well as Bohr himself, ex-ploited wave-particle complementarity for pedagogical and ideologi-cal reasons. The simple thought experiments that supposedly demon-strated the necessity of both the wave and particle descriptions were especially effective for promoting the philosophical "lessons" of quan-tum theory to wider audiences. From the analysis of these experiments, supported by Bohr's doctrine of the indispensability of classical con-cepts, both the "finality" of indeterminism and the "impossibility" of unified objective description followed, leaving seemingly no possibility of other interpretive options.

The accessibility of the explanatory strategies fed the illusion that no technical understanding of the quantum mechanical formalism is needed in order to grasp the essence of the quantum revolution.[3] This illusion was most vigorously sustained by Bohr himself. "Philosophy is but a sophisticated poetry"—this view of philosophy aptly char-acterizes Bohr's voluminous improvisations on the theme of comple-mentarity, filled with affective analogies, subjective associations, and allusions to "harmonies," expressed in "common language." In chap-ter 12, I argue that Bohr's philosophy is best characterized as a richly imaginative, yet ultimately misleading attempt to comprehend the quantum mystery without recourse to the mathematical formalism of quantum theory. There is, in this sense, a vast difference between the complementarity principle and Bohr's correspondence principle, which

3. Many postmodernist critics of science fell prey to the temptations of this strategy of argumentation (Beller 1998).

guided the search for the new quantum theory in the early 1920s, and with which the complementarity principle is sometimes confused. The metaphors of the complementarity principle are vague and arbitrary, in contrast to the more rigorous use of analogies between the macro- and microdomains guided by the correspondence principle. While the correspondence principle was a potent heuristic that led to the discovery of the rigorous quantum formalism, the complementarity principle was a device of legitimation—it led to no new physical knowledge.

In chapter 12, I analyze the ways Bohr, Pauli, and Heisenberg, by their imprecise allusions to quantum "wholeness," spun a metaphorical web of associations that disguise, rather than reveal, quantum entanglement and nonlocality. I argue that these allusions and analogies are fed by classical intuitions and contain nothing quantum about them. Not surprisingly, Bohm's version of quantum theory and its recent variants, which fundamentally incorporate quantum wholeness as a basic principle, are compatible with deterministic description. Bohr's numerous opaque allusions to quantum wholeness contribute to the illusion that his philosophical views were stable, despite the fact that he used this notion differently in different contexts. The notion of wholeness undergoes especially drastic change between Bohr's pre-1935 and his post-1935 writings (as a result of EPR). This ingenious and misleading improvisation on the idea of quantum holism contributes to the deception that a well-defined philosophical framework exists, thus further entrenching the Copenhagen orthodoxy.

Support for the orthodox view comes not only from historians but from philosophers of science as well. The strongest support came from Norwood Hanson and, after him, Thomas Kuhn, who incorporated the Copenhagen ideology into an overarching theory of the growth of scientific knowledge (chapter 14). Hanson and Kuhn canonized the concepts of paradigm and incommensurability into objectified philosophical notions that exclude, in principle, diversity of opinion and legitimate disagreement. Thus opposition is discredited and eliminated in the most radical way—by definition. Kuhn and sociologists of science who follow Kuhn's approach talk in terms of "deviance" and "impermissible aberration" rather than acknowledging reasonable disagreement. Kuhn's theory of incommensurable paradigms and the orthodox narrative of the quantum revolution reinforce each other—the quantum revolution is cited as a prime example supporting Kuhnian philosophy,[4] and the orthodox narrative of the quantum revolution is

4. Even historians who question the adequacy of Kuhnian notions to describe other historical developments (Westman 1994, discussing the Copernican-Newtonian revolution) assume that Kuhn's philosophy adequately describes the case of the quantum revolution.

objectified by a Kuhnian theory of the growth of knowledge (chapter 14). By such circular argumentation, the orthodox perspective is made to appear unassailable.

By incorporating addressivity and disagreement as fundamental notions, the dialogical approach, developed here, presents an alternative to current approaches to the study of science. Dialogical analysis incorporates conversation and communicability both as social realities and as epistemological presuppositions. From the dialogical perspective, a creative scientist cannot, in principle, be isolated—he, or she, is linked fundamentally to the efforts and concerns of others. In the dialogical approach, the notorious question of whether science is "rational" or "social" in nature becomes a pseudoproblem. Science is simultaneously rational and social—the rationality of science is dialogical and communicative.

The view of scientific activity as an ever-changing, open-ended communicative flux fits well scientists' own image of their work. Theoretical physicist David Finkelstein (1987), describing the state of his discipline today, chose the Heraclitean "All is flux" for the title of his paper. David Bohm emphasized the essential communicability of scientific doing: "Communication plays an essential role in the very act of scientific perception. . . . They [scientists] constantly engage in a form of internal dialogue with the whole structure of their particular discipline. . . . When insight occurs, it emerges out of this overall structure of communication and must then be unfolded so that it obtains its full meaning within it" (Bohm and Peat 1987, 67).

In the concluding chapter of this book I provide a tentative outline for a dialogical historiography and philosophy of science. Dialogical historiography reestablishes scientific individuality as the focus for studies of the growth of knowledge. Dialogical analysis demonstrates that scientific theorizing can be both free and nonarbitrary, and that theoretical achievements can be simultaneously well grounded and imaginatively beautiful.

We constantly conduct conversations with others—with living people, with the dead, and even with the yet unborn. The Russian poet Marina Tsvetaeva, in one of her poems, addressed a reader one hundred years in the future—the one who will truly love and understand her, and who will prefer her remains to the flesh of the living. Without unceasing addressivity and communicability, existence and thought, artistic imagination and scientific creativity are inconceivable.

PART ONE

Dialogical Emergence

CHAPTER 2

∽

Matrix Theory in Flux

Antagonistic cooperation is the key.
William Carlos Williams 1936, 177

Introduction

The years 1925–27 were a time of astounding scientific creativity. Dirac (1977) referred to this time as the "exciting era," while Pauli called it a "period of spiritual and human confusion" (1955a, 30). It was both. The "exciting era" refers to the breathtaking advances in the foundations of physics, "human confusion" to the treacherous matter of interpretation, to intensely emotional confrontations, to the slippery philosophical implications of the new conceptual tools. In this period of unprecedented advance the openness and ambiguity of theoretical practice seemed "confusing." But perhaps it was this ambiguity that created conceptual room for such rapid progress. There was no strong "belief" either in indeterminism or positivism, as the received story implies—pronouncements on these issues were uncommitted, fluid, and opportunistic.

The human confusion is reflected in the rich folklore of the history of quantum physics—anecdotes about inflamed passions and wounded egos, about emotional strain and outspoken hostility. These emotions produced not merely heat—they begot light (Beller 1996b).

In this chapter I analyze the initial efforts to interpret the new matrix formalism. I also describe the emotional confrontation between the matrix physicists and Schrödinger, which fused, or confused, issues of foundational preference with that of professional privilege. I argue that the emergence of Born's statistical interpretation of quantum mechanics was devoid of prior philosophical preferences and was embedded in the dialogical context of the "battle" between Schrödinger and the Göttingen-Copenhagen camp.

This chapter and the following one undermine the myth that the matrix and wave approaches were from the beginning conceptually distinct and historically independent. I argue that the strong distinction, and even polarization, of the wave theoretical and matrix approaches developed as a result of the confrontation between the competing groups.[1] This being the case, physicists on both sides were not generous in their papers and later recollections with acknowledgments of mutual debts.

A Revision of the Origins of the Matrix Theory

The still widely accepted scenario for the development of the new quantum mechanics in the 1920s runs as follows. In a period of only a few months (late 1925 to early 1926), two major theories of atomic phenomena emerged: matrix mechanics and wave mechanics. Matrix mechanics originated in Werner Heisenberg's radical reinterpretation of basic physical magnitudes (Heisenberg 1925), and Max Born, Pascual Jordan, and Heisenberg immediately expanded it into a complete and logically consistent theoretical structure in the so-called Dreimännerarbeit (Born and Jordan 1925a; Born, Heisenberg, and Jordan 1926). Wolfgang Pauli (1926b) demonstrated that the abstract formalism of matrix mechanics gave the correct approach to atomic theory by solving the problem of the structure of the hydrogen atom—the central problem of the atomic domain. The basic concept of matrix theory was the particle, or corpuscle.[2]

The basic concept of the rival theory, Erwin Schrödinger's (1926a, 1926b, 1926c, 1926d, 1926e, 1926f), was instead the wave, based on ideas first proposed by Louis de Broglie (1923). Schrödinger's theory combined intuitive and adequately developed mathematical tools with familiar physical concepts, and it was enthusiastically welcomed by the conservative wing of the physical community, which distrusted the revolutionary physical ideas of matrix mechanics and the complicated mathematics involved. Though radically different in their basic assumptions and mathematical treatments, the two theories were soon proved to be equivalent by Schrödinger himself,[3] and the only major problem left was that of physical interpretation. Schrödinger attempted to ascribe to his theory a classical, continuous wave interpretation, but the quantum physicists at Göttingen and Copenhagen demonstrated that these conservative attempts were untenable. The physical interpretation that prevailed instead was a direct continuation of the indeterministic beliefs of Heisenberg, Born, and Niels Bohr.

1. This book analyzes the Göttingen-Copenhagen side of the confrontation. An analogous argument can be made about Schrödinger. It is explored in part in Wessels (1983). I am currently exploring Schrödinger's case as well.

2. See also Darrigol (1992a), Jammer (1966), and Mehra and Rechenberg (1982).

3. Recently, Muller (1997) challenged the "myth of equivalence." He claimed that there was no strict equivalence between Schrödinger's and Heisenberg's versions originally—the two theories became truly equivalent after von Neumann's formalization of transformation theory.

Contrary to this story, I have argued that matrix mechanics was not immediately recognized as a pivotal breakthrough (Beller 1983). A few months before the publication of Schrödinger's mechanics and its use in transformation theory, matrix mechanics was beset by difficulties of such magnitude that no one, including the authors themselves, considered it to be more than a first step on a long path toward the ultimately correct theory. This was the main reason for the enthusiastic acceptance of Schrödinger's theory: acclaim for Schrödinger's theory was not limited to the conservative quarters of the physical community.

Nor was matrix mechanics a theory of corpuscles before Born's probabilistic interpretation: an atom in the matrix approach was endowed with electromagnetic, not kinematic, meaning. It is impossible to understand the genesis of the philosophy of quantum mechanics without closely following the original interpretive announcements of matrix physicists and the modifications they later underwent in dialogical response to Schrödinger's theory. The radical assumption of the matrix approach—within the atom there is no geometry—evolved into the more moderate: within an atom there is only statistics. The classical space-time container, eliminated by the matrix approach, was restored in response to Schrödinger—first by Born, then by Pauli and Jordan, and finally by Heisenberg himself. It is easy to overlook this process because of the speed with which it took place. The resulting quantum philosophy evolved into a hybrid of the original radical matrix program, concepts revived from Bohr's earlier work, and statistical compromises necessitated by the acceptance of Schrödinger's continuous theory. The philosophy of quantum physics became a mixture of quantum and classical concepts, whose inherent difficulties plague physicists and philosophers to the present day.

Problems of Physical Interpretation: The Elimination of Space-Time

Matrix mechanics, designed to avoid the problems of the old quantum theory, implied a radical change in the conventional description of the space-time continuum used in earlier physical theories. Heisenberg's 1925 paper—the turning point in quantum physics—announced the aim of eliminating unobservables and dispensing with visualizable models that relied on continuous space-time pictures. Instead of such unobservable kinematic variables as an electron's position, velocity, and period of revolution, Heisenberg sought to incorporate only observable spectroscopic data into the theoretical framework.[4]

Heisenberg and his colleagues at Göttingen justified their approach

---

4. Although Heisenberg did not rely on it explicitly, the "virtual oscillator" model, which contradicted the usual picture of space-time within the atom, played an important heuristic role (MacKinnon 1977).

by arguing that the difficulties of the old quantum theory stemmed from the fact that the quantum calculation rules operated with unobservable quantities—they were based on impermissible classical mechanical pictures, the use of which made the old theory inadequate for solving any but the simplest cases and led to grave inner contradictions (Heisenberg 1925). The Göttingen program was then to replace the explanatory mode of continuous classical mechanics with a discrete descriptive approach.

Born had made a pivotal contribution toward a "truly discontinuous" theory by inventing a method (used in Heisenberg's reinterpretation paper) of replacing all differential coefficients with the corresponding difference quotients (Born 1924; Born, Heisenberg, and Jordan 1926). This replacement reflected the growing realization that in the new quantum theory each physical quantity should depend on two discrete stationary states, and not on one continuous orbit as in classical mechanics.[5] The old quantum theory chose certain stationary states from all possible mechanical motions by means of quantization rules. The new theory, Born emphasized, was to contain only the permissible values: consistency demanded that forbidden fractional values should have no meaning at all in the theory.

This approach implied the elimination of classical space-time as a container of motion—such a container is as superfluous as the unrealized mechanical motions: "No one has been able to give a method for the determination of the period of an electron in its orbit or even the position of the electron at a given instant. There seems to be no hope that this will ever become possible, for in order to determine lengths or times, measuring rods and clocks are required. The latter, however, consist themselves of atoms and therefore break down in the realm of atomic dimensions" (Born 1926a, 69).

Heisenberg admitted later that he intended to eliminate not only the orbits of bound electrons but even the experimentally observable paths of free electrons, despite Wilson's experimental substantiation of the latter (Heisenberg 1933, 292). The authors of matrix mechanics initially implied not only that it is meaningless to represent electron motion as

5. According to Born, one cannot visualize atomic phenomena as continuous processes in space (for example, as orbital motion) because space itself is not infinitely divisible: there exist ultimate units of matter that cannot be further divided. Born derived this stand that nature has a scale and that one should not expect to find smaller and smaller elements not so much from the presupposition that matter is corpuscular as from the special form of quantum laws, in which only whole numbers appear. If physical laws are expressions of relations between whole numbers, more accurate measurement does not add any new information, and therefore physicists have reached the bottom of nature's scale (Born 1926a, 2).

a change of position in time but also that it is impossible to ascribe position to an electron at a given instant. Space-time exists only in the
macroscopic domain—in the atomic domain "space points in the ordinary sense do not exist" (Born 1926a, 128).

This approach was tantamount to giving up all hope of devising a
visualizable physical interpretation for the new matrix mechanics. As
Jordan put it later, there is no reason to expect that mathematical relations between observable radiation entities are amenable to visualization built on classical habits (1927a, 648). Matrix mechanics is no more
open to visualization than Maxwell's equations—one can only get used
to manipulating both. The only physical interpretation to be expected
from such an approach is the description of the limiting case of transition from the essentially unvisualizable quantum domain to the visualizable macroscopic world.[6]

Even though Heisenberg and his colleagues could dispense with the
demand for visualization on the atomic level, they still needed to comprehend the transition from microdomain to classical macrodomain.
Yet the problem proved unyielding, and Heisenberg complained to
Pauli about his failure to understand how to achieve the transition.[7]
One of the principal reasons for this difficulty was the purely formal
definition of the concepts of time and space in the matrix theory, which
bore no relation to evolution over time in a physical system. Pauli had
mentioned this fact as early as November 1925 in a letter to Bohr. Cor-
nelius Lanczos later argued that an atom in the matrix approach is a
timeless entity (1926, 815).[8] Heisenberg agreed that the concept of time
has no more the usual meaning in the new theory than the concept of
space—both are merely formal symbolic artifacts.

6. "In the further development of the theory, an important task will lie . . . in the manner in which symbolic quantum geometry goes over into visualizable classical geometry"
(Born, Heisenberg, and Jordan 1926, 322).

7. "The worst is that it does not become clear to me how the transition into classical
theory takes place" (Heisenberg to Pauli, 23 October 1925, PC, 251).

8. Although the matrix elements of momentum (p) and position (q) contain the expo-
nent $2\pi i \nu_{ik} t$ (where i is the square root of -1, ν_{ik} is the frequency associated with the
transition from state i to state k, and t is time), nothing changes in the calculations if one
simply writes $\exp(2\pi i \nu_{ik})$, without dependence on time. Actually, Lanczos notes, in a consistent discontinuous theory there is no place for such a continuous parameter, and indeed the dependence of the matrix elements on time is never used. Differentiation with
respect to time is introduced into the theory only in order to preserve the formal analogy
with the Hamiltonian equations of classical mechanics: the differentiation of matrix q
with respect to t gives the result $dq/dt = (2\pi i/h)(Wq - qW)$, where W is the energy
matrix and h Planck's constant. This formula indicates a purely algebraic relation between
matrices, rather than an evolution in time. Pauli presented similar arguments in his letter
to Bohr and speculated that perhaps time could be defined through the concept of energy
(Pauli to Bohr, 17 November 1925, PC, 260).

The meaninglessness of the concept of space-time in the original version of the matrix theory had momentous consequences. Carried to a logical conclusion, it meant eliminating the concept of a particle, or "thinghood," from the atomic domain.

The original matrix program precluded not only a probabilistic interpretation (which presupposed that an electron does indeed occupy a definite position in space and provided formulas for estimating that position's probability) but the type of physical interpretation Heisenberg gave later in his paper on uncertainty relations (Heisenberg 1927b). To doubt the existence of the position of the electron in time, as the original theory did, is more radical than to question the existence of simultaneous position and velocity. Heisenberg's uncertainty interpretation does not limit infinite precision of position when momentum remains undetermined: it was a visualizable (*anschauliche*) interpretation of microphysics, although Heisenberg had previously asserted that such a visualizable interpretation was impossible. In order to understand how this change occurred, we need to take into account the overwhelming success of Schrödinger's theory and the vehement confrontation that developed between Schrödinger and Göttingen.

Was the Original Matrix Theory a Theory of Particles?

The concept of particles was hardly compatible with the original matrix approach. The relations between the matrices q, p, and the like were defined formally so as to retain the Hamiltonian formalism and the analogy with classical mechanics. The matrix theorists considered the positions of electrons themselves to be unobservable: only the frequency, intensity, and polarization of the emitted radiation can be measured. As Born announced, "We therefore take from now on the point of view that the elementary waves are the primary data for the description of atomic processes; all other quantities are to be derived from them" (1926a, 70).[9]

In developing matrix mechanics, the Göttingen physicists tacitly

9. Heisenberg himself pointedly explained the substitution of radiation entities for kinematic entities:

> In the classical theory the specification of frequency, amplitude and phase of all the light waves emitted by the atom would be fully equivalent to specifying its electron path. Since from the amplitude and phase of an emitted wave the coefficients of the appropriate term in the Fourier expansion of the electron path can be derived without ambiguity, the complete electron path therefore can be derived from a knowledge of all amplitudes and phases. Similarly, in quantum mechanics too, the whole complex of amplitudes and phases of the radiation emitted by the atom can be regarded as a complete description of the atomic system, although its interpretation in the sense of an electron path inducing radiation is impossible. (1933, 292)

relied on the "virtual oscillator" model (Bohr, Kramers, and Slater 1924).[10] In order to describe interaction between an atom and radiation in close analogy with classical theory, the authors replaced the atom with a set of "virtual" oscillators whose frequencies corresponded to frequencies of transitions between the stationary states of the atom. They held that in each stationary state the atom continuously radiates "virtual" waves whose frequencies correspond to the possible transitions from this state to all others. The authors attempted to reinterpret from a wave theoretical standpoint all the phenomena successfully explained by light quanta. Thus Bohr, Kramers, and Slater explained the Compton effect, in which X-rays scatter and change frequency upon impact with matter, by ascribing the change in frequency to the Doppler effect. This explanation required them to assume that the center of the waves emitted by the electron did not coincide with its *kinematic* movement. We have here the first intimation of how incompatible a wave theoretical description is with regular kinematics.

In 1925 Heisenberg joined Kramers in a paper that dealt with the dispersion of radiation by atoms and spelled out, in a rigorous mathematical way, the ideas only roughly outlined in the presentation of Bohr, Kramers, and Slater. The authors further pursued the program of eliminating light quanta from physics. The same kinematic curiosity appeared as in the paper by Bohr, Kramers, and Slater, on which it was based: the center of the waves emitted by the atom moved relative to the excited atom (Kramers and Heisenberg 1925).

Faced with this incompatibility between kinematics and the wave theory, the Göttingen-Copenhagen physicists chose the wave theory. Born, lecturing at MIT in the winter of 1925–26, argued that explanations of the Compton effect assuming that light quanta and electrons are corpuscular were less fruitful than explanations relying on the wave approach through the Doppler effect. Calculations showed that the directions of motion of the electron and the wave center did not coincide: "We therefore stand before a new fact which forces us to decide whether the electronic motion or the wave shall be looked upon as the primary act." Because "all theories which postulate the motion have proved unsatisfactory," Born opted for the description based on elementary waves (1926a, 70).

Heisenberg gradually realized how basic is the incompatibility between the usual kinematics and the wave theory of light. Even for a hydrogen atom (a case of one-particle periodic motion), according to

10. Slater's original conception involved light quanta guided by a "virtual" field; Bohr and Kramers disposed of the light quanta. For a discussion of Bohr's opposition to Einstein's idea of light quanta and its influence on the Bohr-Kramers-Slater theory, see Klein (1970).

the wave theory one should obtain equidistant spectral lines, and not lines merging into a continuous limit series (Heisenberg 1926b, 990).[11] One could adhere either to ordinary kinematics or to the wave theory, but not to both; the recurrent confrontation with kinematic peculiarities made Heisenberg fully aware of the need to choose between the usual kinematics and the wave theory. Heisenberg chose the latter and reinterpreted kinematics in such a way that it would lead to the correct spectral, and not the orbital, frequencies of the emitted waves.

This account suggests that the elimination of unobservables was invoked ex post facto—as justification and not as a guiding principle—a point I will elaborate on in chapter 3. The following statement by Heisenberg supports this conclusion: "One could circumvent this difficulty only by giving up altogether the assignment to the electron or to the atom of a definite point in space as a function in time; *for justification* one had to assume that such a point also cannot be directly observable" (1926b, 990, my italics).

The Göttingen physicists regarded the virtual oscillator model as more than a heuristic device. For Born, virtual oscillators were "the real primary thing," and the interaction of electrons in the atom "consists of a mutual influence (irradiation) exerted by virtual resonators on each other" (1924, 190). Heisenberg assumed that "something in the atom must vibrate with the right frequency" (Heisenberg to van der Waerden, 8 October 1963, van der Waerden 1967, 29). Pauli (1926b) described an atom at the time as a collection of harmonic partial vibrations, associated with transitions between different stationary states, and not as a constellation of particles tied kinematically to the occupation of certain stationary states. These vibrations cannot be combined into "orbits."[12] As Heisenberg put it, "In quantum theory it has not been possible to associate the electron with a point in space. . . . However, even in quantum theory it is possible to ascribe to an electron the emission of radiation" (1925, 263).

Thus the original matrix theory was saturated with wave theoretical concepts. Originally, matrix mechanics was a symbolic algebraic theory, and the only concepts connected to the physical situation were wave

11. This is what Heisenberg meant by the remark at the beginning of his reinterpretation paper that "the *Einstein-Bohr* frequency condition . . . already represents such a complete departure from classical mechanics, or rather (using the viewpoint of wave theory) from the kinematics underlying this mechanics, that . . . the validity of classical mechanics simply cannot be maintained" (1925, 261–62).

12. Moreover, virtual oscillators, as opposed to classical theory, could not even be looked upon as charged particles. Their "ghostlike" character was revealed by the fact that for emission oscillators the expression $(e^*)^2/m^*$ (where e^* is the charge of the virtual oscillators and m^* the mass) is a negative number (Kramers 1924).

theoretical (frequency, intensity, and polarization of radiation), not particle kinematic (position in space, or even its probability).

Stationary States and Quantum Jumps

The matrix theorists justified the existence of stationary states by analogy with waves, very much in the spirit of Schrödinger's explanation in terms of vibrations, which they later rejected vigorously. One result the authors of the Dreimännerarbeit were proud of was that in the matrix theory the assumption that stationary states with certain discrete energies existed was not arbitrary, since their existence followed mathematically from the matrix formalism: "The existence of discrete stationary states is just as natural a feature of the new theory as, say, *the existence of discrete vibration frequencies in classical theory*" (Born, Heisenberg, and Jordan 1926, 322, my italics). Originally, matrix mechanics and Schrödinger's theory had more in common than is usually appreciated: both considered some vibration process as primary; but whereas Heisenberg reinterpreted classical space-time through these vibrations, Schrödinger left the usual concepts of space-time unchallenged. Schrödinger, indeed, initially regarded his own work as being in the same vein as the matrix approach. Even though their methods differed, "in its tendency, Heisenberg's attempt stands very near the present one," he asserted (Schrödinger 1926b, 30).

The authors of the matrix method considered one result of the theory particularly valuable: its ability to obtain discrete solutions without assuming a priori that stationary states existed as Bohr had. Heisenberg was especially proud of the independence of the new quantum mechanics from this assumption. The theory is instead derived solely from the commutation relation $pq - qp = (h/2\pi i)\mathbf{1}$ (Heisenberg to Pauli, 18 September 1925, *PC*). The stationary states in the matrix approach were purely formal artifacts, characterized by the numbers n (or $n_1, \ldots, n_f$ in the case of f degrees of freedom), which were associated with the definite values of energy and perhaps of angular momentum. Besides this, it was not clear whether the states could be given any physical interpretation at all, because they were all "interlocked" by the formalism in such a way that one could not assign a varying physical magnitude to a given stationary state (Dirac 1925).

Stationary states in the new theory were different from the original conception.[13] In the old theory the sequence of stationary states was that

13. The authors of the Dreimännerarbeit stated this clearly when discussing the solution of the harmonic oscillator in the new theory. They argued that nothing changes physically if the sequence of quantum numbers 0, 1, 2, 3, . . . is rearranged into a new sequence $n_0, n_1, n_2, n_3, \ldots$ such that certain formulas still hold.

of increasing values of energy, and this order had a fundamental physical significance (of merging continuously into a classical limit). This was not the case in the new theory. "The new mechanics presents itself as an essentially discontinuous theory in that herein there is no question of a sequence of quantum states defined by the physical process, but rather of quantum numbers which are indeed no more than distinguishing indices which can be ordered and normalized according to any practical standpoint whatsoever" (Born and Jordan 1925a, 300–301).

The matrix theorists realized that the theory's inability to incorporate the concept of the state of the system was a serious drawback. It was clear that an atomic system can exist in certain states for definite amounts of time; therefore, a theoretical description of such states was needed. The ability of Schrödinger's theory to give a straightforward definition of a stationary state through a wave function was one of its advantages, as Hendrik Lorentz pointed out in a discussion of the comparative merits of the two theories (Lorentz to Schrödinger, 27 May 1926, Przibram 1967, 44), and as Bohr argued later in the Como lecture (see chapter 6). That superiority originally disposed Born also to regard wave mechanics as physically more significant than the matrix approach.

In order to counter Schrödinger, the Göttingen theorists needed a matrix interpretation of the quantum state. The only way available at the time was to resuscitate Bohr's original concepts of stationary states and jumping electrons. Heisenberg collaborated fully with Bohr for this purpose. As he admitted years later: "It was extremely important for the interpretation to say that the eigenvalues of the Schrödinger equation are not only frequencies—they are actually energies. In this way of course one came back to the idea of quantum jumps from one stationary state to the other. . . . But even when we knew this and accepted the quantum jumps, we did not know what the word 'state' could mean" (1973, 269). Schrödinger was desperate about the reintroduction of "these damned quantum jumps" (quoted in Heisenberg 1967, 103). This weird concept seemed alien not only to his theory but to the matrix approach, originally constructed to dispense with such illustrative notions.

Matrix Mechanics and Determinism

In their recollections, the founders of quantum mechanics described their efforts to construct the new quantum mechanics as guided by a belief in indeterminism. Historians and philosophers of science often follow this lead, seeking the sources of such beliefs in the cultural milieu (Jammer 1966; Forman 1971). Yet we find no strong opinions expressed on the issue of causality during the creative stages of the erection of the new theory. Nor was the new matrix theory initially interpreted in an indeterministic fashion. Instead, matrix mechanics was viewed as a discrete deterministic theory. In this approach the problem was fully determined when the Hamiltonian of the system was known. By a principal axis transformation that diagonalized the matrix H, in principle one obtained the energy values and the matrices p and q. Matrix mechanics was constructed by analogy to be "as close to that of classical theory as could reasonably be hoped" (Born, Heisenberg, Jordan 1926, 322).

What about the fact that the matrix elements represented transition probabilities? The Göttingen theorists made no categorical indeterministic deductions from it; rather they simply did not exclude the possibility that this was a temporary weakness of the theory, very much the conclusion Einstein had reached about transition probabilities in his quantum theory of radiation in 1917 (Einstein 1917, 76). Pauli, for example, considered that the inability of the matrix theory to determine exactly when a transition occurs was a weakness: the time of emission of the photoelectron is definitely observable, and so a satisfactory physical theory should be able to calculate it. A satisfactory physical theory should contain no probability concepts in its fundamental propositions. Pauli was prepared to "pay a high price" to eliminate these probabilities (Pauli to Bohr, 17 November 1925, *PC*, 260). Heisenberg, under Pauli's influence, also considered this question in a letter to Einstein. Should the times of transition be regarded as observable, and would matrix mechanics be able to determine them? The matrix theory was still in such an incomplete *(unfertig)* stage, complained Heisenberg, that he simply did not know what stand to take on these questions (Heisenberg to Einstein, 30 November 1925, Einstein Archive, Hebrew University of Jerusalem).

In accordance with his emphasis on using observable electromagnetic variables, Heisenberg (1926b) presented matrix elements as "radiation magnitudes" *(Strahlungsgrössen)*, and not as transition probabilities, in his *Naturwissenschaften* paper. He seemed originally to suppress the statistical meaning of matrix elements, expressing his pleasure that certain relations formerly deduced from statistical considerations (frequencies of transitions) could be deduced solely by the mathematical manipulation of "radiation-value tables" (matrices; 1926b, 990). Heisenberg was nonetheless open to the possibility that for certain questions quantum theory might not go beyond statistical answers. Einstein himself was not sure whether a deterministic account of microscopic phenomena would be possible, or whether there would always remain a statistical residue (Einstein to Born, 27 January 1920, Born 1971). The answer was unclear to the majority of physicists. The use of statistical methods is not, of course, in itself evidence of a desire to dispose of causality; it is rather a sign that, at least for the time being, statistical methods yield the best scientific results. Heisenberg's preference for indeterminism evolved gradually in complex theoretical and human dialogical contexts (see chapters 4, 5, 9, and 10).

Mathematical Difficulties of the Matrix Approach

The matrix method of solving quantum mechanical problems was formulated in an astonishingly simple and general way. For any pair of values p_0, q_0 that satisfies the basic commutation relation $pq - qp = (h/2\pi i)\mathbf{1}$ (as, for example, in the simple case of the harmonic oscillator), solving the quantum problem (the problem of integrating the canonical equations for the given $H(pq)$) reduces to determining a matrix S that diagonalizes the Hamiltonian: $H(pq) = SH(p_0q_0)S^{-1} = W$, where W is a diagonal matrix. S then gives the solution of the canonical equations $p = Sp_0S^{-1}$ and $q = Sq_0S^{-1}$.

The simplicity of this economical and beautiful formulation was, however, quite misleading: it was practically useless because of the difficulty of calculating the reciprocal matrix S^{-1}. Matrix mechanics could not solve any general case, only, by perturbation methods, those physical problems that could be considered approximations to solved cases. The application of perturbation methods presupposed "that several specially simple systems which are used as starting points in calculus are completely known" (Born 1926a, 99; see also London 1926). The key word here is "completely": the "interconnectedness" of the quantum system in the matrix approach was such that one had to know both the frequencies and the intensities of the unperturbed system in order to calculate any of them for the perturbed system.

The crucial test of the theory was whether it could solve the central problem of the atomic domain—the hydrogen atom. No wonder Heisenberg rejoiced when Pauli succeeded in working out the Balmer formula for the hydrogen spectrum, though Pauli did so with difficulty and at the price of additional assumptions (Pauli 1926a; Heisenberg to Pauli, 3 November 1925, PC). P. A. M. Dirac and Gregor Wentzel also obtained the solution of the hydrogen atom, a success for the matrix approach, but not the grand triumph generally assumed (Dirac 1925, 1926; Wentzel 1926a). In order to proceed from the simple hydrogen atom to more complicated atomic systems by perturbation methods, one needed to know both the frequencies and the intensities of radiation of the hydrogen atom; but Pauli, Dirac, and Wentzel all failed to calculate the intensities. Moreover, it seemed hopeless to obtain this solution by matrix methods.[14] Matrix mechanics was thus at an

---

14. This opinion is expressed by Jordan: "It appeared hopeless to calculate the matrix elements" (1927a, 641).

impasse: the solution of the hydrogen atom could not lead to any further advance of the theory, and consequently the perturbation methods discussed in the Dreimännerarbeit were useless for new physical applications. This serious drawback was duly noted, by Fritz London, for example (London 1926, 921).

The authors of matrix mechanics found themselves in an ironic position. They had erected the new quantum theory primarily in order to approach multiple-electron systems, where the old methods had already failed. Yet the mathematical difficulties of matrix mechanics virtually prohibited its extension beyond simple systems with one moving particle. In fact, it was only after Schrödinger's wave mechanics had appeared that Pauli succeeded in calculating the intensities of the Balmer terms, and only by using Schrödinger's eigenfunctions of the hydrogen atom for his solution (Pauli to Landé, 2 June 1926, *PC*, 327). The superiority of Schrödinger's solution of the hydrogen atom over those of Pauli, Dirac, and Wentzel showed the clear advantage of his theory for solving quantum systems in central fields. Dirac's and Wentzel's calculations did not determine whether the quantum numbers of Balmer energy terms were integers or half-integers, while Schrödinger's calculations resulted unambiguously in half-integers. Both Dirac and Wentzel, and later Heisenberg and Jordan, used a two-dimensional treatment instead of the correct, three-dimensional one (Schrödinger to Lorentz, 6 June 1926, Przibram 1967, 64–65; Heisenberg and Jordan 1926).[15] They reached agreement with the experimental results by tailoring questionable methods to results known in advance: if they had conducted the calculations rigorously, they would have obtained the wrong answer, as Schrödinger suspected and John van Vleck (1973) has since demonstrated. Dirac himself used Schrödinger's solution in *The Principles of Quantum Mechanics* (Dirac 1930), and Pauli scolded Born and Jordan for including his own complicated solution, inferior to Schrödinger's, in their book *Elementare Quantenmechanik* (Born and Jordan 1930). There were other substantial mathematical difficulties as well, including the problem of coordinates (for a full discussion, see Beller 1983).

The matrix approach nevertheless showed promise. It seemed especially well suited to calculations involving angular momentum matrices. Consequently, it allowed straightforward quantization of the angular momentum components (that is, the calculation of magnetic quantum numbers) and the deduction of selection rules without extraneous assumptions (unlike the old quantum theory, where certain orbits, leading to collision between electrons and the nucleus, were excluded). It

---

15. Schrödinger initially obtained half-integers, not integers, because spin was not yet taken into account.

also enabled the intensities in the normal Zeeman effect and the Stark effect to be calculated (Born 1926a; Born, Heisenberg, and Jordan 1926). Moreover, matrix mechanics successfully treated the harmonic and anharmonic oscillators, dispersion, and, though incompletely, the hydrogen atom.

The Emotional Confrontation between the
Matrix Physicists and Schrödinger

The successes of the matrix mechanical method were modest compared with the difficulties encountered. Most of the problems treated using the matrix approach had been solved previously. Their solution was therefore more an encouraging sign of the new theory's possible validity than a guarantee of future success. The most outstanding atomic problems—hydrogen intensities and helium energy terms—awaited solution. They proved to be as resistant to the matrix approach as to the old Bohr-Sommerfeld theory. Even those who favored the matrix approach were reluctant to use it. Thus Arnold Sommerfeld declared: "There was clearly an element of truth in it [matrix mechanics], but its handling is frighteningly abstract" (1927, 231).

There were also grave problems of physical interpretation. "Heisenberg's theory in its present form is not capable of any physical interpretation at all" was the harsh verdict (Campbell 1926, 1115). There was simply no space-time "gravy," to use Hermann Weyl's characterization, in the matrix representation (Sommerfeld 1927, 231). The elimination of unobservables led to the elimination of space-time and of physical reality itself, including the by then familiar electrons. The extreme Machean approach, which eliminated everything but immediate sense perception (intensities, frequencies, and polarization of spectral lines) and a highly abstract uninterpreted formalism, did not seem tenable. Einstein questioned the soundness of it, and London (1926) and Nicholas Rashevsky (1926) argued that it was inconsistent. The ability to understand the transition from the micro- to the macrodomain was questioned. Nor could the matrix approach initially be reconciled with the definition of a quantum state. The lack of visualizability was a heuristic, if not a conceptual, hindrance. To build a truly discontinuous theory, one had to proceed with no suitably developed mathematics and in a virtually complete conceptual vacuum. After the emergence of Schrödinger's theory, it is no wonder the approach was abandoned, first by Born (who proposed the probabilistic interpretation), then by Pauli (who extended Born's approach), and later by Heisenberg (who advanced an interpretation of the indeterminacy relations).

No wonder, too, that the creators of the new mechanics did not seem to have much confidence in their theory in the beginning. Heisenberg's changing moods in the fall of 1925 reflected his doubts about the new theory. Thus, in a letter to Pauli, Heisenberg wrote that he considered the principal axis transformation the most important part of the whole theory (Heisenberg to Pauli, 23 October 1925, *PC*), only to call it "formal rubbish" three weeks later (Heisenberg to Pauli, 16 November 1925, *PC*). Heisenberg's letters to Dirac and Einstein at the time do not reflect a belief in the validity of the matrix approach. "I read your beautiful work with great interest," he wrote to Dirac. "There can be no doubt that all your results are correct, *insofar as one believes in the new theory*" (quoted in Dirac 1977, 124, my italics).[16]

Nor do the original published versions of matrix mechanics convey any feeling of an extraordinary scientific breakthrough. A cautious tone is adopted in the Dreimännerarbeit: even though the authors would like to conclude that the theory might be the correct one because of its mathematical simplicity and unity, they realize nonetheless that it is yet unable to solve the crucial problems of the quantum domain (Born, Heisenberg, and Jordan 1926). Even less optimistic are Born's concluding remarks in his MIT lectures: "Only a further extension of the theory, which in all likelihood will be very laborious, will show whether the principles given above are really sufficient to explain atomic structure. Even if we are inclined to put faith in this possibility, it must be remembered that this is only the first step towards the solution of the riddles of the quantum theory" (Born 1926a, 128).

It is not surprising, therefore, that except for the inventors of the new theory and those under their direct influence, only a very few physicists attempted to employ the matrix method. Most physicists, it seems, decided to "wait and see"—they would not submit to studying complicated mathematical techniques until the new theory proved its worth. Nor were they inclined to direct their students to study matrix mechanics. Felix Bloch, Peter Debye's student, was not even aware of the matrix theory before he learned about it from Schrödinger's publication.[17] Even the biggest success of the matrix approach—the solution of the Balmer terms for the hydrogen atom—lacked both the completeness and the elegance one would expect from a full-fledged theory. Pauli himself was pointing to the limitations of the matrix approach when later referring to his own result as having been derived by an "inconvenient and indirect method" (1932, 602).

In contrast, Schrödinger's theory, mathematically powerful and

16. Heisenberg's insecure and self-critical mood obviously persisted after Pauli's solution of the hydrogen spectrum, which, according to the usual historical accounts, dispelled any doubts about the correctness of the new theory.

17. As he wrote years later: "I did not learn about the matrix formulation of quantum mechanics by Heisenberg, Born and Jordan until I read that paper of Schrödinger, in which he showed the two formulations to lead to the same results" (Bloch 1976, 24).

familiar (as well as physically more accessible), was hailed by the community of theoretical physicists. "It was the most astonishing among all the astonishing discoveries of the 20th century," declared Sommerfeld (quoted in Moore 1989, 2). It is a historical myth that the enthusiasm was limited to the conservative part of the scientific community. It was welcomed, for example, by such unconventional minds as Charles Darwin, Fritz London, and Enrico Fermi.

The success of Schrödinger's theory contributed, in an indirect way, to the dissemination of the matrix approach. Many physicists learned about the existence of matrix mechanics from Schrödinger's publications. The proof of the equivalence of matrix and wave mechanics endowed the unfamiliar and abstract matrix approach with credibility.[18]

Yet the young matrix theorists reacted to Schrödinger's theory with disbelief and hostility (except for Born; see the next section of this chapter). Heisenberg hoped initially that Schrödinger's theory was wrong. Years later Heisenberg explained that he deliberately ignored wave mechanics before Schrödinger's equivalence paper because of the interpretation Schrödinger attached to it (interview with Heisenberg, AHQP). This explanation is not convincing. Schrödinger's initial presentation of his theory was undertaken in "neutral mathematical form" (Schrödinger 1926a, 9). Schrödinger's opinion about the need "to connect the function ψ with some *vibration process* in the atom, which would more nearly approach reality than the electronic orbits, the real existence of which is being very much questioned to-day" (1926a, 9), did not contradict the original inspiration behind Heisenberg's reinterpretation paper. Heisenberg also doubted the reality of electron orbits and tried to detect those internal vibrations that produce radiation in agreement with experiments. There was no substantial reason for Heisenberg's aversion to Schrödinger's theory; indeed, Schrödinger believed initially that his and Heisenberg's work in their "tendency" were very close.[19] Heisenberg's hostility to Schrödinger's theory seems more likely to be connected with his instinctive reluctance to admit anybody else into territory that the ambitious Heisenberg considered his own. This probably was also the reason for Dirac's initial opposition to Schrödinger's theory. As Dirac reported later, he ignored Schrödinger's theory

18. Lorentz wrote to Schrödinger: "I was particularly pleased with the way in which you . . . construct the appropriate matrices and show that these satisfy the equations of motion. This dispels a misgiving that works of Heisenberg, Born and Jordan . . . inspired in me" (Przibram 1967, 43).

19. The hint about a possible "beat" wave model could have aroused Heisenberg's objection because he no longer inclined toward building visualizable realistic models. Yet the interpretive issues in Schrödinger's first paper were so peripheral and understated that it would have been more natural to overlook them instead of the advance Schrödinger had made.

in the beginning because there already was one quantum mechanics, so no other was needed. Dirac admitted that "he definitely had a hostility to Schrödinger's ideas to begin with, which persisted for quite a while" (1977, 131).

Pauli also reacted to Schrödinger's ideas with suspicion—he considered Schrödinger's approach *verrückt* (crazy, foolish; Pauli to Sommerfeld, 9 February 1926, *PC*). Learning of Sommerfeld's regard for Schrödinger's work, however, Pauli took a closer look at it, decided that it belonged among the "most meaningful" recent publications, proved the equivalence between matrix mechanics and wave mechanics, and subsequently strove to elucidate its physical meaning by trying to understand the connection between Schrödinger's formalism and Einstein–de Broglie waves (Pauli to Jordan, 12 April 1926, *PC*).

Heisenberg, who could no longer contend after the equivalence proof that Schrödinger's theory was wrong, declared that its only value was its ability to calculate the matrix elements (Heisenberg to Pauli, 8 June 1926, *PC*). Jordan (1927a) arrived at a similar conclusion: the "meaning of the thing" was clear—eigenfunctions merely provided a new mathematical method for solving the equations of matrix mechanics. Jordan became one of the most militant opponents of Schrödinger's interpretive efforts. Schrödinger should have been satisfied with the mathematical advance and should not have even attempted anything more than a mathematical elaboration, declared Jordan (1927a, 615). The new theory must be "interpreted physically in close analogy with the older notions of stationary states and quantum jumps, and with Heisenberg's theory" (1927b, 567). In a review of Schrödinger's collected papers on wave mechanics, Jordan declared Schrödinger's theory to be devoid of any physical meaning and stated (incorrectly) that such was the opinion prevailing among the majority of physicists (Jordan 1927d). A novice in the field should not be exposed to Schrödinger's work without prior appropriate instruction in physical matters in the Göttingen-Copenhagen spirit, Jordan continued. Schrödinger, understandably, was outraged, and he complained about this review to Born (Schrödinger to Born, 6 May 1927, AHQP). Born admitted that Jordan somewhat exceeded his limits and blamed it on Jordan's youthful temperament (Born to Schrödinger, 16 May 1927, AHQP).

Schrödinger's hope that his theory would have a self-sufficient physical interpretation was not unreasonable. This expectation was consistent with the belief of physicists from Galileo to Einstein, that mathematical simplicity and power are unmistakable signs of a theory's physical significance.[20] Schrödinger attempted initially to elucidate some

20. Schrödinger undoubtedly expressed the feelings of many physicists when he pointed to the conceptual hindrance implied by the elimination of intuitive space-time

"conceivable mechanism" by which microphenomena take place in regular space-time. Physical reality, according to Schrödinger's early attempts at interpretation, was akin to the reality of classical electrodynamics—smeared clouds of electron matter (wave packets) obey wave mechanical equations. This ontology soon proved to be abortive; yet the "intuitiveness" of Schrödinger's approach was not confined to its ability to provide (or not) classical idealized space-time models. Rather, the physical significance of Schrödinger's theory lay in its ability to provide some qualitative handle on the essential aspects of the microworld. Schrödinger's theory deciphered the mystery of quantization; it explained why atoms in stationary states do not radiate. Despite the official line that denied physical significance to Schrödinger's theory, the matrix physicists, even Jordan, could not help but sometimes praise Schrödinger for just these assets.[21]

Heisenberg understood early that there must be some very close connection between Schrödinger's approach and his own. He realized that it was Schrödinger's theory that might prove helpful in elucidating the physical meaning of his own abstract version of quantum mechanics. After Schrödinger's first paper appeared, Heisenberg wrote to Dirac: "A few weeks ago an article by Schrödinger appeared . . . whose contents to my mind should be closely connected with quantum mechanics. Have you considered how far Schrödinger's treatment of the hydrogen atom is connected with the quantum mechanical one? This mathematical problem interests me especially because . . . *one can win from it a great deal for the physical significance of the theory* (Heisenberg to Dirac, 9 April 1926, AHQP, my italics). Pauli agreed, citing the possibility of "look[ing] at the problem from two different sides" (Pauli to Heisenberg, 12 April 1926, *PC*).

Heisenberg fully exploited the interpretive possibilities opened by the wave theory, using Schrödinger's imagery, which he publicly

---

pictures in matrix mechanics. The problems that atomic physics had to treat theoretically were presented, Schrödinger pointed out, in "an eminently intuitive form; as, for example, how two colliding atoms or molecules rebound from one another, or how an electron or α-particle is diverted when it is shot through an atom with a given velocity and with the initial path at a given perpendicular distance from the nucleus." How is one even to begin to treat such problems if one operates only with such abstract ideas as transition probabilities, energy levels, and the like? Darwin's words that "the ultimate theory will be one of space and time again" echoed Schrödinger's hopes (Darwin to Bohr, 1928, AHQP).

21. Jordan noted that Schrödinger's demystification of quantization (quantization follows from the condition of finitude and one-valuedness of wave amplitude) was more akin to "our physical understanding" than quantization imposed a priori. Born explained why there is no radiation in stationary states by using Schrödinger's wave function (Born 1969, 148).

denounced, as a rich source of suggestions for constructing a rival interpretation. Heisenberg's elaboration of the uncertainty relations hinged on a translation of a "wave packet à la Schrödinger" (Heisenberg to Pauli, 23 February 1927, *PC*) into the language of a particle ontology (see the detailed discussion in chapter 4). Bohr understood clearly that the wave theory played an essential role in Heisenberg's considerations, both historically and philosophically.[22] Bohr himself strove to elucidate the physical significance of Schrödinger's wave theory in his Como lecture (see chapter 6). Like the majority of the scientific community (the alleged "conservatives") he preferred the intuitiveness of Schrödinger's version to the formal abstraction of the matrix theory: "This ingenious attack [the matrix approach] upon the problem of the quantum theory makes, however, great demands on our power of abstraction, and the discovery of new artifices [wave mechanics] which, in spite of their formal character, more closely meet our demands for visualization, has therefore been of profound significance in the development and clarification of quantum mechanics" (Bohr 1929a, 110–11).[23]

Nor is the emergence of the Göttingen-Copenhagen statistical ontology in particle terms comprehensible without taking into consideration the selective borrowing of Schrödinger's interpretive ideas and their translation into particle language. Born's statistical interpretation was, to a large degree, a translation into particle language of Schrödinger's idea that the ψ-function determines the electron charge density; Pauli fused Schrödinger's idea of a "weight function" with Born's interpretation, arriving at the interpretation of the wave function as giving the probability that the system is in a specific configuration. Dirac similarly used the suggestiveness of Schrödinger's conceptions in forming his probabilistic ideas.[24]

22. Heisenberg wrote to Dirac about his confrontation with Bohr over the uncertainty paper: "Prof. Bohr says, that one in all those examples sees the very important role, which the *wave*-theory plays in my theory and, of course, he is quite right" (Heisenberg to Dirac, 27 April 1927, AHQP).

23. Bohr expressed similar reservations about the mathematical complexity of matrix mechanics and maintained his preference for the wave mechanical version throughout his life. In 1938 he wrote: "On account of the intricate mathematical operations involved, it was, however, of utmost importance, not only for the practical use of the formalism, but even for the elucidation of essential aspects of its consequences, that the treatment of any quantum-mechanical problem could be shown to be essentially reducible to the solution of a linear differential equation" (Bohr 1939, 387).

24. "Schrödinger's wave representation of quantum mechanics has provided new ways of obtaining physical results from the theory based on the assumption that the square of the amplitude of the wave function can in certain cases be interpreted as probability" (Dirac 1927, 621).

From a purely conceptual point of view, the intense hostility to Schrödinger's ideas (rather than, say, polite disagreement) is puzzling.[25] Partly, as I have noted, the matrix theorists resented Schrödinger's intrusion into their "territory." It did not seem feasible that an outsider could arrive at the long-sought solution of the quantum riddle, and that the solution could be that simple. Schrödinger's success aroused both disbelief and envy. The assessment by Heisenberg that it was "too good to be true" reveals this attitude (1971, 72). So does Born's dismissal: "It would have been beautiful if you [Schrödinger] were right. Something that beautiful happens, unfortunately, seldom in this world" (Born to Schrödinger, 6 November 1926, AHQP).

Yet disbelief and envy, strong as these feelings are, were only a part of the psychosocial setting in which the new quantum mechanics was erected. The Göttingen-Copenhagen physicists saw in Schrödinger's program something capable of extinguishing the matrix approach (interview with Jordan, AHQP). Had Schrödinger succeeded in giving a satisfactory interpretation without the Göttingen-Copenhagen concepts, he might have eliminated the need for the matrix approach altogether, in view of the mathematical equivalence of the matrix and wave theories and the greater manageability and familiarity of the latter. Many physicists found Schrödinger's theory much more attractive and were inclined to regard Schrödinger's physical approach as the more correct of the two candidates (interview with Jordan, AHPQ).

A flood of papers, most following Schrödinger and ignoring the matrix approach, highlighted the reality of the threat. Schrödinger's theory was successfully applied to a great variety of problems unamenable to matrix treatment. Born recalled the successes of Schrödinger's mechanics: "In the meantime, Schrödinger's wave mechanics appeared, and won the approbation of theoretical physicists to such an extent that our own matrix method was completely pushed into the background, particularly after Schrödinger himself had shown the mathematical equivalence of wave and matrix mechanics" (1971, 104). Expressing himself more strongly, Born said during an interview with Thomas Kuhn: "Wave mechanics was considered the real quantum mechanics by everybody, while matrix theory was completely neglected" (interview with Born, AHQP).[26] Born was indeed unhappy at the time about the "world-wide victory" of wave mechanics (Schrödinger to Born, 17 May 1926, AHQP).

This "victory," as Heisenberg clearly understood, was pregnant with

25. Heisenberg used such emotionally charged words as *abscheulich* (repelling, disgusting) in characterizing Schrödinger's approach. The connection between anecdotal and conceptual history, and the formative role of emotions in cognitive endeavors, is analyzed in Beller (1996b).

26. Citation analysis confirms Born's judgment (Kojevnikov and Novik 1989).

far-reaching consequences. It implied no less than the power to "influence . . . the research of the following century" (1952, 60). The "victory" in Bohr's and Schrödinger's characterization similarly meant "to realize one's wishes for the future of physics." [27] The weight of the threat to the Göttingen-Copenhagen version of physics was revealed at the Munich conference held at the end of summer 1926. Most participants there aligned themselves not only with Schrödinger's methods but also with his interpretive aspirations. Even Sommerfeld, or so it seemed to Heisenberg, succumbed to the persuasive force of Schrödinger's mathematics.[28] Heisenberg's critique of Schrödinger's intentions "failed to impress anyone," and Wilhelm Wien crowed to the despairing Heisenberg that his version of quantum mechanics with the nonsensical quantum jumping was "finished" (Heisenberg 1971). Heisenberg immediately reported this alarming situation in a letter to Bohr, and Schrödinger was subsequently invited to Copenhagen for what would become heated discussions.[29] After Schrödinger left Copenhagen, a feverish hunt for an adequate interpretation occupied both Heisenberg and Bohr.

The emerging Göttingen-Copenhagen interpretation did not weaken the "victory" of Schrödinger's methods. Even physicists who had made some initial contributions to the matrix approach turned to the wave theory with relief, Wentzel and London among them. Results obtained in the matrix framework were often translated into wave language, as, for example, William Gordon's wave treatment of the Compton effect, which came after Dirac's work in the matrix framework. Such duplication of scientific results was justified by the argument that "Schrödinger's methods possess the advantage of using the familiar mathematical forms only" (Gordon 1927, 117). When Born tried to show that some results achieved in the wave framework could also be easily obtained by matrix methods, Paul Ehrenfest called the latter "a bad habit" (interview with Oskar Klein, AHQP). Matrix tools could not

27. Schrödinger to Bohr, 23 October 1926, AHQP; reprinted in *BCW*, 6:459–61, translation on 12.

28. Sommerfeld did have some reservations but apparently abstained from expressing them. Sommerfeld reported to Pauli about the Munich conference in a letter: " 'Wave mechanics' is an admirable micromechanics, yet it is still far off from solving the fundamental quantum riddle" (Sommerfeld to Pauli, 26 July 1926, *PC*).

29. Heisenberg's description of these discussions is among the most quoted passages in the literature of the history of quantum theory. Heisenberg's tale is highly dramatic: "Though Bohr was an unusually considerate and obliging person, he was able in such a discussion, which concerned epistemological problems which he considered to be of vital importance, to insist fanatically and with almost terrifying relentlessness on complete clarity in all arguments. . . . He would not give up, even after hours of struggling. . . . It was perhaps from over-exertion that after a few days Schrödinger became ill and had to lie abed as a guest in Bohr's home. Even here it was hard to get Bohr away from Schrödinger's bed" (1967, 103).

compete with wave mechanical ones. By 1929 "the research was dominated by wave mechanics, and matrix mechanics so to speak only came
back through group theoretical arguments" (interview with Hendrik
Casimir, AHQP). Born and Jordan made a last, desperate attempt to
oppose this trend: they wrote a book—*Elementary Quantum Mechanics*
(1930)—relying solely on matrix methods, in which Schrödinger's wave
function did not appear even once. This book "in view of the general
predisposition in favor of Schrödinger . . . was not favorably received"
(interview with Born, AHQP). Pauli (1932) himself wrote a devastating
review:

> What is the "elementary" quantum mechanics of the present vol-
> ume? . . . Elementary is that quantum mechanics which makes use of
> elementary tools, and elementary tools are purely algebraic ones; the
> use of differential equations is . . . avoided as much as possible. . . .
> Many results of quantum theory can indeed not be derived at all with
> the elementary methods defined above, while the others can be de-
> rived only by inconvenient and indirect methods. Among the latter
> results belong, for instance, the derivation of the Balmer terms, which
> is carried out in matrix theory according to an earlier paper of Pauli's
> dealing with it. In this regard, he will not be able to accuse the re-
> viewer that he finds the grapes to be sour because they hang too high
> for him. The restriction to algebraic methods also often inhibits insight
> into the range and the inner logic of the theory. . . . The setup of the
> book as far as printing and paper are concerned is splendid.

Schrödinger's *methods* did indeed win an overwhelming victory. But
the *interpretation* that most physicists seemed to accept was the one given
by Schrödinger's opponents. The Göttingen-Copenhagen line was, of
course, not unassailable—that critical debates on the philosophy of
quantum physics continue to the present day attests most eloquently to
this fact. Yet quite apart from the philosophical question of the validity
of the opposing approaches, we can ask a sociological question concerning the initial distribution of forces. We will not be surprised to find
out who prevailed. At first Schrödinger had the emotional support of
many physicists, some distinguished and some not. Yet he struggled
with the problems of interpretation virtually alone. Those who identified with his aspirations hardly did anything to advance his efforts.
They hoped, as Jordan recalled, that Schrödinger would accomplish his
task unaided (interview with Jordan, AHQP).

The Göttingen-Copenhagen physicists, in contrast, presented a united
front. They cooperated intimately, and each contributed extensively to
the emergence of the new philosophy. Bohr and Born headed the era's
most prestigious schools of theoretical physics, and promising physicists were eager to work with them. Soon Pauli and Heisenberg would

occupy important chairs of their own (Cassidy 1992). Young physicists, who streamed into these centers from all over the world, were exposed automatically to the new philosophy. Because they were more interested in calculating and obtaining definite scientific results than in philosophizing, most of them simply adopted the official interpretation without deep deliberation. Heisenberg was aware of this. In his 1930 book, which he dedicated to the "diffusion of the Copenhagen spirit," Heisenberg conceded that "a physicist more often has a kind of faith in the correctness of the new principles than a clear understanding of them" (1930, preface).

The aging Schrödinger witnessed a remarkable state of affairs: the universal use of his theory coupled with an almost total rejection of his interpretation. Schrödinger's methods proved indispensable. His philosophy did not.

Born's Probabilistic Interpretation: A Case Study of "Concepts in Flux"

Born's probabilistic interpretation of the wave function occupies a central place in the philosophy of quantum mechanics. In the orthodox interpretation it signifies the abandonment of determinism and the introduction to a new kind of reality, abstract and "ghostlike." According to Born's reminiscences, his belief in particles rather than waves, together with Einstein's idea of a connection between the intensity of the electromagnetic field and the density of light quanta, made it "almost self-understood" to interpret $|\psi|^2 dx$ as the probability density of particles (Born 1961). However, scholars have expressed some doubts about the authenticity of Born's recollections.[30]

I have suggested a revision of the history of Born's probabilistic interpretation along the following lines (Beller 1990):

1. Born's probabilistic interpretation was a conceptual contribution that crystallized over a considerable period of time—time during which Born's ideas, as a result of his dialogues with Schrödinger, Heisenberg, Pauli, and other physicists, underwent significant changes.

---

30. Pais (1982) has noted that if Born had really been stimulated by the analogy of light intensity as a quadratic function and Schrödinger's wave function, he would not have been able, however briefly, to suppose initially that ψ rather than $|\psi|^2$ was a measure of probability, as he did in his first collision paper. Stachel (1986) has argued that the source of Born's inspiration was not Einstein's unpublished speculations but rather his work on monatomic gases, in which he suggested a connection between material particles and de Broglie fields. Wessels (1981) has pointed out that in none of Born's original papers on this subject did he interpret $|\psi|^2 dx$ as a probability of position. These valuable insights found additional reinforcement and explanation in my account (Beller 1990).

During the formative stage, all of Born's intellectual pronouncements were fluid, ambiguous, and uncommitted.

2. Born's initial contribution—his first collision paper (Born 1926c)—was not written, as Born later claimed, in opposition to Schrödinger's waves. Initially, Born was not involved in any controversy with Schrödinger at all. In fact, Born was very enthusiastic about Schrödinger's contribution, including its interpretive possibilities. He appears to have had no strong "belief" in particles; Born was undecided about the wave-particle issue.

3. Born's aim in his first collision paper (Born 1926c) was not to contribute to the clarification of interpretive issues but to solve a particular (yet central) scientific problem.

4. The aim of the collision papers (Born 1926b, 1926c) was not to argue the reality of particles and indispensability of indeterminism, as Born's later recollections would lead us to believe, but rather to describe and theoretically substantiate Bohr's concept of "quantum jumps"—discrete energy changes within an atom during collision processes. Most of the disagreement between Born and Schrödinger, as their correspondence confirms, centered not on the wave-particle dilemma or indeterminism but rather on the existence of these quantum jumps.

5. The probabilistic interpretation of Schrödinger's wave function—that ψ gives the probability of position—developed not from the relatively obvious suggestion that ψ describes the motion of free particles but from the pregnant question of how the wave function is to be interpreted for bound systems. Born's interpretation of $|\psi|^2$ as giving the probabilities of the stationary states of an atom was a crucial contribution, around which the issues of indeterminism and the particle-kinematic ontology were elaborated and established.

6. Born's original probabilistic interpretation played a key role in the emergence of the new philosophy of physics not because it was "obviously" correct but because of the ambiguities, difficulties, and paradoxes it raised. The solution to these problems led to the modification of earlier concepts and to the elaboration of new theoretical and philosophical ideas.

Born's probabilistic interpretation can be seen as a "concept in flux" process.[31] The uninterpreted mathematical tools left room for ambiguity

31. The term "concept in flux" was used by Elkana (1970) to denote those vague and unspecific ideas that only during the process of simultaneous formation of a theory and its basic concepts become scientifically legitimate. Thus, according to Elkana (1970), in Helmholtz's case the concept of energy and the law of conservation of energy evolved and were clarified as a single process. The dialogical approach of this book indicates that all concepts at the focus of research are in fact "in flux."

and permitted freedom from binding epistemological and ontological assumptions. In the development of abstract theoretical physics, it is tools (mathematical or experimental), rather than preconceived metaphysical ideas, that constitute the driving force in the growth of knowledge. And if Born was not committed to particles and indeterminism to begin with, he similarly did not commit himself to wave concepts, despite all his enthusiasm for Schrödinger's theory. Born was instead working in the creative conceptual twilight where "mathematics knows better than our intuition" (interview with Born, AHQP).

Born's Collision Papers

The first paper in which Born provided a statistical description of collision processes using Schrödinger's formalism is his "Zur Quantenmechanik der Stossvorgänge" (1926c). In his recollections Born implied that his opposition to Schrödinger's realistic wave concepts preceded the actual solution of the collision problem, yet a careful reading of the paper does not disclose any opposition. The paper glows with Born's emotional enthusiasm for Schrödinger's theory. As Born conceded in this paper, he was unsuccessful in his initial attempts to solve the collision problem within the matrix framework but was finally able to obtain the solution with the help of Schrödinger's formalism. This is the reason, Born declared, that he regarded Schrödinger's formalism as "the deepest formulation of the quantum laws" (1926c, 52).

The problem of aperiodic collisions required inquiry into the evolution of atomic phenomena. Earlier matrix methods had answered only questions of structure (energy spectra). Moreover, because there is a similarity between the behavior of the atom in collision situations and during its exposure to light irradiation, Born hoped that the solution of the collision problem would pave the way to understanding the interaction between matter and radiation as well.[32]

Born's first collision paper clearly asserts the superiority of Schrödinger's theory over Heisenberg's for the solution of this crucial physical problem. Heisenberg's version of quantum mechanics, argued Born, describes only one aspect of quantum problems (stationary states) and says nothing about the occurrence of transitions. In contrast, the one-sentence abstract of Born's paper claims that "quantum mechanics in the Schrödinger form allows one to describe not only stationary states but also quantum jumps" (1926c, 52). Born "intentionally avoids" the

32. As Born recalled many years later: "I wanted to use quantum mechanics because that would give the only way of experimenting with these things. The spectroscopic methods give only terms, energies, nothing more . . . the direct way of measuring is by collisions. Even excitation of light in an atom means collision" (interview with Born, 1963, AHQP).

term "transition probability," using instead Schrödinger's term "amplitudes of vibration." In fact, Born employed Schrödinger and de Broglie's wave concepts quite literally. An atom in stationary state n is a "vibration process" with frequency $(1/h)W_n^0$ spread over the whole space. An electron moving in a straight line is, in particular, such a vibration corresponding to a plane wave. When the atom and electron interact, they produce a complex vibration (*verwickelte Schwingung;* Born 1926c).[33]

The wave field determines the atomic transitions probabilistically, and quite independently of the way in which the wave field of the scattered electron is interpreted—literally, or probabilistically in terms of particles.[34] This is the source of Born's noncommittal reference to a putative corpuscular interpretation. As Born conceded later: "It is true that I considered the collision of particles with other particles as a scattering of waves" (appendix to letter to Einstein, 13 January 1929, Born 1971). There is indeed no indication in the paper itself of the staunch belief in particles that Born professed to have. Rather, Born's mind was open on the wave-particle issue.

There are several reasons for Born's indecision. It was a common practice at the time to introduce theoretical solutions merely as formal schemes, to be filled with physical content in due time. Schrödinger

33. Heisenberg was unhappy about this literal use of wave concepts: "Ein Satz erinnerte mich lebhaft an ein Kapital aus dem Christl[ichen] Glaubensbekenntis: 'Ein Electron ist eine ebene Welle.' . . . Aber ich will ihnen in Lästern keine Konkurrenz machen" (A sentence vividly reminded me of an article of faith from the Christian confession: "An electron is a plane wave." . . . But I am not intending to compete with you in slander) (Heisenberg to Pauli, July 1926, *PC*).

34. To the unperturbed atom with discrete energies Born ascribed eigenfunctions $\psi_1^0(q_k)$, $\psi_2^0(q_k)$, . . . ; to the unperturbed electron whose direction is determined by (α, β, γ) there correspond eigenfunctions of the type $\sin 2\pi/\lambda(\alpha x + \beta y + \gamma z + \delta)$. In case the interaction $V(x, y, z; q_k)$ is taking place, the scattered wave at infinity will be expressed through (Born does not provide the actual calculations in this paper):

$$\psi_{n\tau}^{(1)}(x, y, z; q_k) = \sum_m \iint_{\alpha x+\beta y+\gamma z>0} d\omega \Phi_{n\tau m}(\alpha, \beta, \gamma) \sin k_{n\tau m}(\alpha x + \beta y + \gamma z + \delta) \psi_m^0(q_k).$$

where τ is the energy of the incoming electron, coming from the $+z$-direction, and $\Phi_{n\tau m}$ is a function of the energy $V(x, y, z; q_k)$ of the interaction between the atom and the electron.

Concerning the possibility of interpreting this result in terms of particles rather than waves, Born said simply: "If one translates this result into terms of particles, only one interpretation is possible: $\Phi_{n\tau m}(\alpha, \beta, \gamma)$ gives the probability for the electron, arriving from the z-direction, to be thrown into the direction designated by the angles α, β, and γ with a phase change δ. Here its energy τ has increased by one quantum $h\nu_{nm}^0$ at the cost of the energy of the atom" (1926c, 54).

The calculated function Φ determines the quantum transitions $n \rightarrow m$, albeit only probabilistically. One does not obtain the actual state of the atom after the collision, only the probability of a certain event (quantum jump).

himself introduced his first wave mechanical paper merely as a formal treatment, leaving the question of physical interpretation aside (Schrödinger 1926a). Similarly, de Broglie left the meaning of his wave concepts vague: "The present theory may be considered a formal scheme whose physical content is not yet fully determined, rather than a full-fledged definite doctrine" (quoted in Jammer 1966, 247). It was natural at a time when the foundations of physics were shifting to introduce formal symbolic solutions with minimal interpretive content, especially among mathematical physicists. And Born was indeed "a mathematical method man" (interview with Heisenberg, AHQP). Because there was no elaborate idea of what de Broglie–Schrödinger waves meant, only tentative suggestions, Born preferred to conquer additional territory by mathematical means rather than indulge in a controversy over undefined issues. As Born revealed many years later, his solution to the collision problem did not depend on an interpretation in terms of particles: "I cannot see at all that these purely mathematical objections have anything to do with the question of particles-waves. . . . For if we accept Schrödinger's standpoint that there are no particles, only wavelets, the scattering calculations would be exactly the same as before" (Born 1953b, 148).

Why was Born open at the time to Schrödinger's ideas? Despite the fact that Born recalls James Franck's experiments as a source of his belief in particles, he seemed to consider the wave nature of matter quite seriously, as is clear from a letter to Einstein (Born to Einstein, June 1924, Born 1971).

De Broglie's original insights acquired credibility in Göttingen early in 1925 through Einstein's paper "Quantum Theory of the Monoatomic Ideal Gas" (1924). Just as in his previous treatment of blackbody radiation (Einstein 1905), Einstein obtained two terms: one that seemed to correspond to a fluctuation due to particles and another that seemed due to wave interference. In support of de Broglie's concept, Einstein concluded that "it appears that an undulatory field is connected with every motion [*Bewegungsvorgang*], just as the optical undulatory field is connected with the motion of light quanta. This undulatory field, whose physical nature is for the moment still unclear, must in principle permit its existence to be demonstrated by the corresponding phenomena of its motion [*Bewegungserscheinungen*]" (Einstein 1924, quoted in Stachel 1986, 368).

We know that these "brief, yet infinitely farseeing" remarks stimulated Schrödinger to look for a wave equation that would describe the undulatory field of matter. Born, after reading Einstein's paper, became convinced that "a wave theory of matter can be of great importance" (Born to Einstein, June 1924, Born 1971), and he supported the efforts of Walter Elsasser to attribute the Ramsauer effect to the diffraction of electrons.[35] It is not surprising, therefore, that when Schrödinger's theory appeared, Born immediately recognized its importance and applied it to the problem of collision phenomena, associating a wave with an electron in the spirit of de Broglie and Einstein.

Nor is it superfluous to add here that Jordan, Born's close collaborator, also conceived of Born's initial collision treatment as being a direct continuation of the idea of matter waves in de Broglie's, Schrödinger's, and Elsasser's works (Jordan 1927a). What these matter waves exactly meant was as unclear to Born as to Einstein, Schrödinger, and other participants. Born's elaboration of the corpuscular interpretation, and his later preference for it, as well as Schrödinger's gradual polarization to an overall wave ontology, crystallized through the subsequent dialogue among all the physicists involved.

It is not clear whether Born was aware at the time he wrote his first collision paper that Schrödinger and Pauli had each proved the equivalence of matrix and wave mechanics (Schrödinger 1926f; Pauli to Jordan, 12 April 1926, *PC*). But even if he had been (which is likely), Born could still have considered Schrödinger's version superior. For Schrödinger proved the equivalence between the two theories for bound systems only. Schrödinger also believed that extensions of the theory would be more amenable to his version, proving its superiority. Indeed, he suggested the collision problem as a possible case in point (Schrödinger 1926f). Born's statement in his paper clearly echoes Schrödinger's hopes: "Of the different forms of the theory only Schrödinger's has proved suitable for this process [collisions] . . . and exactly for this reason I might regard it as the deepest formulation of the quantum laws" (Born 1926c, 52).

Born's initial enthusiasm for Schrödinger's ideas is confirmed by the correspondence between the two at the time. Born admitted in a letter to Schrödinger that he was so enraptured by Schrödinger's work that, with "flying banners," he was drawn back again to the clear conceptual structures of classical physics (Born to Schrödinger, 6 November 1926, AHQP). Born was so impressed that he originally considered Schrödinger's theory to have more physical meaning than matrix mechanics (Born to Schrödinger, 16 May 1927, AHQP). This enthusiasm was strongly expressed, according to Born, in his own paper.[36]

35. The Ramsauer effect is a phenomenon in which the scattering of electrons during collisions with atoms of certain gases deviates strikingly from the classical theory (slow-moving electrons are deflected from their paths much less often than faster moving ones, and the angle of deflection is very small). Elsasser (1925) explained this experimental effect as a diffraction phenomenon of de Broglie waves.

36. Born wrote to Schrödinger: "You know that immediately after the appearance of your first works I expressed very strongly my enthusiasm for your conceptions in my

As was first pointed out by Linda Wessels (1981), Born did not initially connect the wave function with the probability of position; the ψ-function controlled the energetic transitions of an atom and the energy and direction of motion of colliding electrons. Doubting the possibility of ascribing position to intraatomic electrons, Born questioned the reality of individual electrons within the atom as well: "Matter can always be visualized as consisting of point masses (electrons, protons), but in many cases the particles are not to be identified as individuals, e.g. when these form an atomic system" (1927b, 9). Even though the presence of particles seemed to be implied by the fact that a "disturbance is propagated along a path away from the atom, and with finite velocity, just as if a particle were being thrown out," their existence should not "be taken too literally" (Born 1927b, 10). (Later, of course, Born would change his attitude: "$|\psi|^2$ denotes [the] probability that the electron will be found in the volume-element dv; this holds in spite of the fact that the experiment, if carried out, would destroy the connection with the atom altogether"; Born 1969, 147.)

Only in his second collision paper (Born 1926b) do we find the beginning of Born's opposition to Schrödinger. This opposition was not yet militant, though it later became so. Although Born found Schrödinger's idea of wave packets unsatisfactory and proposed his own interpretation of the ψ-function as guiding free particles in a "ghostlike spirit" in analogy with Einstein's idea, he nevertheless expressed satisfaction at the retention of the usual concepts of space-time in Schrödinger's theory. Born did not yet foresee the applicability of his conception of the ψ-function to the possible motions of particles in the interior of the atom.[37] In his "adiabatic" paper (Born 1927a), Born still did not regard the $3N$-dimensionality of Schrödinger's waves for N particles—later to be one of the major arguments against Schrödinger's interpretation— as a reason to object to a wave ontology.

Only with the development of transformation theory (which reinforced and axiomatized Born's statistical approach) did Born reach his final stand: "Schrödinger's achievements reduces [sic] to something purely mathematical" (Born to Einstein 1926, quoted in Pais 1982,

---

treatise. Heisenberg from the beginning did not share my opinion that your wave mechanics is more physically meaningful than our quantum mechanics; yet the treatment of the simple phenomena of aperiodic processes (collisions) led me initially to believe in the superiority of your point of view. In the meantime, I found myself again in agreement with Heisenberg's position" (Born to Schrödinger, 6 November 1926, AHQP).

37. "The motion of the particles follows laws of probability, but the probability itself propagates in harmony with the causal law. If three stages of the development of the quantum theory are reviewed, it can be seen that the earliest, that of the periodic processes, is quite unsuitable for testing such an idea. . . . Nothing fundamental in favor of our statement can be obtained as long as we consider periodic processes" (Born 1926b, 208).

1198).[38] Together with Heisenberg and Jordan, and in contrast to his early enthusiasm, Born came to see Schrödinger's wave mechanics as no more than a mathematical appendix to matrix mechanics. To Schrödinger's total dismay, and despite his desperate pleas for Born to remain open to different interpretive options (Schrödinger to Born, 2 November 1926, AHQP), Born allied himself with Schrödinger's adamant opponents. Born's opposition to Schrödinger's ideas became so complete that he indoctrinated younger physicists against them in a "military," albeit "playful" manner.[39] Bohr, however, did not agree with the Göttingen attitude, which depreciated entirely the physical meaning of Schrödinger's theory, and proposed his complementarity as a compromise (see chapter 6).

Born's Probabilistic Interpretation and Quantum Jumps

According to Born's recollections, every collision experiment conducted by his colleague Franck in the adjacent building at Göttingen was a clear demonstration of the corpuscular nature of electrons, thus prompting Born to suggest an interpretation in direct opposition to Schrödinger's ideas (Born 1961; interview with Born, AHQP). I have already argued that this story does not withstand historical scrutiny. Franck's name is indeed essential in another respect: his famous work (with Gustav Hertz) on atom-electron collisions was generally regarded as direct experimental evidence for Bohr's ideas about stationary states and the discontinuous energetic transitions between them—quantum jumps. In fact, the theoretical quantum mechanical description of these jumps was Born's central aim, as is clear from his first and second collision papers. The abstract of the first collision paper clearly states its main purpose: the theoretical description of quantum jumps with the help of Schrödinger's formalism. Born summarized the essence of his collision papers as a "precise interpretation of just these observations which may be regarded as the most immediate proof of the quantized

38. Born was impressed by Pauli's argument that regular position space is not essential, and that by a suitable canonical transformation the description in position space can be converted into a description in momentum space (see Pauli to Heisenberg, 19 October 1926, *PC*; Born to Schrödinger, 6 November 1926, AHQP).

39. According to Jordan: "Einmal in Göttingen bei einem Spaziergang von Born mit Dirac und Oppenheimer und vielleicht mit noch einem oder zwei anderen . . . hat Born also einen Brief von Schrödinger vorgelesen und Oppenheimer erzählte mir scherzhaft, Born hatte so wie ein Feldherr seine Truppen zusammenruft, so hätte er den anderen erklärt, was für falsche Ideen das waren die Schrödinger da hätte" (Once on a walk in Göttingen with Dirac and Oppenheimer and perhaps with one or two persons . . . Born read a letter from Schrödinger aloud, and Oppenheimer told me jokingly that Born, as if he were the commander in chief summoning his army, explained to everybody how false were Schrödinger's ideas at the time) (interview with Jordan, AHQP).

structure of energy, namely the critical potentials that were first observed by Franck and Hertz" (1927b, 11). Born (1927a) further substantiated Bohr's concepts of stationary states and discrete transitions between them by discussing the adiabatic principle in quantum mechanical terms.

In opposition to Schrödinger, who wanted to do away with Bohr's "monstrous" concepts by suggesting an alternative, continuous wave ontology, Born fully defended Bohr's ideas. Born's (1927a, 1927b) central inquiry was: to what extent is it possible to harmonize the concept of quantum jumps, so fruitful from an experimental point of view, with Schrödinger's wave mechanics, so mathematically powerful? As the Born-Schrödinger and Jordan-Schrödinger correspondences reveal, it is around quantum jumps, and not around waves versus particles, that heated controversy evolved. Schrödinger wrote in a letter to Born that he did not rule out the possibility that Born's ideas were correct. Yet Schrödinger felt strongly that Born and his colleagues were too addicted to the old concepts of stationary states and quantum jumps. He could not comprehend why an interpretation that does not dogmatically postulate the discontinuities, as the Göttingen-Copenhagen does, is a priori impossible. His own continuous treatment of resonance between two atoms (Schrödinger 1927b) led to the same consequences as the theory that postulated discontinuous exchanges of energy (Heisenberg 1927a). Does not this indicate, inquired Schrödinger, that one should explore whether all other discontinuities can be similarly deduced rather than dogmatically postulated (Schrödinger to Born, 2 November 1926, AHQP)?

In his reply to Schrödinger, Born conceded that after the appearance of Schrödinger's theory, he was at first disposed to return to the clear conceptual framework of classical physics. But he changed his mind. Of course, Born agreed with Schrödinger, it is desirable to explore all possibilities, yet he himself did not intend to do so. Instead, Born had decided to rely on his own feeling that one cannot dispense with quantum jumps (Born to Schrödinger, 6 November 1926, AHPQ).

In his second collision paper (Born 1926b), Born spelled out the connection between a system's stationary states and Schrödinger's wave function. For a periodic system (an unperturbed atom, say) represented by $\psi(x) = \Sigma C_n \psi_n(q)$ (discrete spectra), Born proposed that $|C_n|^2$ is the "frequency" of the state n in a group of identical noncoupled atoms. In other words, $|C_n|^2$ denotes the probability that the system is in a stationary state described by $\psi_n(q)$.

Born's interpretation of $|C_n|^2$ as giving the probability of the stationary states of the system was a pregnant idea. It shifted the probability considerations one level deeper, from transitions to stationary states, thus making the transition probabilities calculable from the more elementary probabilities of these states. This point would be crucial in the Copenhagen deliberations on indeterminism (see chapter 3).

Born and his contemporaries initially viewed the probabilistic interpretation of stationary states as the crux of his interpretive achievement.[40] Precisely because this hypothesis for atomic states and transitions was farther reaching and more speculative than Born's relatively straightforward, even obvious, suggestion for free particles, he and his contemporaries saw his interpretation of $|C_n|^2$ as a possibly fruitful conjecture, still in need of experimental verification, rather than an "obviously" correct interpretation.[41]

It was Pauli who generalized Born's suggestion concerning the probability of stationary states and endowed this concept with particle-kinematic meaning. Born deemed it meaningful only to talk about energies and angular momenta of stationary states. Pauli, however, defined $|\psi(q_1, \ldots, q_f)|^2 dq_1 \cdots dq_f$ as the probability that, for a definite stationary state, the coordinates q_k of the particles ($k = 1, \ldots, f$) lie between q_k and $q_k + dq_k$ (Pauli to Heisenberg, 19 October 1926, PC). This was actually the first time the wave function had been described as giving the probability of the positions of intraatomic particles. Pauli's definition later appeared in print, in a footnote to a paper on gas degeneracy and paramagnetism (Pauli 1927, 83).

40. Jordan, discussing recent developments in physics, characterized Born's contribution as allowing the determination of the probability that an atom is in a definite stationary state at a definite moment in time:

> In Verfolgung seiner Vorstellungen hat Born die physikalische Bedeutung der Schrödingerschen Wellenfuktion schärfer bestimmen können. Er hat gezeigt, dass man mit ihrer Hilfe ein Maass der Wahrscheinlichkeit dafür angeben kann, dass ein Atom zu einem gewissen Zeitpunkt sich gerade im n-ten Quantenzustande befindet. Diese Deutung der Schrödingerfunktion konnte gestützt werden durch den Nachweis, dass das bekannte Ehrenfestsche Adiabatenprinzip der Quantentheorie im Anschluss an diese Deutung auch vom Standpunkte der neuen Quantenmechanik formuliert und bewiesen werden konnte. (1927a, 646)

> (In further developing his conceptions, Born succeeded in defining more precisely the physical meaning of Schrödinger's wave function. He demonstrated that with its [wave function's] aid, one could determine the value of the probability for an atom to be in the *n*th quantum [stationary] state at a certain moment in time. This interpretation of Schrödinger's wave function was corroborated by demonstrating that Ehrenfest's familiar Adiabatic Principle of quantum theory could also be formulated and proved from the standpoint of the new quantum mechanics in conformity with this [Born's] interpretation.)

41. Born emphasized this point at the time: "It is, of course, still an open question whether these conceptions can in all cases be preserved. . . . Unfortunately, the present state of quantum mechanics only allows a qualitative description of these phenomena [collisions of electrons with atoms—in particular, collisions of electrons in helium]" (1927b, 11–12).

Pauli, who had stood somewhat aloof from the development of matrix mechanics, was not as hostile to Schrödinger as Heisenberg was, nor did Pauli subscribe to the opinion that the positions of intraatomic electrons are unobservable in principle.[42] Thus Pauli favored resuscitating the visualization of the stationary states through the kinematic data of positions of intraatomic particles. Pauli's contribution, which stimulated Heisenberg's own revival of regular space-time through the uncertainty principle (see chapter 4), was indeed a landmark on the way to the establishment of a particle ontology of quantum physics. No such particle ontology had underlain Born's original interpretation.

Born's own calculations for the collisions between electrons and hydrogen atom were carried out only to a first approximation. Therefore, Born concluded, "it would be decisive for the theory if it should prove possible to carry the approximation further" (1927b, 13). In his textbook *Atomic Physics*, written in 1935, Born emphasized the importance of verifying his original hypothesis by subjecting scattering and other phenomena to precise quantum mechanical calculations (Born 1969). Born mentioned in this respect his and Vladimir Fock's work on the excitation of atomic systems, Dirac's work on the excitation of coupled systems, Born's and Mott and Massey's work on collisions, and Wentzel's verification of the Rutherford formula. Such experimental confirmation was, according to Born, more important than philosophical elucidation.

This process of critically evaluating and modifying Born's original ideas is still going on. Eugen Merzbacher (1983) has described advances in ion-atom collisions as, in many cases, substantiating Born's work. Abraham Pais (1982) has described his own success in carrying out a calculation for scattering by a static, spherically symmetrical potential, and his failure to extend Born's method to relativistic field theories. "To this day, proof or disproof of the convergence of the Born expansion in field theory remains an important challenge yet to be met" (Pais 1982, 1198). Recent work on scattering deals with both mathematical improvements and experimental physical difficulties (Daumer 1996; Cushing 1990).

The meaning and the foundational status of the Born-Pauli probabilistic interpretation remain today a subject of lively controversy (Dürr et al. 1992a, 1992b, 1996; Valentini 1996). What appears in textbooks and philosophical writings to be a closed box has never been sealed off at the frontier of research.

---

42. According to Leon Rosenfeld (1971), Pauli did not welcome a complete break between the formalized conceptual scheme of matrix mechanics and the classical notions of space-time and Keplerian orbits that underlay it.

CHAPTER 3

〜

Quantum Philosophy in Flux

The external conditions, which are set . . . by the facts of experience, do not permit him [the scientist] to let himself be too much restricted in the construction of his conceptual world by the adherence to an epistemological system. He therefore must appear to the systematic epistemologist as a type of unscrupulous opportunist.

Albert Einstein 1949b, 684

Introduction

In the previous chapter I argued against the widely accepted misconception that the Göttingen-Copenhagen physicists developed matrix mechanics by implementing a committed indeterministic corpuscular ontology as opposed to Schrödinger's causal wave ontology. I also argued that the two approaches were not conceptually distinct and historically independent. I described the development of the Göttingen-Copenhagen version of quantum mechanics as characterized by vacillation on foundational questions and openness to the wave theoretical perspective. Perhaps the strongest expression of this openness, if not preference, is a statement Max Born made during his lectures in the United States in 1926: Heisenberg's matrix elements are not amplitudes of radiation but "real waves of an atom" (1926a, 70). The polarization and conflict between the wave theoretical and matrix approaches was indeed the result of the threat Schrödinger posed to matrix theorists; acknowledgments of mutual debt were largely eliminated from the public discourse. Yet the development of the Göttingen-Copenhagen version depended, in part, on the de Broglie–Schrödinger ideas—as a stimulus to response and as a resource for selective appropriation. De Broglie's work reinforced Bohr's realization that one had to dispense with particle space-time models in the quantum domain (Bohr 1925); this realization was implemented in Heisenberg's reinterpretation paper (1925). Wave theoretical imagery informed Jordan's early treatment of electrodynamics in matrix mechanics and guided his pioneering work on the theory of quantized matter waves (Darrigol 1986; Kojevnikov 1987).

Another widely held misconception is that a positivist philosophy of "elimination of unobservables" was central to the emergence of the theoretical structure of the new quantum theory. It was this positivist philosophy, so the story goes, that guided Heisenberg's efforts from his reinterpretation paper to his uncertainty paper (Hendry 1984).

I argue that positivist philosophy was less a heuristic principle and more a tool with which theoretical advances could be justified ex post facto. Contrary to the received opinion that Heisenberg's philosophical stand remained stable from the reinterpretation (1925) to the uncertainty paper (1927b), a careful analysis reveals a radical change, if not an about-face. Heisenberg employed two different strategies of justification: the Machean positivist principle of elimination of unobservables in his reinterpretation paper and the operational approach of defining physical concepts through the procedure of their measurement in his uncertainty paper. If in 1925 Heisenberg claimed that directly observable experimental data (frequencies and intensities of emitted radiation) should determine theoretical structure, in 1927 he declared that it is theory that gives meaning to experiment: "It is the theory which decides what we can observe" (1971, 77). No coherent philosophical choice between positivism and realism guided Heisenberg's efforts. A fascinating, ever changing mixture of realist intuition and positivist legitimation characterizes Heisenberg's work leading to, and springing from, the reinterpretation paper.

Positivism in Flux

The myth of the fundamental heuristic role of the elimination of unobservables originated with the matrix physicists themselves. Born declared in his presentation of the new theory in winter 1925–26 (as well as in later writings) that Heisenberg ended the crisis in physics by introducing a fundamental epistemological principle that could be traced back to Ernst Mach. This positivist principle states that concepts and representations that do not directly correspond to observable "facts" are not to be used in a theoretical description of reality (Born 1926a, 68–69).[1] Heisenberg, supposedly, was encouraged by Einstein's example—in 1905 Einstein ended another major crisis in physics using the principle of elimination of unobservables. Heisenberg banished the picture of electron orbits and the concept of "electron position within

1. Instead of such unobservable kinematic variables as the position, velocity, or period of revolution of an electron in an atom, Heisenberg incorporated experimentally observable spectroscopic data (frequencies and intensities of radiation) into the theoretical framework.

the atom," Born claimed, for the same epistemological reasons that Einstein had eliminated the concepts of absolute velocity and absolute simultaneity.

Historical research leads to a different picture (MacKinnon 1977; Mehra and Rechenberg 1982; Beller 1983, 1988; Hendry 1984). During the years preceding Heisenberg's reinterpretation paper, quantum physicists increasingly encountered the inadequacy of the notion of intraatomic orbit. When physicists questioned the adequacy of orbital notions, their doubts had more to do with the theoretical failure of orbits than with their experimental unobservability. Orbital assumptions failed in the domain of the interaction of light with matter; they could not be reconciled with the fact that the dispersion of light occurs with spectroscopic rather than orbital mechanical frequencies. Moreover, in the domain of the constitution of atoms and molecules (with the full price of ad hoc assumptions), all modifications of the orbital mechanical models failed to do justice to the experimental state of affairs (Hendry 1984).

Nor did it help much to attach to these models a formal symbolic, rather than realistic, significance. The orbital model failed for the anomalous Zeeman effect, and for the spectrum of the helium atom. As Born attempted to argue in 1924, all possible orbital assumptions led to equally wrong results (Hendry 1984).[2] Gradually it became clear that there was really no good reason to cling to the failed orbital concepts and the mechanical models built on them. The need now arose, as Pauli understood, for a new, more adequate kinematic law of motion and perhaps too for a new dynamic law responsible for the changes in kinematics. It seemed reasonable at this point to introduce the so-called virtual oscillator model, first applied to the interaction of light and matter, and soon extended to theories of atomic structure. In the domain of the interaction of matter with radiation, this symbolic model proved to be a powerful tool for translating and extending successful classical formulas into the quantum domain via the correspondence principle (Darrigol 1992a). Reaching the conclusion that "reality was not in orbits, but rather in Fourier components, or rather their quantum mechanical analogues" (interview with Heisenberg, AHQP), and encouraged by his previous results (Kramers and Heisenberg 1925), Heisenberg decided in the middle of May 1925 to attack the problem of hydrogen line intensities. Heisenberg's correspondence with Ralph Kronig, whom he kept informed about his progress, clearly reveals that epistemological considerations were far from Heisenberg's mind during his first attempts

2. Today we know that at the price of nonlocality one can retain space-time trajectories within an atom (for example, Bohm's theory).

to tackle the problem.[3] The hydrogen problem was too complicated mathematically, and Heisenberg had to content himself with the simple case of a harmonic oscillator. Even in this case, mathematical difficulties remained, and Heisenberg was not yet sure what physical consequences his new scheme implied: "The physical interpretation of the above mentioned scheme yields very strange points of view" (Heisenberg to Kronig, 5 June 1925, AHQP). Heisenberg confessed to Pauli that his private philosophy was simply "a mixture of all possible moral and aesthetic calculations and rules" through which he did not "find his way anymore" (Heisenberg to Pauli, 24 November 1925, PC). In a letter to Pauli, Heisenberg argued that the interpretation of the experimental formula for the hydrogen spectrum on the basis of kinematic orbital concepts was impossible: "An interpretation of the Rydberg formula in terms of circular and elliptical orbits does not have the slightest physical significance." He added that his efforts were devoted to "killing the concept of an orbit which cannot be observed *anyway*" (Heisenberg to Pauli, 9 June 1925, PC). The order of precedence seems clear: orbits are theoretically inadequate, and they had therefore better be eliminated. One can dispense with orbits without regret because they do not have any observational significance.

A comment on the possible influence of Pauli on Heisenberg is in place here. Most historians who treat this subject (Hendry 1984; Serwer 1977) assume that Pauli's operational attitude called for the elimination of the orbits and positions of intraatomic electrons because they are, as Hendry puts it, "operationally meaningless." Heisenberg's reinterpretation paper is often perceived as an implementation of Pauli's epistemological program. Yet this point of view needs to be qualified.

It is true that, at least after 1919, Pauli advocated in certain theoretical contexts an operational attitude toward physics. For example, he claimed that the strength of electric fields in the interior of an electron is a meaningless concept because the field strength is defined as a force acting on a test body, and there are no smaller test bodies than the electron itself. Similarly, the concept of continuous space-time in the interior of the electron is meaningless. Pauli argued that "one should

---

3. In a letter to Kronig, Heisenberg's program is clearly stated: In classical theory, Heisenberg wrote, knowledge of the Fourier series is sufficient to calculate the radiation completely, not only dipole moments but higher moments as well. In principle, one should be able, by replacing classical variables with their quantum analogues in the equation of motion, to obtain an exact, complete expression for the intensities. Because Heisenberg wanted to take into consideration quadruple and higher moments, he had to immerse himself in the technical mathematical problem of what multiplying X and Y meant, when X and Y were algebraic sets of quantum substitutes for classical position coordinates (Heisenberg to Kronig, 5 June 1925, AHQP).

adhere to introducing only those quantities in physics which are observable in principle" (quoted in Hendry 1984, 19).

However, the key phrase here is "in principle." The question was not whether the force acting on the electron could be experimentally measured but whether the definition of the concept of field strength was consistent with the hypothetical possibility of measurement. Because the electron was assumed to be the smallest test body, it was meaningless (according to the operational approach) to talk about space-time in the electron's interior, but it was not meaningless to talk about the position of this smallest test body in the larger atom. There was nothing inconsistent in assuming that in principle electron position could be measured in the interior of an atom, though the state of experimental techniques did not allow it in practice. One could, for example, devise a γ-ray thought experiment—such as Heisenberg proposed later in his uncertainty paper. And if electron positions are not unobservable in principle, the same holds for electron orbits. Pauli's stand, pushed to its logical conclusion, implied merely that intraatomic positions and trajectories could be determined only with the accuracy of the size of the electron—the smallest particle known at the time. And, in fact, Pauli did emphasize in a letter to Eddington of 20 September 1923 (PC) that the electron's position need not be considered unobservable in principle, and that the question of technical difficulties should not enter into considerations of principle connected with the definition of physical concepts.[1]

As already mentioned, Heisenberg was led to his reinterpretation procedure by trying to solve the problem of hydrogen intensities. His attempt did not succeed. Heisenberg was forced, by technical difficulties, to stop at the programmatic point. Had he solved this problem, Heisenberg's motto "success sanctifies the means" would suffice to justify the procedure of replacing the classical coordinates with a set of quantum theoretical magnitudes. Yet at this programmatic point, Heisenberg needed a more general conceptual justification, and he chose the principle of elimination of unobservables.[5] This principle, sup-

---

4. Not by accident did Pauli fail to mention the principle of elimination of unobservables when presenting the essentials of Heisenberg's approach in his paper on the hydrogen atom (Pauli 1926b). Significantly, it was Pauli himself who later restored the kinematic positions of intraatomic electrons in his extension of Born's probabilistic interpretation.

5. Why did Heisenberg not use Bohr's correspondence principle instead, a principle that had played a most fundamental role in his reinterpretation procedure? The most probable reason is the collapse of the Bohr-Kramers-Slater theory (1924) and its associated research program. In this theory, the correspondence principle, the wave nature of radiation, and nonconservation were tightly connected—and the refutation of the Bohr-Kramers-Slater theory by the Bothe-Geiger experiments seemed to indicate that the cor-

ported by Einstein's authority, was a clever strategic choice. Precisely because it was only invoked a posteriori it played no essential role in the derivation of the formulas in the reinterpretation paper (1925). The reinterpreted formulas held true whether one held the position of electrons to be observable or not. The matrix formalism was not undermined after Heisenberg changed his opinion about the observability of electrons in his uncertainty paper (1927b). The above analysis demonstrates that the elimination of unobservables was, in fact, not a guiding principle, but rather a general justification of a powerful technical method that de facto eliminated classical positions and orbits. The elimination of the space-time container and the loss of visualization were prices to be paid, not goals to be attained. Not surprisingly, the authors of matrix mechanics sometimes would talk not in terms of "observable in principle" but rather in terms of what "can be observed experimentally" or what is "really observable" (Born 1926a, 69).[6]

One of the weapons Göttingen-Copenhagen physicists wielded against Schrödinger's competitive approach was an accusation of anachronistic, "naive" realism. Yet Heisenberg and Schrödinger, despite a difference in intellectual style, were not far apart on the issue of realism. Both employed visualizable intuitions; both willingly sought new "artifices of thought" when old ones failed; both engaged in heated controversies about the "reality" of certain scientific notions (quantum discontinuities, ψ-functions). In fact, heated controversies about unobservable "reality" played a decisive part in the genesis of interpretations of quantum physics. Thus the reality of quantum jumps was the central issue in the interpretive efforts undertaken during the crucial years 1925–27. Indeed, this issue channeled the efforts of the "positivist" group made up of Born, Heisenberg, and Jordan no less than it did those of the "realistically" inclined Schrödinger (see chapters 4 and 6).

When, in his reinterpretation paper, Heisenberg replaced the familiar classical parameters of motion with abstract algebraic constructs, the basic physical intuition behind his innovation was that there must be a correspondence between the internal dynamical "mechanism" of an

---

respondence principle could no longer lead to progress. Indeed, Einstein objected in a letter to Heisenberg that his approach in the reinterpretation paper was too close to the discredited Bohr-Kramers-Slater theory. Serious doubts about the fundamental role of the correspondence principle had been expressed even before the Bothe-Geiger experiments. Sommerfeld, Heisenberg's teacher, declared in 1924 that the correspondence principle was not part of the foundation of quantum theory—it was merely its limiting theorem (Mehra and Rechenberg 1982). Pauli similarly denied the ability of the correspondence principle to be a faithful guide for understanding the structure of atoms (Mehra and Rechenberg 1982).

6. A more detailed analysis of the role of operationalism in the erection of quantum theory is given in Beller (1988).

atom and the radiation it emits. Heisenberg's "realistic" presupposition "that something in the atom must vibrate with the right frequency" (Heisenberg to van der Waerden, 8 October 1963, van der Waerden 1967) prompted him to construct a bizarre noncontinuous space (through a reinterpretation of kinematics) so that electron "motions" in this space would have the correct spectral frequencies. The advantage of gaining a physical sense of the underlying atomic model came at the price of abandoning visualization in regular space-time and was justified by the positivist ideology of elimination of unobservables.[7]

Schrödinger shared Heisenberg's conviction that a connection must exist between electron motion and radiation frequencies, yet he did not have Heisenberg's inclination to operate in nonclassical spaces torn, as Schrödinger called it, by "yawns" and "gaps." Schrödinger announced that "what we cannot comprehend within it [regular space-time] we cannot understand at all" (1926b, 27). He offered his own wave mechanism, which preserved the direct relationship between electron motions and radiation frequencies. The contrast between Heisenberg and Schrödinger is not adequately described as a realist-positivist dichotomy. The difference between them lies elsewhere.

Shifting theoretical and social circumstances forced Heisenberg to abandon the positivist position that he advocated in his reinterpretation paper. In his uncertainty paper, Heisenberg emphasized not the close proximity of basic theoretical terms to direct sense perceptions (what can be closer to Machean sense perceptions than the frequency and intensity of light, or, in other words, light's color and brightness?) but the supremacy of theory over facts (a good theory is not refutable by facts; a good theory is needed to define what the facts are).

Heisenberg's interpretive attempts can hardly be viewed as a smooth evolutionary process, guided by consistent epistemological concerns. What we have witnessed, rather, are changes of opinion on basic issues, trial and error, about-faces. But though extraordinarily agile, Heisenberg was not the only physicist who skillfully adapted himself to changing theoretical opportunities. Pauli, for example, expressed the opinion in 1925 that probability statements should not enter into the basic postulates of a satisfactory physical theory, only to become one of the architects of the probabilistic kinematic interpretation of the wave

7. Heisenberg's physical intuition was obscured by Born and Jordan's chillingly elegant matrix formalism; Heisenberg's original awkward mathematical formulation contained physical heuristics that did not appear in the matrix formulation. This is exactly what Pauli feared when he refused to collaborate with Born, telling him that his "tedious and complicated formalism . . . is only going to spoil Heisenberg's physical ideas" (quoted in van der Waerden 1967, 37).

function two years later. Nor do we find consistency between the epistemological views of quantum physicists and their actual scientific practice. Pauli, who asserted the need to replace orbits in atoms with more satisfactory kinematics two years before Heisenberg's reinterpretation, continued research based on orbital assumptions. Einstein, at least in his younger years, refused to be restrained by an epistemological straitjacket. In his discussion with Pauli in the 1920s, when Pauli argued that the continuum must be abandoned on operational grounds, Einstein replied that the issue must be decided only on the basis of what was theoretically functional (Hendry 1984).

Nor did Niels Bohr consider philosophical analysis heuristically valuable. He did not take seriously considerations of simplicity, elegance, or even consistency (the "epistemic virtues"), holding that "such qualities can only be properly judged after the event" (Rosenfeld 1967, 117). A harsh and crisp verdict came from Paul Dirac: "I feel that philosophy will never lead to important discoveries. It is just a way of talking about discoveries which have already been made" (interview with Dirac, AHQP).

While the above quotes seem to indicate that philosophy is always used only a posteriori, in the "context of justification," as opposed to the "context of discovery," such an assertion would be too categorical. Quite apart from the fact that a meaningful distinction between the context of discovery and the context of justification cannot always be sustained, the sources of scientific creativity can be found in diverse fields of human activity—philosophy is not the most unlikely of them. Philosophical ideas might be suggestive in particular theoretical settings: the idealist German philosopher Fichte might have been a surprising source of Heisenberg's idea of the reduction of a wave packet (see chapter 4). Historians of science have described extensively the rich cultural and political milieu in which the development of quantum mechanics took place (Faye 1991; Forman 1971, 1979; Heilbron 1987; Holton 1970; Krips 1996; Wise 1987). Yet these contexts served more as resources and less as influences. The point is not that philosophy cannot influence science. Creative scientists might adopt a certain foundational stand (sometimes indistinguishable from a traditional philosophical stand) in order to pursue a definite line of research. Such a philosophical orientation is, however, local and provisional. The longevity of philosophical "commitment" is coterminous with its usefulness in solving the problem at hand.[8]

8. A scientist can develop a preferred philosophical position, yet such a position crystallizes only after an entire career, when certain basic characteristics of successful theories that the scientist has developed are internalized and "entheorized." For the notion of "entheorizing," see Fine (1986) and chapter 9.

Indeterminism in Flux

I argued in chapter 2 that one should not ascribe to Born any committed opinions about particle ontology or about the physical significance of Schrödinger's theory; these were issues that Born initially left open. My claim applies similarly to the problem of indeterminism. Heisenberg, I have argued (chapter 2), was open on the question of determinism versus indeterminism. I will argue, as well, that Born did not have a committed opinion about causality, only tentative, uncommitted suggestions.[9] My analysis of Born's stand will focus on two issues: the nature of positivist statements in quantum mechanics and the problem of elementary probabilities.

Born's indeterminism in his first collision paper was rather hesitant: "I myself am inclined to give up determinism in the world of atoms. But that is a philosophical question for which physical arguments alone are not decisive" (1926c, 54). He wavered even more in his second collision paper (1926b): even though Born himself inclines toward indeterminism, he fully realizes that attempts to look for "hidden" internal parameters, which determine individual processes, are consistent with the prevailing state of quantum theory. And even later, when Born did commit himself publicly to indeterminism, he confessed in a letter to Einstein that the issue is undecidable in principle (Born to Einstein, 13 January 1928, Born 1971).

Born's main concern at the time was neither deep conceptual analysis nor philosophical deliberation, but the solution of the collision problem. As Born indicated in an interview with Kuhn, he did not think his solution involved any far-reaching implications for the issue of indeterminism: "I did not consider it very philosophical. I thought giving up the description in space and time and replacing it by symbolic description was much deeper and much more philosophical. And to find a way of expressing it in simple terms—probability—seemed to me not so very important. . . . We were so accustomed to making statistical considerations, and to shift it one layer deeper, seemed to us not so very important" (interview with Born, AHQP).

The description of collision phenomena that Born provided was statistical: one does not get an answer to the question "What is the state after collision?" but rather to "How probable is a specified outcome of the collision?" (1926c, 54). Because the necessity of indeterminism was

9. For a different opinion, see Forman (1979). I use the words "causality" and "determinism" interchangeably here, following the usage of quantum physicists. In a careful philosophical discussion these notions need to be dissociated. For a wide-ranging analysis of the complexity of the notion of determinism, and of the application of this notion in different branches of modern physics, see Earman (1986).

far from obvious, Born had to face a natural question: was his solution only tentative and a more complete, deterministic description possible in the future, or was his contribution fundamental and not susceptible to any substantial modification? In short, Born was looking for arguments to legitimate his contribution, as is often done in theoretical physics.[10] Moreover, Born had a special need for such justification because, in his treatment, the statistical conclusions followed from exact data: even though the momentum and energy of the incoming electron, as well as the energy of the stationary state of the atom, were given, his description of the result of the collision was probabilistic. This situation was clearly unsettling after the Bothe-Geiger experiments, which seemed to indicate that individual atomic phenomena are determinate (that conservation laws apply strictly; Bothe and Geiger 1925a, 1925b).

Born tackled this problem from two different angles. He denied an analogy between his solution and the Bohr-Kramers-Slater theory—discredited by the Bothe-Geiger experiments—despite a very close resemblance between them (Born 1927b, 10 n; Born to Schrödinger, 6 November 1926, AHQP).[11] The conservation laws in collisions are enforced, Born claimed, by the basic formalism of quantum theory (1927b, 10 n). He tried to justify his statistical solution by positivist arguments: because formal quantum mechanics renounced "internal atomic motions," there is "no reason to believe that there are some inner properties of the atom which condition a definite outcome for the collision" (1926c, 54).

Yet Born must have realized at the time that the renunciation of space-time as a container of motion was far from necessary. As I have mentioned, Heisenberg's claims that electron position within an atom is in principle unobservable were faulty: intraatomic position is unobservable not in principle but merely in practice. And because positivist arguments for the nonexistence of internal atomic parameters, or "hidden variables," were inconclusive, similarly inconclusive were assertions about indeterminism based on such arguments. This is the reason

10. Discussing quantum mechanical and semiclassical solutions in collision processes, Merzbacher (1983) remarked: "The virtues or disadvantages of a partially, or even fully, classical approach are frequently contrasted with the quantum mechanical approach, but since convenience often dictates the choice, such comparisons are usually biased by the desire to justify a particular calculation after its completion."

11. The analogy between the Bohr-Kramers-Slater theory and Born's collision theory is in fact very close. The electronic "phantom" ϕ-field probabilistically controls the energetic transitions of atoms in collision, in the same way that the virtual radiation field probabilistically controls atomic transitions in the Bohr-Kramers-Slater theory. The action of the ϕ-field is twofold: according to Einstein's idea of light quanta in the case of free particles and in analogy to the Bohr-Kramers-Slater theory in the case of bound particles (an atom).

Born made, not decisive statements about the necessity of indeterminism, but rather mild arguments concerning the agreement of theoretical indeterminism with an existing practical statistical approach. Actual experiments with collisions deal de facto only with statistics: "Ought we to hope later to discover such properties (like phases of the internal motions) and determine them in individual cases? Or ought we to believe that the agreement of theory and experiment as to the impossibility of prescribing conditions for a causal evolution is a preestablished harmony founded on the nonexistence of such conditions?" (Born 1926c, 54).[12] "Preestablished harmony" is not a forceful argument, but Born had no better—yet. The difference between classical and quantum probabilities was yet to be realized, probabilistic transformation theory was yet to be established, Heisenberg's uncertainty paper was yet to be written.

In a nontechnical paper that summarized Born's position at the time, we find no strong arguments for indeterminism: "The quantum theory is in agreement with the experimentalists, for whom microscopic coordinates are also out of reach, and who therefore only count instances and indulge in statistics. It is not forbidden to believe in the existence of these coordinates, but they will only be of physical significance when methods have been devised for their experimental observation" (1927b, 11). As long as such a possibility was not ruled out in principle, Born should have remained (and did in fact remain) undecided on the issue of indeterminism.

There are additional, theoretically deeper reasons for the indecision among physicists on the issue of indeterminism. It is often supposed in historical accounts that because quantum theory was de facto statistical, the creators of quantum theory had some prior indeterministic beliefs, or at least strong predispositions. The typical story runs as follows: Bohr talked about probabilities of spontaneous emission in the 1920s; therefore, he and the Göttingen-Copenhagen group already believed in indeterminism at that time. Only the conservative Einstein (1917) regarded the introduction of a priori transition probabilities as a weakness of the theory, because he could not accustom himself to a "dice-playing God."

But this account does not hold up against the historical record. I know of no quantum physicist before 1927 who did commit himself to indeterminism. Physicists at the time thought statistically *along classical lines* (the uniqueness of quantum probabilities was recognized only

---

12. That such a "preestablished harmony" existed between quantum theory and the possibility of experiment was in fact argued by Heisenberg (1927b) in his uncertainty paper, though without that particular term being used.

after Born's interpretation). In classical statistical theory one starts with an assumption about equally probable cases (elementary probabilities) and derives from them more complicated probabilities (Born 1927b; Jordan 1927b). In quantum theory, however, complex probabilities (such as transition probabilities) were introduced a priori, without being reduced to elementary probabilities. As long as no consistent theory of probabilities was developed within quantum physics, there could be no verdict about indeterminism. For Born, Jordan, and Pauli, this reasoning was an undercurrent of their struggles with the issue of indeterminism.

Born, despite his openness to indeterminism, was equally open to the possibility, echoing Einstein, that a priori transition probabilities could be viewed as a flaw in quantum theory: "It is notable that here [in quantum mechanics], even historically, the concept of *a priori* probability has played a part that could not be thrown back on equally probable cases, for example, in the transition probabilities for emission. Of course, this might be merely a weakness of the theory" (1927b, 9). Pauli similarly argued against the current form of quantum mechanics, claiming that probability assumptions should not figure in the fundamental propositions of a theory (Pauli to Bohr, 17 November 1925, *PC*). A searching appraisal of the quantum theoretical situation was made by Jordan: "We must reduce the quantum theoretical probabilities to independent probabilities. Only then can we say that we really understand the laws. Only then can we know exactly what is causally determined, and what is left to chance." And even more eloquent: "The circumstance that quantum laws are laws of averages, and can only be applied statistically to specific elementary processes, is not a conclusive proof that the elementary laws themselves can only be put in terms of probability (Jordan 1927b, 569).

Because statistical calculations are not in themselves indications of indeterminism, it is clear why none of the quantum physicists, though acknowledging the possibility of indeterminism, were ready to take any forceful stand on the issue. Born's probabilistic interpretation of the stationary states of an atom was significant precisely because it pointed to the possibility of resolving this crucial issue. Born's interpretation changed the status of transition probabilities from a priori to calculable from the more elementary probabilities of the stationary states.[13] The

13. "Born ist mit diesen Aufstellungen in einem wesentlichen Punkte über den ursprünglichen Heisenbergschen Gedankenkreis hinausgegangen: Während bei Heisenberg als statistische Begriffe nur die Übergangswahrscheinlichkeiten erschienen, hat Born als primären Begriff die Zustandwahrscheinlichkeit eingeführt, aus denen sich die Übergangswahrscheinlichkeiten erst sekundär ergeben (durch Bestimmung der zeitlichen Änderungen der Zustandswahrscheinlichkeiten)" (By using these ideas Born transcended

probabilistic interpretation by Born and its extension by Pauli indicated the possibility of building a consistent quantum probabilistic calculus, and thus the possibility of reaching a final verdict on indeterminism: "It is thus very significant that in Pauli's above-mentioned formulation nothing is said about a probability of a transition—for we saw that this could not lead to independent probabilities. What the theory does specify, is the probability that the system-point be at a given place in the configuration space. One might therefore hope that these considerations would lead us to independent elementary physical probabilities." In any case, "a trustworthy decision will only be possible after a further analysis of quantum mechanics on the lines laid down by Born and Pauli" (Jordan 1927b, 569).

Born's probabilistic treatment had many difficulties. I have already mentioned one of them: the statistical nature of Born's solution in contrast to the strict causality of individual atomic processes (supported by the Bothe-Geiger experiments). Several people, Schrödinger and Heisenberg among them, began to realize the problem of interference of probabilities (Schrödinger to Joos, 17 November 1926, AHQP; Heisenberg to Jordan, 7 March 1927, AHQP). Moreover, Born's electron, if a particle, was a strange creature indeed. Represented by a plane wave, it had a precise momentum, but no definite position at all (Schrödinger to Born, 2 November 1926, AHQP)! It was about these crucial problems that quantum physicists addressed each other in their correspondence. And their wide-ranging dialogues resulted in new theoretical and interpretive developments. Born's probabilistic interpretation was a focal point of these exchanges—between Born and Pauli, Schrödinger and Born, Pauli and Heisenberg—precisely because it was crying out for elucidation, modification, and confirmation, not because it was an obvious, uncontroversial advance.

The development of transformation theory by Jordan (1927f) and Dirac (1927), and especially Heisenberg's uncertainty paper, signified a breakthrough on the issue of indeterminism. Heisenberg, impressed by Jordan's discussion, announced that he had solved definitively *(definitiv festgestellst)* the philosophical problem of indeterminism: Quantum theory is essentially a probabilistic theory, not because quantum laws, as opposed to classical laws, are statistical in Born's sense (this possibility is contradicted by the Bothe-Geiger results), but because one needs

---

Heisenberg's conception on an essential point: Whereas for Heisenberg only transition probabilities appeared as statistical concepts, Born introduced probability of a state as the fundamental concept from which the probabilities of transitions are deducible merely as secondary concepts [through determination of the time variation of the probabilities of states]) (Jordan 1927a, 646).

probabilities to describe fully the state of the system. The future cannot be known in all its detail, not because quantum laws are statistical, but because the present cannot be known in all its detail and therefore must be described probabilistically (Heisenberg 1927b).

Heisenberg's uncertainty paper made a forceful public statement in favor of indeterminism, in contrast to the earlier tentative employment of statistical considerations. In addition, Heisenberg's paper restored particle ontology within the atom, in opposition to Schrödinger's waves. Despite disagreements among Heisenberg, Dirac, Jordan, Bohr, and Born on the ultimate source of acausality, acausality itself, from this point forward, became a pillar of the quantum orthodoxy: an umbrella under which different persuasions were united, a sword with which opponents were chastised as reactionaries. In historical accounts written by the orthodox, indeterminism became a "natural interpretation" in Feyerabend's (1975) sense: indeterminism was present in quantum theory from the very beginning; one simply had to "realize" fully what one had actually known all along.[14]

14. See, for example, the discussion of classical indeterminism in Born (1955a).

~

CHAPTER 4

The Dialogical Emergence of Heisenberg's Uncertainty Paper

Let's get physical.

Singer Olivia Newton-John, Max Born's granddaughter

Introduction

Less than two years passed between Heisenberg's reinterpretation paper (1925) and his formulation of the uncertainty principle (Heisenberg 1927b).[1] During this time, Heisenberg often contemplated interpretive issues related to the new quantum theory. How is one to interpret the new formalism, when space-time imagery no longer seems to fit? What can the notion of "interpretation" in such circumstances mean? Should one give up space and time notions, as the original formulation of the matrix philosophy implied, or can one determine, in a systematic way, the limits of application of the old, classical concepts? What is the connection between quantum theory and classical mechanics, when Bohr's correspondence principle is no longer necessary as an independent postulate? Are the discontinuous and acausal features of the new theory derivative, or fundamental? Does the apparent incompatibility between the new formalism and experiment (precise times of energetic transitions and continuous trajectories of an electron in a Wilson cloud chamber that the new formalism is unable to describe) imply that quantum theory is in need of further elaboration and modification? Or should one accept the formalism as complete and final and undertake a suitable reinterpretation of experiments and experimental terms?

Heisenberg's elaboration of these issues took place in a context of intense dialogue. In this chapter I describe how Heisenberg's opinion on these central issues emerged gradually in his dialogues with Schrödinger, Pauli, Dirac, Jordan, Campbell, and Sentfleben. I analyze Heisenberg's dialogue with Bohr in chapter 6. The emergence of Heisenberg's

---

1. Heisenberg (1927b) is quoted in translation from Wheeler and Zurek (1983). Page references are to Wheeler and Zurek. I have retranslated the opening sentence of the paper (see pages 69 and 109 below) to render it more faithful to the original German text.

interpretation, with the uncertainty principle as its main focus, was intimately connected with his growing confidence in the finality of the new quantum theory. This confidence crystallized as he communicated with Dirac, Pauli, and Jordan and was intensified by his emotional reaction to Schrödinger's competitive interpretive attempts. It would underpin his resolution of other issues, notably those of acausality and discontinuity.

Heisenberg's treatment of the space-time problem in the quantum domain similarly took place in the midst of many overlapping dialogues. Schrödinger's version of quantum theory provided both a strong incentive and a rich conceptual resource for the restoration of space-time imagery in the quantum domain. Pauli urged Heisenberg to find a particle-kinematic interpretation, drawing Heisenberg's attention to the problem of the joint inapplicability of conjugate variables. Dirac singled out the basic commutation relation $pq - qp = (h/2\pi i)\mathbf{1}$ as the focus of a space-time analysis, despite the earlier stand taken by matrix physicists that this formula is unamenable to visualizable interpretation. Jordan, Campbell, and Sentfleben, each in his own way, analyzed the connection between the issues of space-time and causality in quantum theory. Their treatment triggered Heisenberg's creative dialogical response, culminating in the formulation of the uncertainty principle. Heisenberg's decision to focus his analysis on the operational meaning of experiments emerged only in the last stages of his interpretive efforts, through dialogues with Jordan, Campbell, and Dirac. In this context Einstein's dictum—"It is the theory which decides what we can observe"—became crucial.

According to the usual accounts, Heisenberg's challenge in 1927 was to demonstrate compatibility between the limited degree of precision with which conjugate variables could be simultaneously determined and the limited accuracy with which such variables could be measured in experimental arrangements (Jammer 1974, 61). Yet the idea of such a one-dimensional comparison of theory and experiment, no matter how adequate to describe the underlying logic of an argument, can hardly do justice to the contrived process of Heisenberg's creative efforts. Nor does it suffice to elucidate the text of the uncertainty paper (see chapter 5).

The analysis in this chapter demonstrates the indispensability of "lesser" scientists in the dialogical process. Two crucial components of Heisenberg's uncertainty paper were provided by "marginal" physicists—H. A. Sentfleben and Norman Campbell. Sentfleben provided the formulation for Heisenberg's "definitive" resolution of the causality issue: causality does not hold because the present cannot be known precisely and therefore the future cannot be predicted exactly. Campbell, before Heisenberg, suggested that the path of an intraatomic electron should be regarded as an irregular sequence of discrete points. Jammer has identified the two claims as the most startling and novel in Heisenberg's uncertainty paper (1974, 85).

Heisenberg's least expected interlocutor was the German philosopher Johann Fichte. One of the most pregnant ideas in the uncertainty paper was Heisenberg's reduction of wave packets during measurement. Such measurement selects a definite value for a measured observable "from the totality of possibilities and limits the options for all subsequent measurements" (Heisenberg 1927b, 74). With this idea, Heisenberg inaugurated the notorious measurement problem of quantum mechanics, which plagues physicists and philosophers of quantum physics to this day. The source of this idea, as Heisenberg pointed out just a few years later, in 1932, during his lecture at the Academy of Science in Saxony, was Fichte's philosophy of self-limitation of the ego: "The observation of nature by man shows here a close analogy to the individual act of perception which one can, like Fichte, accept as a process of the *Selbst-Beschränkung des Ich* (self-limitation of the ego)." Heisenberg explicated Fichte's idea in the following way: "It means that in every act of perception we select one of the infinite number of possibilities and thus we also limit the number of possibilities for the future" (1952, 28). These words are almost identical with the concluding lines of the uncertainty paper: "Everything observed is a selection from a plenitude of possibilities and a limitation on what is possible in the future" (Heisenberg 1927b, 74).

Dialogue with Schrödinger

Heisenberg's need to counter the threat posed by Erwin Schrödinger to the Göttingen-Copenhagen achievement, and the theoretical possibilities opened by Schrödinger's work, form the background needed to understand the emergence of Heisenberg's uncertainty paper. In reaction to Schrödinger's challenge, Heisenberg resurrected *Anschaulichkeit* and the regular space-time container, eliminated by the original philosophy of matrix mechanics. Heisenberg thus restored the particle-kinematic interpretation of quantum theory. In an emotional confrontation with Schrödinger, Heisenberg moved gradually, if not to a commitment, at least to a forceful emphasis on the discontinuity and acausality of quantum description. It was the threat implicit in Schrödinger's work that pushed Heisenberg toward the assertion that the new quantum mechanical formalism was final. While publicly denying any physical significance to Schrödinger's wave mechanics, Heisenberg, in his efforts to create an alternative to Schrödinger, relied substantially on the "enemy's" theoretical resources and physical imagery. We often hear Heisenberg speak in conflicting voices on the same issue: while resurrecting *Anschaulichkeit*, Heisenberg admired *Unanschaulichkeit*; while asserting the physical equivalence of the different versions of quantum theory, Heisenberg attempted to argue that the matrix version was nevertheless more significant ontologically.

Restoration of Anschaulichkeit *and of Space-Time*

The uncertainty paper represents a striking epistemological about-face to the original philosophical interpretation of matrix mechanics. If in the reinterpretation paper (1925) Heisenberg claimed that observable magnitudes dictate the structure of a theory, his uncertainty paper stated that the theory determines what can be observed. In 1925 Heisenberg eliminated classical space-time and *Anschaulichkeit*; in 1927 he restored them.[2] In 1925 Heisenberg considered the position of the electron unobservable; in 1927 he described a thought experiment by which electron position within an atom could be measured. Heisenberg's reversal was a direct response to Schrödinger's wave mechanics.

One of Schrödinger's arguments for the superiority of his approach over the matrix approach was the familiarity and intuitiveness of wave theoretical notions. Heisenberg initially dismissed such declarations as "rubbish" (Heisenberg to Pauli, 8 June 1926, *PC*). Yet toward the winter of 1926, Schrödinger's success was becoming manifest—a deluge of papers following Schrödinger's approach and ignoring the matrix method swelled the scientific literature. The familiar tools of wave mechanics were being successfully applied to a great variety of problems unamenable to matrix treatment.

Jordan (1927a) explained the overwhelming success of Schrödinger's theory by citing not only its mathematical convenience but also its wide intuitive appeal. Schrödinger (1952a) indeed hoped to provide an intuitive interpretation that would relegate Bohr's quantum jumps to the status of Ptolemaic epicycles. No wonder neither Bohr himself nor the authors of matrix mechanics (erected on the basis of Bohr's concepts) sympathized with Schrödinger's aspirations.

The success and popularity of Schrödinger's theory very much annoyed the ambitious Heisenberg. He called disgusting not only Schrö-

---

2. The matrices p and q initially had no kinematic meaning: they were, as Heisenberg admitted in a letter to Pauli, simply formal symbolic artifacts. Precisely because the formulas had only symbolic meaning, originally no attempts were made to draw any kinematic conclusions from them. For example, the basic commutation relation $pq - qp = (h/2\pi i)\mathbf{1}$, if interpreted kinematically means that p and q cannot simultaneously have definite values. This meaning was contained in the theory from the very beginning, yet it was only a year later that physicists drew this conclusion.

dinger's theory but also work by other physicists that substantiated the wave theoretical point of view. Thus Heisenberg found Darwin's and Wentzel's work *abscheulich* (Heisenberg to Pauli, 9 March 1927, *PC*). An irritated Heisenberg complained to Pauli that physicists should also learn matrix methods. Significantly, Heisenberg informed Pauli about his interpretative efforts immediately after expressing his discontent with the wave theory's success (Heisenberg to Pauli, 4 November 1926 and 5 February 1927, *PC*). This success, Heisenberg realized, posed a direct threat to his achievement, and Heisenberg decided to counter Schrödinger on his own terms.

Even though the uncertainty paper is known for its refutation of classical determinism, Heisenberg's original concern was with the "intuitive content of quantum theoretical kinematics and mechanics." This is, in fact, the title of the uncertainty paper. Heisenberg began this paper with a definition of the *Anschaulichkeit* (intuitiveness) of physical theory: "We believe that we intuitively understand physical theory when we can think qualitatively about individual experimental consequences and at the same time we know that that application of the theory never contains internal contradictions" (Heisenberg 1927b, 62). By analyzing the measurement process with the help of a γ-ray microscope, Heisenberg claimed to have demonstrated that it supported the basic commutation relation $pq - qp = -i\hbar$ (1927b, 64). According to his own definition, Heisenberg succeeded in providing a visualizable interpretation of the matrix formalism: "All the concepts which can be used in the classical theory for the description of a mechanical system can also be defined exactly for atomic processes in analogy to the classical concepts" (1927b, 68). In this way definability is not restricted by indeterminacy relations, because each physical variable, if measured alone, can be measured with arbitrary accuracy: it is the *simultaneous* measurement of physical variables that sets limits on the precision of measurement.

Heisenberg discussed other experiments and concluded: "As we can think through qualitatively the experimental consequences of the theory in all simple cases; we will no longer have to look at quantum mechanics as unphysical and abstract" (1927b, 82). In order to dispel any doubt as to whose accusations he was refuting, Heisenberg appended a footnote to this statement, reminding the reader of Schrödinger's remark about the "repulsive" and "frightening" nonintuitiveness of the matrix theory. It is not the *Unanschaulichkeit* of Heisenberg's theory but the misleading *Anschaulichkeit* of Schrödinger's that hinders the progress of physics, Heisenberg said in effect.

Despite these statements, Heisenberg still considered lack of visualizability a positive feature. In a letter to Pauli, Heisenberg explained that he preferred Dirac and Jordan's theory to wave mechanics because it was "less visualizable [*unanschaulicher*] and more general" (Heisenberg to Pauli, 31 May 1927, *PC*). Heisenberg's effort to render matrix mechanics visualizable was undertaken less to satisfy his own intellectual preferences and more to contribute to the matrix theory's wider acceptance.[3]

Against Waves

In order to comprehend Heisenberg's reasoning in the uncertainty paper, we have to take into account his hostility to Schrödinger's wave theory as a competing system, and his resulting aversion to wave theoretical concepts. Heisenberg himself regarded as "the most important result of the uncertainty paper" his critique of Schrödinger's idea of the wave packet (Heisenberg to Pauli, 23 February 1927, *PC*).

One of the problems Heisenberg treated in his uncertainty paper was the transition from micro- to macromechanics. A description of this transition was essential for obtaining a physical interpretation of the quantum formalism. Yet using the matrix theory, Heisenberg failed repeatedly to provide such a description, and Schrödinger seized on this failure to proclaim the superiority of his theory over the matrix version. Schrödinger attempted to identify microphenomena with wave vibrations and succeeded in constructing the example of the harmonic oscillator, in which superposed vibrations constitute a well-defined wave packet (a configuration of waves that does not spread in time, and that therefore keeps its identity; Schrödinger 1926e). Transition to the macrolevel is achieved by noticing that this wave packet can be identified with an oscillating particle. Schrödinger also hoped to construct wave packets that would move along Keplerian ellipses and therefore represent electrons moving in the atom. Schrödinger realized early, however, that such a direct intuitive interpretation of the ψ-function would not be feasible. He subsequently interpreted $\psi(r)$ as a weight function and stated clearly that the ψ-function "cannot and may not be interpreted directly in terms of 3-dimensional space—however much the one-electron problem tends to mislead us on this point" (1926d, 120).

Thus Schrödinger abandoned his first, "naive" interpretation and declared this in print a few months before Heisenberg's uncertainty paper was published. That Heisenberg found it appropriate to attack a position his opponent no longer held is indicative of his eagerness to dis-

3. *Anschaulichkeit*, as Forman (1979) and Miller (1978) have argued, was a pregnant notion in the German cultural milieu. In the German neo-Kantian philosophical tradition, it was important as a cultural value. Forman and Miller have interpreted Heisenberg's restoration of *Anschaulichkeit* as a submission to the pressure of a wider intellectual milieu. Yet Heisenberg showed little desire to tackle the issue of *Anschaulichkeit*, until subjected to concrete scientific pressure—by Schrödinger.

credit the wave theoretical approach. As Heisenberg wrote to Pauli, considerations of the transition from micro- to macromechanics were the most important of all: first he would like to demonstrate how this transition *does not* occur (Heisenberg to Pauli, 23 February 1927, *PC*). It does not occur as Schrödinger originally intended, through wave packets that keep their identity.[4] The spreading of the wave packet, Heisenberg pointed out, constitutes a conclusive refutation of the possibility of a micro-macro transition along the lines outlined by Schrödinger. Nor can this difficulty be avoided through considerations of natural radiation width *(Strahlungsbreite):* the transition from quantum to classical mechanics must be intelligible without the use of electrodynamics (Heisenberg 1927b).

Heisenberg's change of heart was complete. If in 1925 he attempted to reinterpret the mechanical behavior of an electron by replacing kinematic concepts with electrodynamic ones (elementary waves of emitted radiation), now he refused to employ electrodynamics in the physical interpretation of quantum mechanics. Upon coming to work with Heisenberg, Felix Bloch had proposed to explore the possibility of keeping wave packets together through radiation damping. Heisenberg, as Bloch recalled, was strongly opposed to this initiative. "As to my hopes for keeping wave packets together by radiation damping, he only smiled and said that, if anything, it could of course make them spread even more" (Bloch 1976, 25; interview with Bloch, AHQP).

There are other reasons for regarding Heisenberg's paper as a direct attack against Schrödinger's competitive efforts. Only under such an assumption do some puzzling features of Heisenberg's paper, as well as the story of Heisenberg's clash with Bohr, become intelligible (see chapter 6). In his description of the γ-ray thought experiment, Heisenberg had committed a trivial error, which both Bohr and Dirac (who was in Copenhagen at the time) brought to his attention (Heisenberg to Pauli, 16 May 1927, *PC*). Heisenberg had treated both photons and electrons as regular point particles and argued that at the time of their collision a photon transfers to an electron a discrete and uncontrollable amount of momentum (Compton recoil). The more precisely the position of the electron is determined, the greater the uncertainty of the discontinuous change in the electron's momentum (Heisenberg 1927b, 64). Yet Compton recoil does not lead to indeterminacy but rather to exactly calculable momentum changes. From the conservation laws for

4. Were the wave packets to carry the periodic motions of classical electrons, then the atom would send out radiation that could be expanded in Fourier series, for which the frequencies of the harmonics are integral multiples of the fundamental frequency. "The frequencies [of] the spectral lines sent out by the atom are, however, according to quantum mechanics, never integer multiples of the basic frequency—except in the special case of the harmonic oscillator" (Heisenberg 1927b, 73).

energy and momentum, and knowing the energy and momentum of the colliding photon, one can calculate the momentum change of the electron exactly. There is no way to transcend the classical deterministic framework once it is assumed that photons and electrons are point particles obeying conservation laws (see diagram).

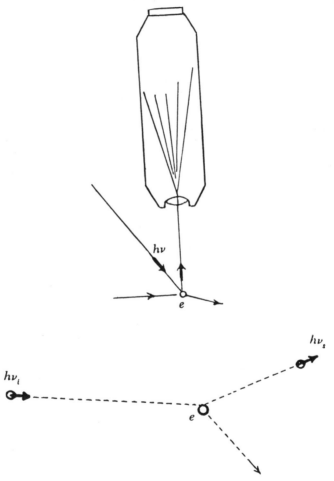

In his initial mistaken analysis of a γ-ray microscope, Heisenberg treated both photons and electrons as particles without taking into account their wave attributes. This analysis is equivalent to the scattering of light quanta by intraatomic electrons in the Compton effect (*bottom*), and it leads not to indeterminacy but to exactly calculable changes. Heisenberg's misleading analysis is sometimes repeated in popular expositions of uncertainty principle.

That a physicist of Heisenberg's stature would make such a mistake is odd enough, but his refusal to correct the mistake, despite powerful criticism from Bohr, becomes incomprehensible unless he had a vested interest in preferring a misleading description to the correct one. Schrödinger sought to argue that reality consists of waves, and waves only; Heisenberg was now taking the opposite stand—that reality consists of particles, and particles only. This sudden insistence on the exclusively corpuscular nature of matter is all the more surprising if we recall that only three months earlier Heisenberg (1926b) asserted that recent developments in physics (matrix theory and the Bose-Einstein statistics) undermined the reality of particles. It was the particle characteristics of radiation and matter that Heisenberg now wanted to emphasize, at the expense of their wave characteristics. Heisenberg's letter to Pauli clearly identifies the focus of his disagreement with Bohr: "I argue with Bohr about how far the relation $\Delta p \Delta q \sim h$ has its origin in the wave or discontinuity aspect of quantum mechanics. Bohr emphasizes that the diffraction of waves is essential in the γ-ray microscope; I emphasize that the light quantum theory and even [the] Bothe-Geiger experiment[s] are essential" (Heisenberg to Pauli, 4 April 1927, PC).

There is an important phrase in the passage above: "discontinuity aspect of quantum mechanics." Heisenberg came to regard discontinuity as the most essential feature of the quantum world. The matrix theory, and not wave mechanics, was best suited to express this basic discontinuous aspect (Heisenberg to Pauli, 16 May 1927, PC). After completing the uncertainty paper, Heisenberg wrote to Jordan that "finally" he felt the discontinuous ground firmly under his feet (Heisenberg to Jordan, 29 September 1926, AHQP). Clearly, replacing his own description based on the discontinuous Compton recoil with Bohr's explanation based on the continuous wave aspect of radiation did not suit Heisenberg's purposes. As Heisenberg wrote to Pauli, discontinuity, as the hallmark of the quantum world, could not be overemphasized; he was therefore happy with his paper, despite the mistakes (Heisenberg to Pauli, 16 May 1927, PC).

Bohr, as one gathers from Heisenberg's postscript to the uncertainty paper, also pointed out to Heisenberg that Compton recoil applies rigorously only to free and not to bound electrons. And for free particles Bohr's own derivation of the uncertainty relations, from the wave theoretical point of view, can be considered more natural. Yet Heisenberg preferred the Dirac-Jordan quantum theory, which was more suited to express discontinuities. This was the reason he ended up "in the camp of matrices and against waves" (Heisenberg to Pauli, 31 May 1927, PC).

Oskar Klein, a witness to Heisenberg's clash with Bohr, offered the following commentary: "It was so much in the beginning, . . . although Heisenberg must have known in principle, if one had asked him, that there couldn't be much difference [between particle and wave theoretical descriptions]. . . . Still, I think, he was not so familiar with the wave point of view at the time, and he was a bit partial to the matrix point of view which he started" (interview with Klein, AHQP). Heisenberg, as he revealed in an interview with Kuhn (AHQP), was inspired by William Duane's (1923) results, which managed to explain the diffraction of light by a grating assuming the corpuscular nature of light (light quanta) and avoiding wave theoretical presuppositions (see chapter 11). Heisenberg insisted in the uncertainty paper that the powerful transformation theory allowed a purely corpuscular interpretation in the spirit of the generalization of Born's statistical interpretation. He therefore rejected not only Schrödinger's attempts at a pure wave ontology but also Bohr's conciliatory solution in terms of wave-particle duality (see chapters 6 and 11).

Nevertheless, Heisenberg supported Bohr's position on complementarity in the postscript to the uncertainty paper. In order to offer a unified explanation, capable of countering the opposition, differences of opinion were to be suppressed. The easiest way to deduce the uncertainty relations is, in fact, by identifying a particle with a limited wave field, as Bohr argued in his Como lecture. Heisenberg, who considered the uncertainty formula fundamental to the self-consistency of the quantum formalism, was willing to employ Bohr's accessible derivation for wider audiences and for "pedagogical reasons." He used it because "it did not do any harm" to his own explanation, yet he did not believe "it was necessary" (interview with Heisenberg, AHQP).

Discontinuities and Quantum Jumps

In the heated debate between Schrödinger and the matrix physicists, each side proclaimed its own approach superior for describing atomic phenomena. Schrödinger claimed that wave mechanics was better because it operated with familiar and manageable mathematical notions and promised to eliminate the irrationalities of quantum jumping. The matrix physicists considered these "advantages" superfluous and misleading—it was the matrix theory and not Schrödinger's, they argued, that expressed the "inevitable," irreducible discontinuities of the quantum world.

The confrontation between Schrödinger and his opponents is one of the most fascinating chapters in the history of physics. Although emotions ran high (or maybe because they did), the dialogue that resulted produced significant contributions by all participants. It led to the Born-Pauli probabilistic interpretation, the Dirac-Jordan generalized mathematical version of quantum mechanics, the Heisenberg uncertainty principle, and Bohr complementarity.

Yet on positivist grounds, the issue was meaningless. If the two theories were fully equivalent, as the participants seemed to hold, what was the meaning of the claim that one theory was superior to the other on physical grounds? How could anyone defend the position, as Heisenberg did, that his theory was able, though Schrödinger's equivalent theory was not, to deduce certain physical results, such as Planck's law? And what meaning could one attach to a "real existence," and instantaneity, of the unobservable quantum jumps, the reality of which Heisenberg was eager to demonstrate? The whole controversy gains intelligibility only when we assume that not only Schrödinger but also Heisenberg had some very strong reason to ask how unobservable processes *really* occur in nature.

Before the wave theory appeared in print, Heisenberg spoke with two different voices on the issue of continuity versus discontinuity. Heisenberg incorporated the guidelines of Born's approach to building "a truly discontinuous theory" in his reinterpretation paper, by integrating the concepts of discrete energy levels and transitions between them into the basis of the matrix mathematical formalism. At the time Heisenberg did not claim (as he would later, after the appearance of continuous wave mechanics) an irreducible, a priori status for discontinuities. In fact, Heisenberg was elated (as Schrödinger would later be) when he learned that his new mechanics did not have to postulate discontinuous energy levels—their existence simply followed from the mathematical formalism. As Heisenberg noted, nothing in the matrix theory depends on assuming a priori that discrete stationary states exist—both the discrete and the continuous solutions follow equally from the basic quantum equation: $pq - qp = (h/2\pi i)\mathbf{1}$ (Heisenberg to Pauli, 18 September 1925, *PC*).

Similarly, Heisenberg, Born, and Jordan argued (as Schrödinger later would about his theory) that "a notable advantage" of the new matrix theory "consists in the fact that the existence of discrete stationary states is just as natural a feature of the new theory as, say, the existence of discrete vibrations frequencies in classical theory" (Born, Heisenberg, Jordan 1926, 322). Although Bohr's concepts played a heuristic role on the way to the development of the matrix theory, the matrix theory itself undermined the fundamental role of a priori stationary states and "irreducible" quantum jumps (chapter 2).

These concepts were resuscitated in response to the threat from Schrödinger. Matrix physicists, unable to develop a comprehensive and satisfying interpretation of quantum physics, and faced with Schrödinger's rival theory, chose the only effective strategy available—claiming that an interpretation dissociated from the original Göttingen-Copenhagen approach was impossible, that quantum discontinuities could not be eliminated.

Heisenberg and Schrödinger first clashed over the interpretation of quantum mechanics at the Munich conference in July 1926, where Schrödinger was presenting his new theory. Most of the participants were captivated by Schrödinger's mathematics and in sympathy with his interpretative aspirations—at least according to Heisenberg's assessment. The next confrontation occurred in Copenhagen in the fall of 1926, when Bohr joined forces with Heisenberg to combat Schrödinger's attempt to do away with quantum jumps.

In order to counter Schrödinger, Heisenberg had to argue that discontinuities are irrevocable. A natural candidate for this task offered itself—fluctuation phenomena. Einstein had analyzed fluctuation phenomena to deduce the existence of discontinuities on the microlevel. He ascribed Brownian fluctuations to the atomistic structure of matter, while the fluctuations of energy and momentum of gas molecules led to his idea of light quanta. Heisenberg chose to treat fluctuations precisely because he viewed them as an appropriate tool with which to emphasize discontinuities.

In his "fluctuation" paper (1927a), Heisenberg considered two similar atoms having energies E_n and E_m. When these atoms are coupled there is "a resonance"—each atom periodically changes its state from E_n to E_m and vice versa. The exchange of energy between these atoms could be viewed either (1) as due to energy jumps $E_n - E_m$ via photons or (2) as a classical interaction between two oscillators. By translating the second alternative into matrix language and pointing out that for such a system only the time average of each individual atom is observable, Heisenberg obtained the value $\frac{1}{2}(E_n + E_m)$ for each atom, describing the fluctuation of an arbitrary function of energy as $f(E(t)) = \frac{1}{2}(f(E_n) + f(E_m))$. The only way this formula can be correct for all f and E, argued Heisenberg, is if the energy changes discontinuously, spending equal amounts of time at the values E_n and E_m.

Schrödinger did not wait long to meet Heisenberg's challenge. Soon he published a response, in which he treated the same problem on his own terms, without assuming discrete energy levels, and using instead only the language of frequencies (Schrödinger 1927b). According to Schrödinger, the interaction is controlled by a law of resonance that requires the difference between two eigenfrequencies in one system to be equal to the difference in the other system. Consequently, he succeeded in showing that the interaction can be adequately described as a continuous change in the amplitudes of the four corresponding characteristic vibrations.

In Heisenberg's fluctuation paper (1927a), the formulas are correct, but Heisenberg's logic is faulty. If we designate by **D**, discontinuous jumps, **M**, the matrix version of quantum mechanics, and **F**, the fluctuation formulas, then Heisenberg's incorrect reasoning runs as follows: **D** → **F** and **M** → **F**; therefore **M** → **D**.

Because in using matrix mechanics Heisenberg restricted himself to only the noncontroversial assumptions of that approach—namely, that a diagonal element of a matrix represents the time average of the corresponding physical variable in the stationary state considered—he felt he had proved conclusively the inevitability of quantum jumps. In order to do so, however, Heisenberg had to present a much stronger case. Heisenberg had to prove that every possible continuous description would lead inevitably to predictions incompatible with experimental results or to inner contradictions. Heisenberg did not even attempt this impossible task. (And, in fact, Bohm has recently offered a continuous description of quantum jumps; Bohm and Hiley 1993). What Heisenberg did show, not surprisingly, is that the assumption of quantum jumps is admirably compatible with the matrix formalism. Heisenberg argued in a letter to Jordan that it was the matrix formalism, and not Schrödinger's wave mechanics, that was best able to treat quantum discontinuous phenomena (Heisenberg to Jordan, 29 October 1926, AHQP).

This letter to Jordan demonstrates how Heisenberg's preference for discontinuity emerged from his opposition to Schrödinger. According to Heisenberg, the progress achieved in his fluctuation paper had nothing to do with his calculation of energy fluctuations, which was a simple problem that could easily have been solved by thermodynamic considerations on the assumption of quantum jumps (Heisenberg to Jordan, 29 October 1926, AHQP). The progress rather consisted in the ability of matrix mechanics to calculate these fluctuations without the explicit assumption of quantum jumps.[5] The important conclusion was not that quantum jumps are irreducible but rather that the matrix theory was the formalism most appropriate to representing discontinuities and quantum jumps. The matrix formalism and the existence of discontinuities, reinforcing each other, undermined, according to Heisenberg, Schrödinger's claims for the wave theory.

Heisenberg's contention that he could prove the inevitability of discontinuities in the domain where the matrix and wave mechanical methods were equivalent was unreasonable. On this issue, Schrödinger's

---

5. "The progress consists rather in the fact that quantum mechanics allows the calculation of these fluctuations without an explicit assumption of quantum jumps, but on the basis of relations between q, q' and so on" (Heisenberg to Jordan, 29 October 1926, AHQP).

position was more sensible: Schrödinger assumed the superiority of his method because of what he saw as its greater suitability to extension beyond the domain where the two methods proved equivalent (Schrödinger 1926f).

Heisenberg's attempt to prove the inevitability of discontinuities in his fluctuation paper was intended less as a genuine contribution to science and more as a weapon to be used against Schrödinger's theory. In a letter to Pauli, Heisenberg apologized that he had submitted for publication a paper that contained nothing new. He explained that the purpose of the paper was purely "pedagogical"—it was directed against those who favored the continuous approach (Heisenberg to Pauli, 4 November 1926, PC).[6]

Shortly after Heisenberg's fluctuation paper, Jordan published a paper along similar lines (Jordan 1927e). Jordan, more consistently, did not claim that he had proved the inevitability, or even the preferential status, of discontinuities. Neither did he assert the superiority of matrix mechanics over Schrödinger's continuous theory. He merely attempted to demonstrate that the Göttingen-Copenhagen approach, based on quantum jumps, was not inferior to Schrödinger's intuitive continuous framework. Jordan arrived at a sound conclusion: "Both theories and their conflicting interpretations, despite their great difference, give the same results for all empirically testable claims" (1927e, 661).[7]

Dirac and Jordan's transformation theory was based on probabilistic axioms that presuppose discontinuity. This probabilistic basis served Heisenberg directly in his derivation of the uncertainty principle. Discontinuous quantum jumps and acausality became entrenched both in the quantum formalism and in its interpretation to such an extent that they appeared to be "inevitable," "hard" facts of nature.

6. "The enclosed note about fluctuation phenomena I am sending to you, not without a long list of excuses—that I will publish such a thing at all that actually contains nothing new for you or other enlightened physicists, that I myself do not know what to do with it, either, etc. The reason that I wrote it is really only a pedagogical one against the Lords of continuums theory" (Heisenberg to Pauli, 4 November 1926, PC).

7. This conciliatory mood did not last long—Jordan soon became the most militant opponent of Schrödinger's interpretive efforts and an ardent defender of quantum jumping. In his fluctuation paper Heisenberg was eager to convince himself and his audience that he had proved the inevitability of discontinuities and the "real existence" of quantum jumps. Years later, Heisenberg would remember his emotional need more vividly than Jordan's initial balanced appraisal: "The result decided clearly in favor of the quantum jumps and against the continuous change" (1977, 4). Similarly, Dirac, who at the time was still hostile to Schrödinger's theory, also confused the desired with the proven. In an important paper, stimulated by Heisenberg's fluctuation considerations, Dirac wrote: "Heisenberg had shown that these calculated means are just what one would expect from the assumption that the energy changes discontinuously from one quantized value to another. . . . The theory can thus be considered to show that the energy actually does change discontinuously" (1927, 622).

Dialogue with Pauli

Wolfgang Pauli was Heisenberg's closest scientific collaborator and critic. Heisenberg's major breakthroughs—both the reinterpretation and the uncertainty papers—would be unimaginable without Pauli's prior penetrating analysis of the issues fundamental to quantum theory at the time. It was not merely that Pauli was generous with his insights in his correspondence with Heisenberg; the very formulation of the issues occurred when Heisenberg and Pauli addressed each other. Pauli and Heisenberg complemented each other. It was Pauli, unusually gifted at elucidating fundamental issues, who often provided the essential pieces of the puzzle. It was the brilliant "unphilosophical" Heisenberg (Pauli to Bohr, 11 February 1924, PC) who, cheerfully sustaining conflicting inner voices, provided the epoch-making solutions to the puzzles Pauli defined. In his later years Pauli regretted his own "conservatism," feeling perhaps that Heisenberg's intellectual boldness (if not recklessness) reaped a richer scientific harvest and wider scientific fame (Mehra 1976).

Pauli's impact in his dialogue with Heisenberg over the issues of the uncertainty paper was crucial in several respects. It was Pauli who insisted, and eventually succeeded in convincing Heisenberg, that the physical interpretation of quantum mechanics must be undertaken in particle-kinematic terms, as opposed to the initial philosophy of matrix mechanics. Pauli formulated the issue of the impossibility of the joint determination of position and momentum as the focus of interpretive efforts. Pauli also provided Heisenberg with the example of the time-energy uncertainty relation—an example that Heisenberg repeated and reanalyzed in his uncertainty paper. By analyzing Born's collision treatment, Pauli translated Born's results into matrix language, thus undermining claims by Schrödinger and Born that wave mechanics was superior to matrix mechanics. This analysis was a crucial contribution to the elucidation of the physical interconnection between the wave and matrix approaches, preparing the ground for the generalized version of quantum mechanics proposed by Dirac and Jordan. In direct dialogue with Pauli, Heisenberg's preference for discontinuities was strengthened, the product $\Delta p \Delta q$ was defined, and the causality issue became more focused.

Particle-Kinematic Interpretation

With the benefit of hindsight, one can argue that the uncertainty principle was contained in the new quantum mechanics from the beginning, implied by the basic commutation relation $pq - qp = (h/2\pi i)1$. The reason this fact was not mentioned and elaborated from the start is that originally the matrices p, q, and the like were considered formal symbolic artifacts, which had no connection with the usual meaning of position and momentum. It seemed a marvel, as Lorentz wrote to Schrödinger, that the equations of motion "can be satisfied when one interprets these symbols as things that have quite another meaning, and only remotely recall those coordinates and momenta" (Przibram 1967). Let us recall that in the original matrix approach, the matrix q was endowed with electromagnetic, not kinematic, meaning.

In Dirac's version of q-numbers, constructed in close analogy with classical mechanics, it is appropriate to ask what happens when one substitutes a c-number for a q-number (q-numbers are symbols that satisfy quantum theoretical equations; c-numbers are ordinary numbers). One can subsequently see that it is impossible to assign c-numbers to both position and momentum simultaneously without violating the equation $pq - qp = (h/2\pi i)\mathbf{1}$.[8] The fact that one cannot talk about the position and the momentum of a given atomic object simultaneously, and that one cannot define a path of a particle, was also implicit in Schrödinger's version of quantum mechanics. Schrödinger stressed that quantum mechanical phenomena cannot be described by image points, only by wave processes, and sought an interpretation of wave mechanics analogous to the interrelation between wave optics and geometrical optics. Born's collision treatment, in agreement with this approach, presented the moving electron by a plane wave e^{-ikx}, where the value of the momentum $p = \hbar k$ is determined exactly while the position of the moving corpuscle is completely undetermined.

Pauli, who opposed Schrödinger's interpretive ambitions and sought to endow the quantum formalism with particle-kinematic meaning by generalizing Born's collision arguments, tried to understand why in cases when momentum is known, positions is completely undetermined. Pauli discussed this issue in a remarkable letter to Heisenberg (Pauli to Heisenberg, 19 October 1926, PC). In this letter Pauli elaborated other cases of quantum peculiarities that were incomprehensible from the classical point of view.[9] He also explored the question of the observability of electron motion. On more than one occasion Pauli had

8. "The general question of classical mechanics can be formulated as follows: What is the value of any constant of integration g of a given dynamical system for any given initial conditions, specified by numerical values q'_{r0}, p'_{r0}, say, for the initial coordinates and momenta q_{r0}, p_{r0}? . . . [As opposed to classical mechanics] on the quantum theory one can also obtain an expression for g as function of the q_{r0}, p_{r0} but the q_{r0} and p_{r0} do not satisfy the commutative law of multiplication, so that if one substituted numerical values for them the result would in general depend on the order in which they were previously arranged. One can thus give no unambiguous answer to the question on the quantum theory" (Dirac 1927, 623).

9. For example, if a mass with energy E is directed toward a finite potential barrier with energy $V(x)$, when $E > V_{max}$ the classical particle will always run over the barrier while the quantum particle will sometimes be reflected, with a calculable probability (Pauli to Heisenberg, 19 October 1926, PC).

expressed the opinion that one must not only eliminate theoretically unobservable entities but also endow experimentally observable entities with the appropriate theoretical representation. At the same time Pauli, a lifelong critic of field theories, sought to transform the original electromagnetic underpinnings of the matrix theory into a particle-kinematic foundation. His aim was to base the interpretation of matrix elements on kinematic attributes of particles rather than on elementary waves of emitted radiation. Pauli realized that he could partly achieve this objective by interpreting Schrödinger's ψ-function as denoting the probability of position.

Pauli defined $|\psi(q_1, \ldots, q_f)|^2 dq_1 \cdots dq_f$ as the probability that for a system in a definite stationary state, the coordinates q_k of the particles of this system ($k = 1, \ldots, f$) lie between q_k and $q_k + dq_k$. Pauli subsequently interpreted the diagonal elements of the matrix of an arbitrary function F such that $F_{nm} = \int F(q_k)|\psi_n(q_k)|^2 dq_1 \cdots dq_f$ as the average value of F in the state n (Pauli to Heisenberg, 19 October 1926, PC).

According to Pauli, the probability $|\psi(q_1, \ldots, q_f)|^2 dq_1 \cdots dq_f$ should be regarded as observable in principle. Despite the fact that, historically, matrix elements were connected with the properties of emitted waves, Pauli became convinced with "all the fervor of his heart" that matrix elements must be connected with "kinematical (perhaps statistical) data" for particles (Pauli to Heisenberg, 19 October 1926, PC).[10]

Pauli, as I noted earlier, after initially ignoring Schrödinger's theory, recognized its fundamental importance. Because of his aversion to continuous field theories, Pauli had no use for Schrödinger's interpretive aims, but he was not prepared to overlook Schrödinger's physical insights. Thus Pauli's kinematic interpretation to a considerable degree simply translated into particle language Schrödinger's interpretation of ψ as a measure of charge density. The fact that Schrödinger's ψ-function operated in a multidimensional configuration space strengthened Pauli's predisposition to ascribe to it a particle-kinematic meaning (Pauli to Heisenberg, 19 October 1926, PC).

At first, Heisenberg vacillated. He was by that time ready to accept that matrix elements could be defined kinematically without recourse to emitted electromagnetic radiation. Yet he disagreed with Pauli's attempt to connect such a definition with a single stationary state. The kinematic variables, suggested Heisenberg, should be tied to two different stationary states. In this way, the question of position would intertwine naturally with the possibilities of transitions (Heisenberg to Pauli, 28 October 1926, PC).

10. That the phrase "perhaps statistical" is in parenthesis is not accidental: Pauli did not yet hold statistical concepts to be fundamental and therefore did not intentionally look for a statistical interpretation; it was the kinematic aspect of the problem that was of prime importance to him.

Pauli's arguments for the equivalence of description in momentum space and that in position space made a great impression on Heisenberg (Pauli to Heisenberg, 19 October 1926, *PC*).[11] Waves in momentum space, Heisenberg concluded, have as great a physical significance as waves in position space (Heisenberg to Pauli, 28 October 1926, *PC*). The equation $qp - pq = hi$ thus always corresponds in the wave presentation to the fact that it is impossible to speak of a monochromatic wave at a fixed point in time. But if one makes the line not too sharp, and the time interval not too short, then the idea begins to make sense. Analogously, one cannot talk about the position of a particle moving with a definite velocity. Yet when one does not apply these concepts too rigorously, it may well make sense. One catches a glimpse therefore of why in the macrodomain these concepts can be used as approximations (Heisenberg to Pauli, 28 October 1926, *PC*). At this stage, however, Heisenberg did not seek to interpret this insight in terms of the limit of precision of kinematic concepts as applied to an individual particle, suggesting instead a statistical meaning for space-time.[12]

Heisenberg's discussion was, in fact, a corpuscular improvisation on the idea of the wave packet—an idea advanced initially by de Broglie and Schrödinger, and used extensively by Bohr in his Como lecture (see chapter 6). The Göttingen-Copenhagen physicists obtained important insights by turning the rival wave imagery to their own ends. At this point in Heisenberg's letter, Pauli commented in the margin: "In a time-interval, short as compared with the period (of revolution), it is also meaningless to speak of a precise energy value" (*PC*, 352). It is this point that led Heisenberg to the formulation of the time-energy uncertainty relation. In this endeavor, Heisenberg relied substantially on Pauli's (1926a) "encyclopedia" article, where the whole issue was comprehensively analyzed.

Time-Energy Relation

According to Klein, Bohr posited "the first instance of the indeterminacy principle—that a stationary state cannot be better defined than h

---

11. These considerations also made a great impression on Born, chilling his initial preference for Schrödinger's visualizable interpretation. In a letter to Schrödinger (6 November 1926, AHQP), Born informed him of Pauli's calculations in the case of a two-dimensional rotator. Its solution can be written in two ways: first, viewing x, y as regular coordinates and p_x, p_y as operators; second, viewing p_x, p_y as variables and x, y as operators. Both solutions lead to the same result. The possibility of substituting momentum space for coordinate space (through an appropriate canonical transformation) reveals— Pauli, Born, and Heisenberg concluded—that there was no reason to suppose that q-space has any conceptual preference and that the essence of quantum phenomena depends on the visualization in q-space (as Schrödinger believed).

12. On this point, see the discussion of Heisenberg's dialogue with Campbell, later in this chapter.

divided by the life-time" (interview with Klein, AHQP). In formulas, this means $\Delta E \Delta t \sim \hbar$. The connection between the breadth of spectral lines and the lifetime of stationary states was at the heart of the Bohr-Kramers-Slater theory. According to John Slater's memoirs (1975b), he was the first to raise this point in connection with Bohr's assumption about radiation emitted during instantaneous transitions from one stationary state to another. This assumption was disposed of in the Bohr-Kramers-Slater theory: here, as in classical physics, radiation (albeit virtual) was emitted continuously during the entire lifetime of an excited state, and not during transitions. This theoretical change took place, according to Slater, because he had insisted that the assumption of radiation width during extremely short transitions is inconsistent with experiment.[13]

Heisenberg was familiar with the Bohr-Kramers-Slater theory, and with its underlying rationale. There was nothing exceptional in the fact that the frequency spectrum of a finite wave train depends on the time during which the train is emitted. Only when, after the uncertainty paper, Δt signified, not simply a "finite precise interval of time," but rather "the breadth of imprecision in evaluation of a point of time," did $\Delta E \Delta t \sim \hbar$, like $\Delta p \Delta q \sim \hbar$, become a relation that signified a break with the past. As Slater mentioned: "It was only some years later, in 1927, that Heisenberg published his paper on [the] uncertainty principle, and that he and Bohr made this such a great feature of their version of wave mechanics" (1975b, 19). Not surprisingly, Heisenberg was afraid that his considerations in the uncertainty paper might be regarded as "an old snow" (*alter Schnee*; Heisenberg to Pauli, 23 February 1927, PC).

It is clear that the relation $\Delta E \Delta t \sim \hbar$ was obtained before $\Delta p \Delta q \sim \hbar$. Heisenberg translated $\Delta E \Delta t \sim \hbar$ into $\Delta p \Delta q \sim \hbar$. He daringly introduced wave theoretical peculiarities exactly where they seemed out of place: the particle-kinematic framework. Pauli's influence was crucial. It was Pauli who insisted on the insufficiency of field concepts, and who sought to reinterpret the matrix formalism in particle-kinematic terms. In this connection Pauli struggled with the statistical meaning of the "path" of a particle, preparing the ground for Heisenberg's discussion

13. As Slater recalls: "The difficulty [with Bohr's theory] was with the assumption that the radiation was emitted in the form of a photon at the instant the atom jumped from the state E_2 to the state E_1. Any student of physics or mathematics knows that a wave train of finite length has a frequency spectrum which is not strictly monochromatic, but which instead has a frequency breadth $\Delta \nu$ which is of the order of magnitude of $1/T$, where T is the length of time during which the train is emitted. . . . If this period of time is much longer than the period of oscillation, which is $1/\nu$, the breadth $\Delta \nu$ will be very small compared to the frequency ν. The observed sharpness of the spectral lines shows that this must be the case. The experiments are consistent with emitted wave trains which have perhaps the order of 10^5 or more waves on the train. . . . Surely it must have taken long enough for 10^5 waves to be emitted" (1975b, 10).

of this problem in the uncertainty paper. Pauli explained his approach in this way: if at time t the particle's position is q_0, what is the probability that at time $t + \tau$ its coordinate will be q? The kinematic answers, argued Pauli, should be statistical, so one would not be able to talk about the definite "path" of a particle, quite in accordance with the impossibility of endowing a particle simultaneously with exact position and momentum (Pauli to Heisenberg, 19 October 1926, PC).

In his uncertainty paper Heisenberg discussed several cases of the time-energy uncertainty relation, including the case of the Stern-Gerlach experiment and that of transitions from one stationary state to another.[14] These examples were treated extensively in Pauli's (1926a) important encyclopedia article, and Heisenberg dutifully appended a footnote referring the reader to Pauli's work: "Quantum mechanics has changed only slightly the formulation of these problems as given by Pauli" (1927b, 63 n).

In his encyclopedia article Pauli discussed the problem of energetic transition from one stationary state to another, referring to Einstein's treatment of elementary processes of absorption and emission in probabilistic terms. Pauli emphasized that this treatment is mute about the exact times of transition. Does this fact indicate a fundamental, irreducible acausality, or is it merely a sign of the temporary incompleteness of contemporary theoretical knowledge? Pauli (1926a, 11) refused to take a stand: "This is very much debated, yet still an unsolved issue." Heisenberg (1927b), in response to this challenge, proclaimed a "definitive" solution of the causality problem.

Another fundamental problem, according to Pauli (1926a), was the duration of transition processes. Any attempt to introduce a detailed description of the transition process into quantum theory proved unfruitful. Certain theoretical considerations lead to the conclusion that these processes must be very short compared with the lifetime of stationary states. Nevertheless, Pauli continued to assume, in the spirit of Bohr's work, that stationary states are characterized by exact values of energy, electric and magnetic moment, and so forth, and that these parameters change discontinuously during sharp and precise moments of time. Pauli suggested that the precision limit of the time of transition is perhaps of the same order of magnitude as the period of emitted light. Pauli admitted that he could not offer a more precise analysis.

Heisenberg immediately took up Pauli's challenge to specify the precise limits on the determination of transition times: "According to the physical interpretation aimed at here, the time of transitions or 'quan-

14. Other examples Heisenberg referred to (both due to Pauli) were of "weak quantization," when quantized periodic motion is interrupted by quantum jumps, and of a rotator in the form of a gear wheel.

tum jumps' must be as concrete and determinable by measurement as, say, energies in stationary states." The imprecision within which the instant of transition is specifiable, argued Heisenberg, is given by Δt in $\Delta E \Delta t \sim h$, where ΔE indicates the change of energy in a quantum jump (1927b, 76). Analyzing the transition from a stationary state $n = 2$ into $n = 1$ and using a thought experiment similar to the Stern-Gerlach experiment for energy measurement, Heisenberg argued that in order to be able to make a distinction between the two states, Δt cannot be smaller than $h / \Delta E$. The instant of transition, Heisenberg concluded, can only be determined within a Δt spread (1927b, 77). Heisenberg interpreted Δt as characterizing the imprecision of the definition of time, rather than as a finite, precise interval of time. Discussing the Stern-Gerlach experiment, Heisenberg demonstrated that the "accuracy of the energy measurement decreases as we shorten the time during which the atom is under the influence of the deflecting field" (1927b, 67).[15] Heisenberg interpreted t_1 (during which the atoms are under the influence of the deflecting field) as "uncertainty in time," rather than as a short, precisely defined interval of time. In this way, Heisenberg claimed, the relation $\Delta E \Delta t \sim h$ is analogous to the relation $\Delta p \Delta q \sim h$.

Yet the uncertainty relations of time-energy and position-momentum are not analogous. The latter is a straightforward deduction from the mathematical formalism of the quantum theory; the former is not. Heisenberg legitimated the time-energy relation merely by discussing several thought experiments—no wonder a complex controversy developed around the validity and the meaning of the time-energy uncertainty formula, in contrast to the wide acceptance of the position-momentum relation.[16] Heisenberg's imprecise treatment of the time-energy relation contained the seeds of future difficulties. In the uncertainty paper, Heisenberg (1927b) presented what he called the "familiar" equation $Et - tE = -i\hbar$ (as written there by Heisenberg).

This time-energy formula is misleading; time, as many physicists point out (see Aharonov and Bohm 1961), enters into the Schrödinger equation as a parameter, not as an operator—the source of the basic difference between the time-energy and position-momentum uncertainty formulas.

Dialogue with Dirac

Heisenberg's dialogue with Paul Dirac was essential to his formulation and interpretation of the uncertainty principle. Dirac's work provided

---

15. Here Heisenberg again refers the reader to Pauli's discussion in his encyclopedia article.

16. This controversy is discussed comprehensively in Jammer (1974).

the mathematical tools of transformation theory for the derivation of the uncertainty principle. His achievement dramatically accelerated Heisenberg's conviction that the new quantum theory was final, or complete. This conviction (or at least strongly held assumption) was crucial for Heisenberg's reinterpretation of experience (his definitions of the position and the path of a particle in operational terms). Dirac's powerful generalization of Heisenberg's fluctuation paper transformed Heisenberg's preference for discontinuity into a passionate commitment. Dirac's stand regarding the essentially nonstatistical nature of quantum mechanics, and his claim that statistics is merely introduced by experiments, opened the way for Heisenberg's legitimation and interpretation of quantum mechanics through analysis of thought experiments. Dirac's impact on Heisenberg was important in other ways as well: it reinforced his kinematic explorations, and it stimulated further his search for physical insight through the analysis of Schrödinger's wave function. Dirac's work strengthened Heisenberg's growing preference for the abstract matrix formalism and *Unanschaulichkeit*, despite Heisenberg's proclamation of an "intuitive" interpretation.

Centrality of Noncommutativity

The most nonclassical result of the original matrix theory was the noncommutativity of "position" and "momentum" in the basic equation $pq - qp = (h/2\pi i)\mathbf{1}$. Heisenberg at first regarded this noncommutativity as a hindrance or defect that might be eliminated as the theory developed further. For Dirac this noncommutativity was not a hindrance but a sign of a fundamental advance. As Dirac explained years later: someone (like Heisenberg) who takes a very daring and unconventional step can only advance up to a certain point—a natural anxiety (about being wrong) prevents the author from going further. Others therefore step in and carry on (interview with Dirac, AHQP).

Dirac's version of quantum mechanics was built on symbols, q-numbers, by analogy with the Poisson brackets of classical mechanics. Dirac therefore viewed the noncommutativity in the basic relation as being a natural counterpart of the noncommutativity in the Poisson bracket algebra. Rather than shying away from noncommutativity, he conceived quantum mechanics as the "general theory of all quantities that do not satisfy the commutative law of multiplication" (Dirac 1929, 716).[17]

Impressed by Dirac's transformation theory, Heisenberg arrived at the same point of view: every mathematical scheme that fulfills the basic relation $pq - qp = (h/2\pi i)\mathbf{1}$ is correct and physically meaningful

---

17. My description of the Heisenberg-Dirac dialogue relies heavily on the interpretation of Dirac's work in the studies by Darrigol (1992a) and Kragh (1990).

(Heisenberg to Pauli, 4 November 1926, *PC*). The problem of interpreting quantum mechanics thus became organically connected with that of deciphering the basic commutation relation.

In contrast to the authors of matrix mechanics, Dirac did not consider noncommutativity a sign that a visualizable, geometrical interpretation was fundamentally impossible in the quantum domain. In Dirac's view, noncommutativity was not counterintuitive, because it could be apprehended in geometrical forms. Dirac was familiar with "non-Pascalian" geometries, in which the "coordinates" of a point did not commute (Darrigol 1992a; Mehra and Rechenberg 1982). He consequently believed that the abstract relations between *q*-numbers, including the basic commutation formula, could be given a geometric interpretation (though not an interpretation continuous in space-time). Dirac made noncommutativity a central part of his version of quantum theory.

Heisenberg adopted Dirac's point of view. The geometric (kinematic) interpretation of the relation $pq - qp = (h/2\pi i)\mathbf{1}$ is the centerpiece of Heisenberg's uncertainty paper. The general interpretive problem in the paper is stated along lines laid down by Dirac: which *c*-numbers (coordinates in regular space-time) correspond to *q*-numbers (general abstract operators) that satisfy the fundamental quantum equation? Heisenberg's analysis of actual and imaginary experiments in his uncertainty paper was intended to demonstrate that experimental situations are consistent with the basic commutation relation of quantum mechanics.

Finality and Discontinuity

In a letter written in the fall of 1926, Heisenberg informed Pauli of Dirac's "extremely broad" generalization of Heisenberg's own fluctuation paper. Heisenberg presented Dirac's reasoning to Pauli as follows: The general dynamical problem was formulated by Dirac in fundamentally probabilistic terms. If p and q are canonical variables, what can one say physically about a function $f(p, q)$? If q is taken to be a specific *c*-number, say, $q = 10$, classically one can calculate $f(p, 10)$. In quantum mechanics this cannot be done—one can only specify the range of p for which f lies between the *c*-numbers f and $f + df$. Dirac succeeded in defining the probability function in the form of a general matrix S, which contained all the physically meaningful statements that could be made in quantum mechanics at the time, including statements about Born's collision processes and Jordan's canonical transformations. Heisenberg concluded enthusiastically, "I hold Dirac's work to be an extraordinary advance" (Heisenberg to Pauli, 23 November 1926, *PC*).

Dirac's generalized version of quantum mechanics provided clear guidelines for moving from one representation in terms of canonical variables to another. Incorporating work by Pauli and Schrödinger,

Dirac demonstrated that the wave function ψ could be viewed as a transformation from a scheme in which position is diagonal to one in which energy is diagonal.[18] Because of the freedom contained in the abstract conception of q-numbers, Dirac did not have to confine himself, as the matrix theorists initially had, to matrix schemes in which the energy matrix is diagonal. This freedom proved potent, especially in Dirac's able hands (Kragh 1990; Jammer 1966; Darrigol 1992a).

Heisenberg elaborated on this feature of Dirac's theory in his uncertainty paper, demonstrating how transformation theory supports the uncertainty relations and enhances the superiority of the matrix point of view over all others (see chapter 5). For Dirac the existence of a general, elegant, powerful set of transformations in quantum theory suggested that his version of quantum mechanics compared favorably with the most developed parts of physics—Hamiltonian mechanics and relativity theory (Darrigol 1992a, 348). Transformation theory gave Dirac strong incentive to consider quantum mechanics complete.

Heisenberg became an eager advocate of the completeness, or finality, of quantum mechanics. He also made a crucial step toward discontinuity: "I hold it now more than ever as completely out of the question that the world is continuous" (Heisenberg to Pauli, 23 November 1926, PC).[19] His enthusiasms for the finality and for the discontinuity of quantum mechanics reinforced each other. The fact that Dirac's powerful generalization was based on Heisenberg's discontinuity considerations further motivated, at the emotional level, Heisenberg's embrace both of discontinuity and completeness.

Yet at first this reinforced belief in discontinuity did not make matters easy. Heisenberg did not see how one could possibly obtain a physical interpretation in terms of familiar concepts in the quantum domain. In this discontinuous world waves and particles seemed completely alien: "But as soon as it [the world] is discontinuous, in all of our words that we use for the description of a fact, there are too many c-numbers. What the word 'wave' or 'corpuscle' means, one no longer knows" (Heisenberg to Pauli, 23 November 1926, PC).

Heisenberg's only consolation at the time was his intuitive "understanding" of why p and q cannot have exact values simultaneously. If space-time is discontinuous, the velocity at a definite point cannot have any meaning, because in order to *define* velocity at a point, one needs a second point infinitely close to the first: impossible in a discontinuous world (Heisenberg to Pauli, 23 November 1926, PC).

---

18. Jordan developed a similar theory, in which he studied transformations from one canonical pair of variables to another pair in a given representation.

19. "Dass die Welt kontinuierlich sei, halte ich mehr denn je für gänzlich indiskutabel."

We see that initially Heisenberg struggled with physical interpretation in terms of a geometric analysis of concepts, and not in (operational) terms of the process of measurement. This approach did not lead far: "When we talk about position or velocity," he wrote, "we always need words that are obviously not defined at all in this discontinuous world" (Heisenberg to Pauli, 23 November 1926, *PC*).

But Heisenberg *needed* a physical interpretation of the microdomain that used familiar kinematic concepts, both because of the theoretical advances of Pauli, Jordan, and Dirac and because of Schrödinger's contention that matrix mechanics was basically nonintuitive. Pauli provided the first clues: the position of a particle should be considered observable in principle, momentum and position cannot be determined simultaneously, the path of a particle should be described in statistical terms. Heisenberg exploited these insights fully in the uncertainty paper. Yet Pauli did not suggest, as far as the available evidence reveals, that the particle-kinematic interpretation of quantum physics be elaborated in operational terms. This decisive step was Heisenberg's, and it was taken, as I will shortly argue, under the influence of Dirac's and Jordan's work.

Both discontinuity and finality were crucial in Heisenberg's reasoning in the uncertainty paper. Discontinuity entered fundamentally (though incorrectly) into his analysis of the "uncontrollable discontinuous" Compton recoil (disturbance) during an idealized position measurement. From this point on, Heisenberg considered discontinuity the most essential feature of quantum mechanics and claimed that its importance could not be overemphasized. This emotional attachment to discontinuity, canonized in the misleading notion of disturbance, is probably the reason for Heisenberg's lifelong use of disturbance imagery. The appealing imagery of the disturbance concept, backed by Heisenberg's authority, would confuse many students and interpreters of the philosophy of quantum theory.[20]

Belief in the finality of quantum mechanics, strengthened in dialogue with Dirac, shaped Heisenberg's interpretative efforts. Without this belief, on encountering a discrepancy between "nature" (the continuous path of a particle) and "formalism" (an inability to describe space-time trajectories), Heisenberg would seek to improve the formalism rather than to reinterpret nature. As I will argue, Heisenberg's opinion that quantum mechanics was complete determined the direction of his concluding interpretive steps in the dialogue with Jordan.

---

20. Heisenberg continued to rely on disturbance imagery after Bohr's explicit critique of this notion. Let us mention that Bohr, despite his critique of the disturbance notion, nevertheless invoked it because it had "intuitive" appeal (see chapter 12).

However, this stand (both by Dirac and by Heisenberg) was premature; formidable problems remained to be solved (Darrigol 1992a, 338). For Dirac, the criterion of "mathematical beauty" reinforced his belief in the finality of quantum mechanics. For Heisenberg, whose philosophy was an opportunistic mixture of mathematical Göttingen and physical Copenhagen, the embrace of finality and discontinuity was spurred by his desire to rout Schrödinger's waves once and for all.

Appeal to Experiment

Heisenberg's initial deliberations on discontinuity did not contain any analysis of experiments (actual or idealized) as a possible solution to the interpretive problem. Nor did Heisenberg, who was at the time with Dirac in Copenhagen, mention the need to analyze experiments while outlining to Pauli the essence of Dirac's advances. Nevertheless, quantum theory as conceived by Dirac contained a puzzle: it was built on a statistical foundation, even as most recent experiments—Bothe-Geiger, for example—seemed to rule against fundamental indeterminism. This conflict had to be faced.

In the fundamental paper that presented the essentials of transformation theory, Dirac (1927) also tackled the question of physical interpretation in general, and the causality issue in particular. Dirac, unlike Born, did not consider quantum theory essentially statistical. The statistical element, Dirac argued, is only introduced when experiments are performed and questions are formulated using the information obtained: "The notion of probabilities does not enter into the ultimate description of mechanical processes; only when one is given some information that involves a probability . . . can one deduce results that involve probabilities" (Dirac 1927, 641). In letters to Pauli and Jordan, as well as in his uncertainty paper, Heisenberg sided with Dirac on this issue (Heisenberg to Pauli, 23 February 1927, *PC*).[21] Might not analysis of experiments prove helpful for understanding other nonclassical peculiarities of quantum theory? Heisenberg could naturally have inquired.

Dirac's attitude toward experiments was influenced by his opinion that there is an analogy between the generalized version of quantum theory and relativity theory (the freedom to move from one scheme to another). This analogy suggested the introduction of the notion of experiment, or even of an observer, into the interpretive search. As Darrigol (1992a) points out, Dirac considered the existence of the group of quantum mechanical transformations, relating different "points of

---

21. "As Dirac does, so I believe that . . . all statistics is only introduced through our experiments" (Heisenberg to Pauli, 23 February 1927, *PC*).

view," very important. By analogy with relativity, it implied that some variables—position, for example—are frame dependent (in the position representation, positions have an exact value, while in the momentum representation, they are completely undefined). The need for an analysis of observational possibilities was then a natural conclusion. Dirac's somewhat later statement sums up this trend of thought in a strong form: "This state of affairs . . . [implies] an increasing recognition of the part played by the observer in himself introducing the regularities that appear in his observations" (1930, v).

Heisenberg's reasoning in the uncertainty paper echoed Dirac's deliberations. In the abstract of the paper Heisenberg stated that the uncertainty formula "is the real basis for the occurrence of statistical relations in quantum mechanics" (1927b, 62). Heisenberg transformed Dirac's question "How is statistics introduced by experiments?" into "How is uncertainty introduced by experiments?"[22] His answer inaugurated the thorny "disturbance" explanation: In experiments used for the operational definition of concepts, microscopic particles "suffer an indeterminacy introduced *purely by the observational procedure we use* when we ask of them simultaneous determination of two canonically conjugate quantities" (Heisenberg 1927b, 68, my italics). Heisenberg identified, in the disturbance concept, the principal distinction between classical and quantum mechanics. In classical mechanics (as in quantum mechanics) one can only give the probability of a definite position of an electron if one does not know the phases of motion, yet in the classical case one can always conceive of determining phases precisely, through suitably chosen experiments. In quantum mechanics, however, such determination "is impossible, because every experiment for determination of phase *perturbs or changes the atom*" (my italics). In a definite stationary state, the phases are in principle undetermined, as follows from the basic equations $Et - tE = -i\hbar$ and $Jw - wJ = -i\hbar$ (Heisenberg 1927b, 66).

Dialogue with Jordan

Heisenberg's dialogue with Pascual Jordan was crucial in the last phase of his search for an adequate interpretation of quantum mechanics.[23] It greatly reinforced Heisenberg's growing tendency to bring the analysis of experiments into the interpretive picture. In direct response to Jordan, Heisenberg formed a structure of argumentation that unified previously fragmented insights. Heisenberg, as his correspondence with

22. "One can say, if one will, with Dirac, that the statistics are brought in by our experiments" (Heisenberg 1927b, 66).

23. My description of the Heisenberg-Jordan dialogue follows Beller (1985) closely.

Pauli reveals, struggled with the issues raised by Jordan (Heisenberg to Pauli, 5 February 1927, *PC*). But so great was the inspiration that within two weeks, he had a response: the core of the uncertainty paper (Heisenberg to Pauli, 23 February 1927, *PC*).

Jordan's Discussion of the Causality Issue

Jordan's paper (1927b) was devoted to exploring the relation between the foundations of quantum mechanics and the problem of causality. Jordan stated that causality is not an a priori necessity of thought but rather a question that should be settled by experiment. Physicists cannot be satisfied by an approximate metaphysical determinism: to arrive at a definition of determinism, the physicist must specify conditions under which its existence can be verified. The definition of determinism must therefore change in accordance with theories and experimental methods. Determinism in classical physics means both complete knowledge of initial conditions (in biology, for example, such knowledge is unattainable) and the special character of physical laws (second-order partial differential equations in a four-dimensional manifold, with one dimension, time, being imaginary). In the classical case, the present will uniquely determine the future course of events, because in the classical world, quantities are continuously propagated through space and all physical motions are continuous.

In the quantum domain one must be prepared to sacrifice this kind of determinism because of the existence of elementary discontinuities—quantum jumps. Quantum laws are indeed at present statistical; in general they say nothing about a single atom but yield only mean properties of an assembly of similar atoms. If, however, one directs one's attention not to discontinuities but to probabilities that evolve according to Schrödinger's differential equations, one can conclude that determinism (for these probabilities) holds by analogy with the classical principles (Jordan 1927b).

What is the connection between the statistical character of quantum laws and experiments? Initially, one is tempted to conclude that experiments would give nothing but average values. This conclusion could be based on the results of Frits Zernike (1926) and Gustaf Ising (1926), who demonstrated the existence of impassable limits of accuracy in measurements, due to Brownian movement. Ising's and Zernike's work was triggered by the attempt of two experimentalists, Erich Moll and Herman Burger (1925), to improve the accuracy and sensitivity of measuring devices. By constructing a "thermal relay" device they achieved an unprecedented improvement in the sensitivity of galvanometers. This device allowed the detection of the origin of various disturbances during measurement, and consequently, the disturbances could be eliminated (for details, see Beller 1988). Yet there always remained a conspicuous disturbance of the zero point.

Ising (1926) analyzed Moll and Burger's data and demonstrated that the perpetual fluctuations of the zero point were simply Brownian fluctuations. Ising's paper, which implied limits on the accuracy of all measurements, did not cause any conceptual stir. The widespread assumption was that by reducing the temperature of the measuring device one could reduce the amplitude of the fluctuations, which should approach zero as the temperature approaches 0°K. Yet Zernike (1926) soon showed that even in this case the fluctuations would persist. Although the oscillations of a freely swinging galvanometer can be reduced by lowering the temperature, there are fluctuations in the charge of the external circuit to which the galvanometer is connected. A galvanometer will exhibit these fluctuations even if its own temperature is 0°K.

What is the connection between quantum indeterminacy and the limits of actual measurement? "When we remember that this is the case with all our apparatus, and that it all 'rattles about' in this way, we may be tempted to think that the *experimentalist is quite as incapable of observing elementary processes as the quantum theorist is of predicting them*" argued Jordan (1927b, 568, my italics). Yet one could avoid these experimental uncertainties, he continued, by making experiments at absolute zero or, more comfortably, by working with particles with a vast store of energy. In such cases, one could observe in Wilson's cloud chamber the trajectory of a single particle and determine exactly the moment when the trajectory ended in a quantum jump.

Thus, Jordan stated, the time of a quantum jump is in certain cases exactly determined. How does this fact compare with available theory? Because quantum theory is statistical, one might conclude that it can only provide the probability of a quantum jump, and that therefore the time of transition is undetermined. Yet this conclusion does not follow from the foundations of the theory; it was an additional hypothesis of the Bohr-Kramers-Slater theory, which was disproved by experiment. In this respect, it is significant that Pauli's assumptions say nothing about the probability of a transition (they refer to the probability that a system will be at a given place in configuration space). In order to deal with this issue, one has to reduce the probabilities of quantum theory to elementary probabilities. "Only then can we say that we really understand the laws; and only then can we tell under what conditions the time of transition is determined. Only then can we know exactly what is causally determined, and what is left to chance." The question of determinism in the quantum domain remains open (Jordan 1927b, 569).

Heisenberg's Response to Jordan's Discussion

It is convenient to summarize Jordan's theses as items 1 through 5 below, and then to follow with Heisenberg's direct responses to them, designated items 1' through 5', in a tacit dialogue with Jordan.

> 1. There are experimental situations that cannot be described by the formalism of quantum mechanics (the trajectory of a particle, the determinate time of a quantum jump in certain cases). Thus quantum theory is as yet incomplete.

> 2. One may think initially that experiments, like theory, contain essential indeterminacies due to Brownian motion. Yet these experimental indeterminacies can be avoided by choosing experimental conditions properly.

> 3. Quantum theory is essentially a statistical theory. Yet it has not been reduced to independent probabilities. Therefore, quantum theory is as yet incomplete.

> 4. The question of determinism is one of both the completeness with which the initial conditions are known and the special form of the physical laws.

> 5. Because of point 3, one cannot yet give a conclusive answer to the question of determinism in the quantum domain.

We know from Heisenberg's correspondence and his recollections that he regarded quantum theory at that time as final, and its mathematical formalism as complete. Thus Heisenberg's responses to points 1 through 5 run as follows:

> 1'. Because quantum theory is already complete, there can be no experimental situations that cannot be described by its formalism. If one finds such situations (such as the path of a particle), they must be reinterpreted so as to fit the formalism.

> 2'. One must be able to demonstrate the existence of unavoidable experimental indeterminacies, in order to have agreement with theoretical indeterminacies.

> 3'. Quantum theory is complete. Therefore, it is not essentially a statistical theory.

> 4'. Because quantum theory is not essentially statistical, the question of indeterminism reduces to the question of the completeness of knowledge of initial conditions. Because of point 2', this knowledge is essentially incomplete.

5'. From points 3', 4, and 4' it follows that one can give a conclusive answer to the question of determinism: there is no determinism in the quantum domain.

Points 1' through 5' are the central theses of Heisenberg's uncertainty paper. Heisenberg was familiar with Jordan's paper, having read it with *grossen Genuss* (*Genuss* means both "joy" and "profit"). Heisenberg mentioned Jordan's work in connection with his attempt to understand the logical foundations of the $pq - qp$ relation, and inspired by Jordan's explorations, he began contemplating what the position of an electron means. In his paper, Jordan had inquired about the "probability of finding an electron in a certain place." Heisenberg found this expression unrigorous. On the basis of his previous stand that an electron's position is unobservable, Heisenberg did not consider "the place of an electron" to be a defined concept: "What, for example, does 'probability for an electron to be in a definite point' mean when the concept 'position of an electron' is not appropriately defined?" (Heisenberg to Pauli, 5 February 1927, *PC*). Just two weeks later, Heisenberg answered this question in another letter to Pauli, in which he laid out all the essentials of his uncertainty paper (Heisenberg to Pauli, 23 February 1927, *PC*). In the spirit of points 1' through 5', the question about an electron's position was answered in operational terms.

Heisenberg's discussion of the limits of measurement was directly suggested by Jordan's paper. Heisenberg sought quantum mechanical limits on measurements by analogy with classical limits in thermodynamic terms. Bohr, familiar with Heisenberg's efforts, was later explicit on this point: "It is true, as emphasized by Heisenberg, that an instructive analogy to the quantum theoretical point of view is obtained by comparing the uncertainty in the observations of atomistic (microscopic) phenomena with the uncertainty inherently contained in any observation, due to imperfect measurements, as considered in the ordinary description" (1927c, 123). Perhaps it is worthwhile to add here that Zernike's paper (1926) about limits on the accuracy of measurements was published in the same issue of *Zeitschrift für Physik* where Jordan's transformation theory was elaborated (Jordan 1927f). Heisenberg, of course, read this paper of Jordan's (as he wrote to Pauli, 5 February 1927, *PC*). It is unlikely that he did not at least glance through Zernike's work, after it had been called to his attention by Jordan's discussion. Thus that Ising and Zernike had an impact on Heisenberg's thought is highly plausible. The realization that there are inevitable limits on the accuracy of real measurements provided Heisenberg with the crucial clue: he would pursue an intuitive interpretation of quantum mechanics through an analysis of the limits of measurement in his thought experiments.

Dialogues with "Lesser" Scientists

The names Zernike and Ising do not appear in the usual historical works dealing with the genesis of Heisenberg's uncertainty principle. Nor do we find the names Duane, Campbell, or Sentfleben.[24] Yet all these scientists had prominent places within the dialogical web that formed as Heisenberg groped toward the formulation of the uncertainty principle. By burying such names in the annals of history, the usual accounts suppress the dialogical nature of reasoning and enhance the "hero worship" tradition. In "revolutionary" accounts the steps taken by the central players seem more discontinuous (and irrational) than they actually were. In linear "rational reconstruction" accounts, the complexity and ingenuity of each scientist's complex reasoning is hardly apparent. A dialogical analysis of the achievements of our scientific heroes puts into proper light their outstanding creativity. Such an analysis also resurrects the "lesser scientists," bringing back to life the relevancy, ingenuity, and indispensability of their thought.

Dialogue with Campbell

The revolutionary realization that time-space concepts do not apply in quantum mechanics is usually ascribed to Bohr, and the ingenious implementation of this realization in the mathematical formalism of quantum mechanics to Heisenberg. Yet many other scientists before Bohr and Heisenberg contemplated the overthrow of mechanics and the modification of space-time concepts: Planck, Poincaré, Richardson, Sommerfeld, Campbell, and Sentfleben, among others.

After the appearance of matrix mechanics, Norman Campbell (1926) repeated and extended his earlier call (Campbell 1921) for a modification of space-time concepts in the quantum domain. Campbell's papers are significant because they concentrate on the necessity of modifying the concept of time and redefining the notion of the path of an intraatomic electron. Both insights were used in Heisenberg's uncertainty paper. Campbell (1926) connected the existence of paradoxes within the quantum description with the uncritical adoption of the time concept in matrix mechanics. Although matrix mechanics arose from Heisenberg's daring reinterpretation of the space container, Heisenberg assumed, for no good reason, that the classical concept of time applied without alteration to atomic systems. Such a concept of time underlies, for example, Heisenberg's statement that frequencies of radiation belong to entities observable in principle. But this statement is controversial: the determination of frequencies demands two types of experiments—

24. As far as I know, only Hendry (1984) noticed the connection between Sentfleben's work and Heisenberg's.

interference experiments, where no temporal component is involved, and experiments on the measurement of the velocity of light, in which temporal magnitudes are associated not with atomic systems but with macroscopic bodies, such as planets or revolving mirrors. It is surely a hasty step "to say that we have observed in 'principle' the frequency belonging to the latter when we have actually measured a frequency belonging to the former" (Campbell 1926, 1115).

Contrary to Heisenberg, Campbell (1926) suggested considering time as a purely statistical concept and, consequently, such notions as velocity or frequency as applicable only to statistical aggregates. Campbell supported his proposal with an analysis of measurement by radioactive clocks. In the case of the radioactive clock, there is no possibility of defining a time interval by observing a single atom—only a large collection of atoms allows measurements of temporal intervals.

In his discussion Campbell raised a number of fundamental issues that were close to Heisenberg's interpretive concerns:

1. Campbell's analysis of the measurement of frequencies implied that such observables are theory laden. Such an analysis could have reinforced Einstein's famous dictum that "it is the theory which decides what we can observe," which, according to Heisenberg's recollections, was crucial to Heisenberg's discovery of the uncertainty principle (1971, 63).

2. Campbell's suggestion that time is statistical in nature became a focus of Heisenberg's deliberations about the statistical nature of quantum theory. Just a few months before his formulation of the uncertainty principle, Heisenberg wrote to Pauli that perhaps "space and time are in reality only statistical concepts, such as temperature, pressure, etc. in a gas. I mean that the space-time concepts for an individual corpuscle are meaningless" (Heisenberg to Pauli, 26 October 1926, PC). Heisenberg's words are remarkably close to Campbell's assertion: "The view I want to suggest here is that time, like temperature, is a purely statistical conception, having no meaning except as applied to statistical aggregates" (1926, 1107). Heisenberg eventually abandoned this approach and began to explore the limits of applicability of space-time concepts for an individual atomic particle. Yet Campbell's insistence that the analysis of space-time concepts and the problem of causality were closely related—"time and chance are merely two aspects of the same thing" and "time and chance are inevitably associated" (1926, 1107, 1108)—reverberates through Heisenberg's uncertainty paper. Campbell's discussion could have suggested to Heisenberg the formulation of the time-energy uncertainty relation in probabilistic terms. Campbell's discussion of the interrelation of space-time and causality was also a trigger for Bohr's discussion of the complementarity of space-time and causality (see chapter 6).

3. There is another point in Campbell's analysis that resonates in Heisenberg's interpretive attempts. Campbell suggested abandoning the usual temporal concepts, yet he clearly realized that the concept of time is deeply embedded in our thought and language and that without it "we could not even state adequately the problems we desire to solve" (Campbell 1926, 1110). Is it possible to proceed in such a case? Yes, Campbell replied to his own question, but only if one abandons formal consistency. By using temporal conceptions freely in our constructive efforts while simultaneously denying their validity, we might eventually realize the direction of a future consistent solution.

How can one explore Campbell's suggestion that the concept of time does not apply to an individual system? Campbell's answer is experiment oriented: "The value of these ideas would be proved decisively if it were possible to produce a theory . . . involving no temporal conceptions . . . and at the same time to show that all the experiments on which the prevailing temporal conceptions are based can be described in terms of statistics" (1926, 1110). As follows from Heisenberg's letters to Pauli, Heisenberg repeatedly hit dead-ends when analyzing such kinematic concepts as "velocity" in the "discontinuous world." Campbell's suggestion that one could clarify the application of the concept of time through an examination of experiments, without achieving prior analytical clarification of the concept, was used in Heisenberg's uncertainty paper.

Another fundamental point, which was common to Campbell's and Heisenberg's discussions, is the reinterpretation of the concept of the orbit of an intraatomic electron. Both Campbell and Heisenberg conceived of such orbits as sequences of discrete points. One of the formidable problems Heisenberg struggled with was how to introduce space-time concepts for an individual particle without simultaneously resurrecting the continuous intraatomic path. While Heisenberg's solution goes beyond Campbell's, Campbell's deliberations nevertheless offered a new way of conceiving the path of an electron as an irregular series of discrete points. According to Campbell, a particle that is moving along an orbit is actually undergoing irregular transitions, the locus of which is a closed curve. Campbell perceived "stationary states as consisting of transitions, just as much as the changes from one stationary state to another" (1926, 1114). While quantum jumps relate to transitions between states of different mean energy, stationary states are transitions between states of the same mean energy. Campbell considered his suggestion to be in the spirit of Heisenberg's, Born's, and Jordan's matrix theory: "Their analysis depends on the use of double-affix quantities, such as $\gamma_{m,n}$ which are associated with transitions, and in the formal treatment they make no essential distinction between quantities in which the affixes are different and those in which they are the same, but these latter can be associated only with transitions within the same stationary state" (1926, 1115).

Campbell does not offer any concrete physical theory—only guidelines for the principles on which the future quantum theory should be built. He apologized for "offering such a collection of incomplete and incoherent suggestions" and then concluded prophetically: "If anyone can make any use of them, he will be much better entitled than myself to be regarded as their author" (1926, 1117).

Dialogue with Sentfleben

While peripheral in the context of the issues discussed in the uncertainty paper, it is Heisenberg's verdict on causality that attracted the attention of philosophers and philosophically minded scientists. Heisenberg announced that his discussion established "the final failure of causality" (1927b, 83). He explained confidently: "In the strong formulation of the causal law, 'If we know exactly the present, we can predict the future' [it] is not the conclusion but rather the premise, which is false" (1927b, 83). Since we cannot know the present precisely, we cannot predict the future with certainty.

This solution was a novel one. In classical physics it was assumed that in principle (though not in practice) one can know the present exactly (or at least Laplace's demon can). According to the philosopher Moritz Schlick, Heisenberg's solution was a great surprise for modern philosophy "since even the mere possibility of such a solution had never been anticipated," despite frequent discussion of this problem through the ages (quoted in Jammer 1974, 75).

Schlick's assessment was wrong: H. A. Sentfleben, a physicist whose name does not appear in the usual accounts of the development of quantum physics, reached a similar conclusion four years before Heisenberg (Sentfleben 1923). Sentfleben argued that in principle the present cannot be precisely defined; therefore, the future can only be known in statistical terms. Sentfleben claimed that space-time must be discontinuous because the size of Planck's constant h sets the limits for the precision with which space-time concepts can be defined. According to to Sentfleben, one need not give up space-time concepts (as the original matrix theory later did), but only determine the limits of their applicability.[25]

25. The realization that the introduction of h demands a modification of space-time concepts was understood as early as 1911 by Planck, who at the first Solvay conference called for a modification of classical mechanics. Planck argued that in order for the energy of an oscillator to be an integral multiple of the energy element $h\nu$, phase space must consist of finite elementary areas h. This issue is discussed in Jammer (1966, 52–56).

Sentfleben's thoughts on the modification of space-time were connected with his agenda to obtain a unified foundation for physics. Sentfleben, like Einstein before him, aimed at avoiding the dualism of discrete atomistic conceptions and continuous field theoretical ones. Sentfleben identified the source of the contradictions in the old quantum theory as an illegitimate extension of continuous field theoretical notions into the discrete microdomain. In order to build a consistent microphysics, one has to reinterpret the notions of space and time. Because description in terms of space-time notions is always connected with phenomena in which h plays a substantial role, Sentfleben introduced the following principle: "The possibility to describe space-time events with arbitrary precision is limited in principle, where Planck's constant h is the factor which determines the limits of precision by which a coincidence can be set between a matter-point and a space-time point" (1923, 131).

According to Sentfleben (1923), the world is discontinuous in principle (let us recall how Heisenberg struggled to define kinematic concepts in such a discontinuous world). For small distances and times, clocks and rods become useless because of their own characteristic oscillations. Anticipating Heisenberg's analysis, Sentfleben argued that the problem of the applicability of space-time notions is intimately connected with the analysis of limits on measurements. In particular, due to the molecular fluctuations of measuring clocks, the time of an event cannot be defined with unlimited precision.

A new discontinuous quantum theory would clearly be opposed to field theoretical macroscopic theories, in which strong determinism holds, argued Sentfleben. This new theory must be formulated in statistical terms. Because Planck's constant h limits in principle the possibility of describing a space-time process with arbitrary accuracy, Sentfleben concluded that the usual formulation of the law of causality ("Given a space-time situation A, we can determine a later situation B") is inadequate: "We simply cannot have a precisely defined situation A" (quoted in Hendry 1984, 33). The similarity with Heisenberg's discussion in the uncertainty paper is striking.

The idea that h determines the structure of space-time is fully expressed in the division of phase space into finite elementary areas of size h (cells). The division of phase space into quantum cells of size h was a common procedure, used in many quantum theoretical papers at the time. Schrödinger understood the direct link between Heisenberg's uncertainty principle and Planck's quantum cells. In a letter to Planck, Schrödinger wrote: "The partition of phase space in cells which you [Planck] introduced for statistical purposes, already contained the fundamental uncertainty relation, which was only much later established *expressis verbis* on the ground of newest developments" (quoted in Darrigol 1992a, 264). Both Heisenberg and Pauli realized the relevance of such a division to their attempts to understand the kinematic meaning of the quantum mechanical formalism (Heisenberg to Pauli, 15 November 1926, *PC*). In this regard, Heisenberg commented in the uncertainty paper: "Equation (1) $[p_1 q_1 \sim h]$ is a precise expression for the facts which one earlier sought to describe by the division of phase space into cells of magnitude h" (1927b, 65). But while Pauli and Dirac reasoned in terms of the precise determination of one conjugate variable at the price of the complete indeterminacy of another, Heisenberg formulated the problem in terms of the degree of precision of a joint specification of p and q (in terms of the size of cells h in phase space). It is this psychological leap from the world partly known into the world that is imprecise in principle that distinguishes Heisenberg's deliberations from those of Pauli and of Dirac. In this respect, there is a strong similarity between Heisenberg's and Sentfleben's approaches. Throughout the uncertainty paper, Heisenberg, as did Sentfleben, used the word *Ungenauigkeit* (imprecision) rather than *Unbestimmtheit* (indeterminacy) or *Unsicherheit* (uncertainty). After Heisenberg formulated the problem along these lines, the uncertainty formula immediately followed: for a microparticle in a cell of size h, the joint specification of p and q must obey $\Delta p \Delta q \sim h$. Knowing the desired conclusion beforehand, Heisenberg could easily deduce it from the mathematical formalism by elementary use of transformation theory.

CHAPTER 5

The Polyphony of Heisenberg's Uncertainty Paper

He [Heisenberg] is very unphilosophical.

Wolfgang Pauli to Niels Bohr, 11 February 1924, *PC*, 143

The one who insists on never uttering an error must remain silent.

Werner Heisenberg 1958, 86

Introduction

Heisenberg's reasoning in the uncertainty paper emerged from a complex network of dialogues, intellectually challenging and emotionally charged. The gradual forming of preferences, of choosing one intellectual option over another, of shaping opinions, occurred as insights gained during many different dialogues coalesced. I argued in the previous chapter that Heisenberg's preference for acausality formed in reaction to Schrödinger's space-time causal program, and in dialogue with Pauli's deliberations about the problem of determining the times of intraatomic transitions. Heisenberg's elaboration of the causality issue received further impetus from Campbell's discussion of chance and space-time in quantum theory, from Sentfleben's discussion of causality, and from Jordan's doubts about the finality (and therefore about the inherently statistical character) of quantum theory. Heisenberg's opinion that the quantum mechanical formalism was final crystallized gradually. Crucial in this respect was the perceived need to insulate matrix mechanics from Schrödinger's competing attempts, and the strong support from Dirac's theoretical advances. Heisenberg's decision to undertake an analysis of thought experiments as the focus of his interpretive attempts emerged in the last stages of the creative process, in dialogues with Dirac, Campbell, and Jordan. Not the magisterial unfolding of a single argument, but the creative coalescence of different arguments, each reinforcing and illuminating the others, resulted in Heisenberg's monumental contribution to physics. In the process of Heisenberg's discovery, the communicative nature of thought was fundamental.

It would be surprising if the final product arising from such a plurality of issues, challenges, and doubts were unambiguous and unequivocal. It would be surprising if the polyphony of the creative act could be systematized into a neat conceptual scheme. Heisenberg's uncertainty paper contains in fact traces of past struggles, vacillations on central issues, subdued, but continuing doubt about the fundamentals.[1] Conflicting voices permeate the text. And the equivocations and contradictions of the uncertainty paper reverberate in subsequent writings by Heisenberg and other Göttingen-Copenhagen physicists.

The fundamentally new formula of quantum mechanics in the matrix formulation is the basic commutation relation $pq - qp = (h/2\pi i)\mathbf{1}$. We saw that initially the authors of matrix mechanics deduced nothing interesting or physically meaningful from this formula:[2] outside a dialogical context formulas are mute. Heisenberg's reasoning in the uncertainty paper was aimed at connecting the uncertainty relations with the commutation formula—yet how exactly they are related remained unclear. Heisenberg was "seeing double" on issues of *Anschaulichkeit*, on the role of classical concepts in the quantum domain, and on the causality issue.[3] He was even unsure about what it meant to provide an interpretation, vacillating between an operational and a model theoretical approach. It was in this context that Heisenberg introduced the thorny disturbance problem, and the ensuing confusion between inevitability and consistency that plague the Copenhagen philosophy. Similarly, the contradictory voices of positivism (operationalism), model theoretical realism (the invariant features of a successful scientific theory refer to genuine aspects of reality), and conventionalism (physicists can freely choose the basic axioms of a theory, worrying only about its consistency and empirical adequacy) are all present in Heisenberg's paper.

The voices in Heisenberg's paper are not neutral or disembodied. They retain traces of the high emotions that characterized Heisenberg's confrontation with the wave theoretical approach. We hear triumph: after Heisenberg's own ingenious interpretation matrix mechanics should no longer be considered nonintuitive, refuting Schrödinger's accusations (Heisenberg 1927b, 82). We hear arrogant but good-tempered sarcasm: Heisenberg reminded Schrödinger of the simple consideration

---

1. See also Holton (1991) on conflicting voices in Einstein's early work in physics.

2. Except declarations about the symbolic and nonintuitive character of the matrix theory.

3. "Seeing double" is Shapin and Schaffer's (1985) term, relating to Hobbes's philosophy.

that wave packets disperse, "only because in recent times they seem to be forgotten" (1927b, 73).

Heisenberg's paper is not a systematic monologue, in which a reader follows the unfolding of a central argument to its end. We will see that Heisenberg stopped his deliberations abruptly in order to send the paper for publication. It is contingent, I will argue, which voices Heisenberg chose to emphasize and which voices he decided to withhold. Some of these voices, suppressed at the last moment, reemerged in Heisenberg's future publications; other voices perhaps were lost.

Looking at Heisenberg's paper in this way, we cannot sustain the illusion that Heisenberg wrote the uncertainty paper in order to convey an unequivocal message, the meaning of which was clear to Heisenberg himself, and that therefore the job of historians and philosophers is to decipher Heisenberg's original message. Heisenberg, who was wavering on many issues, hoped that the very act of writing, together with Pauli's response, would help him to "get some sense of his own considerations" (Heisenberg to Pauli, 27 February 1927, *PC*). These considerations Heisenberg laid out in his letter to Pauli, and the text of this letter was taken almost verbatim into the published version. Pauli's response (which is lost) seems to have served at this stage more as reinforcement than as clarification.

The plurality of voices in a scientific paper should not be confused with stable elements in the Gestalt model of creativity. According to Wertheimer (1959), the creative act consists in the recognition of a restructuring idea, which resolves the tension that exists between separate facts, or elements. The resulting new Gestalt eliminates contradictions, achieves unity, and provides insight. Using this model, Wertheimer, and after him Miller (1984), described Einstein's discovery of the special theory of relativity. From the dialogical perspective, there simply are no final, stable elements, or facts—everything can be questioned and doubted. Nor is there a total, final resolution of tension—conflicting voices coexist, or are temporarily put aside. Rarely are they completely extinguished.

A paradoxical tension exists between the openness of a scientific text, addressed to the future, and its solid roots in the past. This tension is connected with another tension—that between innovation and continuity. Heisenberg ended his letter to Pauli with an apology, and an anxiety, that all his deliberations about uncertainty contained nothing new—they were merely an *alter Schnee* (old snow; Heisenberg to Pauli, 27 February 1927, *PC*). Almost all of us are familiar with such fears. They are connected not only with natural doubts about the worth of our contributions but also with an inner knowledge of their dialogical connection with the work of others. When we are aware that our ideas are rejoinders to those of others, how can we be sure that we not merely repeat but creatively respond? This is the reason, I suggest, for such frequent vacillation in the evaluation of our own work between two extremes—"deeply profound" and "utterly trivial."

The dialogical nature of creativity explains why reinforcement, support from another voice, is so important. The more novel the idea, the stronger the need for reinforcement. For how can we know, except through validation by another's voice, whether our ideas are daring or reckless, profound or absurd?

The Polyphony of the Notion of Interpretation

Heisenberg's uncertainty paper was aimed at providing a satisfactory interpretation of the new quantum theory. Today there are many interpretations of quantum theory—the Copenhagen, the many worlds (Everett 1957; De Witt and Graham 1973), the realistic statistical (Ballentine 1970), the (nonlocal) hidden variables (Bohm 1952; Vigier 1982), the modal (van Fraassen 1981, 1991; Kochen 1985; Dieks 1994; Bub 1992; Healey 1989), the quantum logical (Finkelstein 1965; Putnam 1968; Friedman and Putnam 1978).[4] Yet there is no agreement on the basic question: what does it mean to interpret a mathematical-physical theory? Nor is there an unequivocal answer to another basic question: what is a theory?

The legacy of logical positivism implies that a physical theory is a partially interpreted formal system consisting of an abstract formalism F and a set of rules, or bridge principles, or correspondence rules, R that connect the nonlogical terms of the formalism with observable phenomena.[5] This notion of a scientific theory survives in what Redhead (1987) called the "minimal instrumentalist" interpretation of quantum

---

4. "It sometimes seems as if there are as many different attempts [to understand quantum mechanics] as there are people who have seriously made the attempt" (Healey 1989, 2). In fact, it is difficult even to list all the exciting interpretive work being done nowadays as an alternative to the Copenhagen interpretation. The recent volume on Bohmian–de Broglie alternatives edited by Cushing, Fine, and Goldstein (1996) includes such notable examples as Dürr, Goldstein, and Zanghi (1996) and Valentini (1996). For an overview of this interpretive work, see Redhead (1987) and Cushing (1996).

5. There is no need to rehearse here numerous objections to this approach, the difficulty of distinguishing between theoretical and observational terms being among the most prominent. Suffice it to say that all historically sensitive philosophy of science, starting with Kuhn and Feyerabend, sprang from a critique of this positivist model for a scientific theory. It is worthwhile perhaps to add that in the case of quantum theory, even the notion of an abstract formalism is problematized: philosophers disagree about the best way to formalize quantum theory. There is no unique formal system whose semantics is open (Healey 1989, 5).

theory. This interpretation contains a quantization algorithm for any given observable (it determines what measurement results, or values, are possible) and a statistical algorithm (it gives the probability of a possible measurement result). According to this approach, the quantum mechanical formalism is indeed no more than a tool for the description and prediction of measurement results, as Bohr claimed.

Yet the minimal instrumentalist interpretation serves more as a refuge when one is pressed by paradoxes and contradictions (especially in the case of the Copenhagen interpretation) than a consistent, committed interpretive stand. Most physicists, Bohr and Heisenberg included, wanted more: some feeling of understanding, of illuminating, of explaining the kind of world the quantum formalism describes. The need for this kind of metaphysical grasp is not merely psychological but social as well—the power of a successful explanation and the power of the effective legitimation and dissemination of a theory are connected. This is one of the reasons that Bohr, despite his minimal instrumentalist pronouncements, nevertheless invoked his principle of complementarity to "explain" the quantum mystery.

One need not be a naive realist to appreciate the need for an interpretation more satisfying than the minimal instrumentalist stand. Van Fraassen formulated the issue of interpretation in the following way: "What would it be like for this theory to be true, and how could the world possibly be the way this theory says it is?" (van Fraassen 1991, 242). Heisenberg had his own succinct definition of interpretation: to understand a theory means *daraus klug zu werden* (to get clever)—to get sense out of a theory, to get a feeling of enlightenment and insight.

Heisenberg's definition shows why the question of interpretation is a controversial one—in which circumstances one gets a feeling of understanding is personally, professionally, and culturally conditioned. Those scientists who feel at ease with mathematical formal structures often gain enlightenment from the analysis of a mathematical formalism. Often scientists are willing to make bold ontological predictions relying on the features of mathematical structures.[6] Physicists especially are inclined to ascribe realistic meaning to the invariant features of a theory, as is demonstrated by the writings of Einstein, Dirac, and Born. And indeed most recent attempts to interpret quantum mechanics rely on the mathematical formalism in a significant way. Those who shy away from mathematical formalism try to gain enlightenment by linguistic ingenuity. As medieval theologians, such as Maimonides and Aquinas, ascribed apparent contradictions in the discourse about God's

---

6. Leverrier's prediction of Uranus, Dirac's of the positron, and Gell-Mann's of the quark are characteristic examples.

attributes to the limits of ordinary discourse, so Bohr resolved the contradictions of quantum theory by analyzing the limits of ordinary language (MacKinnon 1994). Heisenberg moved freely between these two options, choosing his approach according to the type of audience he addressed and the type of problem he encountered.

In his letters preceding the formulation of the uncertainty principle, Heisenberg expressed his desire "to get clever" more than once. Yet the sense of this expression changes: originally Heisenberg's dominant voice was that of denial—denial that a quantum ontology was possible (the only enlightenment one could hope for was to understand the transition from the macro- to the microdomain); after Schrödinger's challenge, Heisenberg, more ambitiously, began the search for deeper understanding. In a broad sense, the Göttingen-Copenhagen physicists sought to understand the "naturalness" of agreement between theory and experiment. For Born, a Leibnitzian preestablished harmony existed between the world of mathematical symbols and that of facts observed in the laboratory. Born, as I have argued, used the notion of preestablished harmony to justify his own statistical interpretation. This notion was retained in Bohr's Como lecture, which was woven around the "harmony" between definition and observation (see chapter 6). Physicists outside the mainstream, such as Campbell, also suggested that to substantiate a radically new theory, one should discuss experiments (thought experiments and actual experiments) and demonstrate how the theory works (how the theoretical concepts are used in experimental situations; chapter 4).

Heisenberg's letter to Pauli is built around "making sense" out of the uncertainty formula in numerous thought experiments (Heisenberg to Pauli, 27 February 1927, PC).[7] Heisenberg's discussion of these experiments in the letter convinced him that he was able to offer a satisfying interpretation of quantum physics. Despite his continuing doubts, the discussion had "eased" his conscience (mein Gewissen erleichtert). Indeed, Heisenberg had opened the letter to Pauli by expressing a hope that by the very act of writing he would clarify the interpretive issues to himself (dadurch hoff' ich auch selbst drüber klar zu werden). Yet the feeling of relief and insight that Heisenberg achieved was limited: he concluded the letter by saying that his deliberations might be merely vague speculations, that they probably contained nothing new, and that they were unclear on many points. In order to achieve clarity, Heisenberg revealed, he would need further communication with Pauli.[8] Yet Hei-

7. The existing literature on the history of quantum mechanics usually only mentions one of these thought experiments—the γ-ray experiment.

8. "To make anything clearer, I must write to you about it" (Heisenberg to Pauli, 27 February 1927, PC).

senberg did not wait to achieve the desired clarity and understanding. Driven by ambition, relieved by his letter to Pauli, and encouraged by Pauli's positive response, Heisenberg sent the text of the uncertainty paper to print. This text is, in its main points, identical with the letter. No wonder the uncertainty paper is shot through with doubts and contradictions.

Heisenberg had no definite notion of an interpretation, nor did he need one. All he intended, as he wrote in the abstract to the uncertainty paper, was "to show how microscopic processes can be understood" through quantum mechanics (1927b, 62). Heisenberg's stated goal was to "illustrate" quantum theory by discussing thought experiments. Yet in order to disseminate the new theory effectively and to make it palatable, Heisenberg had to offer not simply a satisfying "illustration" but one that was intuitive (anschauliche). How does one claim intuitiveness for such an abstract, nonintuitive formalism? At the very end of his letter to Pauli, Heisenberg made a hesitant suggestion: perhaps one can say that one understands physical laws in an intuitive way when one can immediately say in each experimental case what should occur—in this way the sense of understanding is advanced (Heisenberg to Pauli, 27 February 1927, PC). After Pauli provided his support, this hesitant suggestion was transformed into the confident redefinition of intuitiveness with which Heisenberg opened the uncertainty paper: "We believe that we intuitively understand physical theory when we can think qualitatively about individual experimental consequences and at the same time we know that that application of the theory never contains internal contradictions"(1927b, 62). The justification for this new definition, as is clear from the Heisenberg-Pauli correspondence, is that Heisenberg's qualitative discussion "eased his conscience" and simultaneously provided an effective way to counter Schrödinger's criticisms. Still, as we will shortly see, Heisenberg was not completely satisfied with this definition, offering, in addition, a visualizable interpretation of the uncertainty formula in terms of the directions of tensors (matrices) in Hilbert space, thus endowing these abstract entities with ontological meaning. These two conflicting notions of *Anschaulichkeit* run through the uncertainty paper.

We can conclude that Heisenberg did not start with a clear conception of what a satisfying interpretation should be; rather, the decision about the adequacy of his notion of interpretation was conditioned by social pressure (for *Anschaulichkeit*) and Heisenberg's own feeling of relief (that his "conscience" was easier). Pauli, whose "relentless criticism" Heisenberg requested (Heisenberg to Pauli, 27 February 1927, PC), had a more demanding conception of an interpretation. If Heisenberg put the uncertainty formula in the center of his interpretive enterprise, Pauli expected that the uncertainty formula should illumi-nate directly the nature of the basic quantitative laws of quantum me-chanics. Heisenberg quite confidently disagreed: Pauli was asking too much (Heisenberg to Pauli, 9 March 1927, PC). One encounters here, according to Heisenberg, the same theoretical freedom as in all other physical theories—quantum theory is no exception. In relativity theory, Heisenberg argued, one cannot substantiate the principle of the con-stant velocity of light—it is merely the simplest assumption if one ac-cepts Einstein's definition of simultaneity. Similarly, if one knows that p and q are not regular numbers but that the relation $p_1 q_1 \sim (h/2\pi i)$ holds, then the assumption that p and q are matrices is the simplest.[9]

Heisenberg, despite his confident tone, was aware that such an in-terpretation is not self-evident: "Naturally, I clearly see that this for-mulation can appear unsatisfactory" (Heisenberg to Pauli, 9 March 1927, PC).[10] Still, he retained his arguments, he informed Pauli, as a con-clusion of the uncertainty paper. This conclusion, Heisenberg admitted, is too dubious, too uncertain—he might yet write "ten times" (!) some-thing completely different, or he might delete these arguments from the final version altogether. As it happens Heisenberg did not delete these arguments. So much for the story of his strong "belief" and "commit-ment" to the arguments in the uncertainty paper.

The Contingency of Acausality

Heisenberg's uncertainty paper has won a permanent place in the an-nals of science and in the history of philosophy because of its radical message of the "final" overthrow of the law of causality. In Wheeler and Zurek's (1983) anthology it is reprinted under the heading "Principle of Indeterminism."

Yet Heisenberg must have introduced his "final" overthrow of cau-sality at the very last stage of writing the paper, because the causality issue is mentioned neither in the outline of the paper contained in the 27 February letter to Pauli, nor in two subsequent letters, which further discussed issues in the paper. Nor is the acausality problem mentioned in the abstract.

9. Heisenberg repeats this argument in the uncertainty paper without mentioning Pauli: "Of course, we would also like to be able to derive, if possible, the quantitative laws of quantum mechanics directly from the physical foundation—that is, essentially, from relation (1) [$p_1 q_1 \sim h$]. We believe, rather, that for the time being the quantitative laws can be derived out of the physical foundations only by the use of maximum simplicity. If, for example, the X-coordinate of the electron is no longer a number . . . , then the simplest assumption conceivable [that does not contradict relation (1)] is that this X-coordinate is a diagonal term of a matrix" (1927b, 82).

10. "Natürlich, ich seh' gut, dass diese Formulierung unbefriedigend scheinen kann."

We can only speculate about what caused Heisenberg to turn to this issue at the last moment. Did he fully realize the consequences of his dialogue with Jordan, along the lines of the analysis in chapter 4? Did he suddenly remember the challenge of his teacher Sommerfeld (1927), who awaited a new Kant who would overthrow classical causality, as Einstein had disposed of Newtonian space-time? Did Heisenberg realize that an essentially statistical theory, which is mute on questions of individual phenomena, is always open to a charge of incompleteness, thus weakening any claim for the finality of the Göttingen-Copenhagen achievement (an achievement to which Heisenberg had contributed so decisively)? Did Heisenberg read, hear, or see at the last moment something of the acausal spirit prevalent in his cultural milieu? We probably will never know the answer. Yet realizing how contingent the introduction of the acausality issue into the uncertainty paper was, and how closely tied to the original content of the paper it later became, is very instructive.

What is the ultimate reason for Heisenberg's "final" overthrow of causality? The mere fact of employing a statistical formalism for the derivation of the uncertainty principle does not, of course, signify any predilection for "essential" acausality—this is the reason Born, Jordan, and Pauli all considered the problem of causality open before Heisenberg's uncertainty paper. Double-voicedness characterizes Heisenberg's discussion of the causality issue in the middle of the paper—it is because Heisenberg added the verdict on causality at the end of the paper that the contrast between the voices is so strong. Is quantum mechanics "characteristically" a statistical theory, as Born and Jordan claim? Or is statistics only "brought in by experiments," as Dirac maintained? No preference is spelled out initially.[11] In a letter to Pauli, Heisenberg had mentioned noncommittally that Dirac's stand seemed "deeper" to him (Heisenberg to Pauli, 27 February 1927, PC). At the end of the uncertainty paper Heisenberg confidently ruled out the "essentially statistical" nature of quantum theory (citing the Bothe-Geiger experiments against it), when "only statistical conclusions can be drawn from precise initial data" (1927b, 83). Heisenberg's resolution was to bring into full view the impossibility of knowing the present, and therefore of predicting the future, thus disposing of any hopes of determinism.

How can one argue that the present cannot be known exactly? One can point out, of course, that it follows from the uncertainty formula,

11. "In the fact that in quantum theory only the probability distribution of the position of the electrons can be given for a definite state, such as 1s, one can recognize, with Born and Jordan, a characteristically statistical feature of quantum theory as contrasted to classical theory. However, one can say, if one will, with Dirac, that the statistics are brought in by our experiments" (Heisenberg 1927b, 66).

which is a mathematical consequence of quantum mechanics. The persuasiveness of the overthrow of causality thus depends on the status of the "finality" of quantum theory. In 1927, when this highly abstract and unconventional theory had yet to prove its mettle, such an argument would convince only believers.

Heisenberg therefore needed arguments that would show "how the world is the way quantum mechanics says it is" without direct appeal to the formalism. This is the reason Heisenberg preferred Dirac's stand, seeking to understand how "statistics is introduced by experiments." Heisenberg's search resulted in the introduction of the (misleading) disturbance imagery—with the help of disturbance imagery Heisenberg argued that there are no continuous orbits in an atom, just as the quantum mechanical formalism implies. One could deduce this fact "without knowledge of recent theories, simply from experimental possibilities" (1927b, 65). Consequently, the impossibility of a causal space-time description seemed to follow directly and inevitably from the conditions of experience, as Bohr later claimed.

Two voices characterize Heisenberg's discussion of causality, *Anschaulichkeit,* and other issues in the uncertainty paper. These conflicting voices persevere in later writings by the Göttingen-Copenhagen physicists, often changing their pitch but never completely fading away. One voice is that of "inevitability": through simple analyses of experiments one can see that the state of affairs claimed by quantum theory is unavoidable. The other voice is that of "consistency": analyses of experiments demonstrate the consistency of quantum theory because the results of such analyses do not contradict the quantum mechanical formalism. While inevitability arguments do not withstand close scrutiny (see chapter 9 and Beller 1993), they served as a powerful tool for legitimating and disseminating the Copenhagen orthodoxy. Inevitability arguments are often mixed with those of consistency. Heisenberg's concluding words in the uncertainty paper are uttered in the more sober, consistency voice. Inevitable acausality follows from the correctness of quantum mechanics (which covers the measurement process as well): "Because all experiments are subject to the laws of quantum mechanics and therefore to equation (1) $[p_1 q_1 \sim h]$, it follows that quantum mechanics establishes the final failure of causality" (1927b, 83).

Heisenberg also employed another strategy in establishing the inevitability of acausality. He claimed that the indeterminacy relation is "the real basis for the occurrence of statistical relations in quantum mechanics" (1927b, 62). He promised to demonstrate that the uncertainty formula "is a straightforward mathematical consequence of the rule $pq - qp = -i\hbar$" (1927b, 65). Yet Heisenberg's promise was not kept. What Heisenberg was able to offer was merely a qualitative discussion of how the uncertainty formula "creates room" for the validity of the quantum mechanical commutation relation: if position is determined precisely, momentum is undetermined. This qualitative "making sense" is, of course, different from straightforward mathematical derivation. Heisenberg derived the uncertainty relation while supplementing the original matrix formalism with the statistical assumptions of the Dirac-Jordan transformation theory—no wonder he gets a statistical result! This was one of numerous strategies used, intentionally or not, to present the Göttingen-Copenhagen perspective as unassailable.

Anschaulichkeit and the Status of Classical Concepts

As I have mentioned, Heisenberg was torn between his desire to render matrix mechanics intuitively appealing to a wide audience and his infatuation with Dirac's abstract version of quantum theory. This conflict was not resolved: it permeates the uncertainty paper. Heisenberg vacillated between his attempts to "intuit," or "understand," quantum theory by discussing simple thought experiments and his confident deduction of philosophical consequences from the mathematical formalism of the theory.

How does one endow the behavior of microscopic objects with intuitiveness, when one rejects their causal space-time description? What happens to objects located in space-time? Eager to resurrect Anschaulichkeit, Heisenberg found a brilliant solution: the microobject, he declared, is nothing but the totality of all its interactions.[12] In this way, one can define the attributes of atomic objects by their interactions, or measurements—this is in fact an operational stand that precedes Bohr's relational definition of concepts. In this way Heisenberg defined the position of an electron through measurements by a γ-ray microscope and redefined the path of a particle as a sequence of measurements. Similarly, Heisenberg defined the "size" of an electron as the difference in the electron's position when two very fast particles hit it one after the other within a very short time interval Δt (1927b, 65).

Heisenberg was considerably more proficient than Bohr in working with the mathematical tools of the quantum mechanical formalism. Heisenberg had no reason therefore to give up his attempt to explore the ontological significance of the quantum formalism. Consequently, two voices permeate Heisenberg's writings—the positivist, identifying

12. "In order to be able to follow the quantum-mechanical behavior of any object one has to know the mass of this object and its interactions with any fields and other objects. . . . About the 'Gestalt' (construction) of the object any further assumption is unnecessary; one most usefully employs the word 'Gestalt' to designate the totality of these interactions" (Heisenberg 1927b, 64).

the meaning of a concept with the procedure for its verification (in real or imagined physical interactions), and the realist, deducing the genuine features of the quantum world from characteristics of the mathematical formalism. The invariant features of this formalism, according to Heisenberg, explain why the uncertainty relations must hold. In order to obtain physical results from Dirac's formulation of quantum theory, one has to associate ordinary numbers with the q-numbers (matrices, or tensors, in multidimensional space). A definite experimental setup prescribes a definite direction in this multidimensional space—the question of the "value" of the matrix in this direction has a well-defined meaning only when the given direction coincides with the direction of a principal axis of the matrix. When these two directions differ slightly from one another, one can still talk about the "value" within a certain statistical error. A definite experiment therefore cannot give exact information about all quantum theoretical quantities—rather it divides the physical quantities into "known" and "unknown," as is exemplified in the uncertainty formula.[13] In this way Heisenberg "makes sense" out of the abstract quantum formalism.

Heisenberg's fearless exploration of quantum reality was intended not merely to gain deeper physical insight into the workings of the quantum world. Heisenberg seemed never to be free from a desire to argue—again and again—the superiority of matrix mechanics to wave mechanics. He offered in this respect a peculiar mixture of positivist (using disturbance imagery) and realist (using ontological discussion) argumentation. Thus Heisenberg claimed: "The X-coordinate of the electron is no longer a 'number,' as can be concluded experimentally," and in this case, the "simplest assumption conceivable . . . is that the X-coordinate is a diagonal term of a matrix" (1927b, 82). Heisenberg did not show why this is the "simplest assumption conceivable"—he immediately proceeded to defend this new and abstract kind of reality, countering a potential critical interlocutor, perhaps Schrödinger: "The prediction that, say, the velocity in the X-direction is 'in reality' not a number but the diagonal term of the matrix, is perhaps no more abstract and no more unvisualizable than the statement that the electric field strengths are 'in reality' the time part of an antisymmetric tensor located in space-time. The phrase 'in reality' here is as much and as little justified as it is in any mathematical description of natural processes" (1927b, 82).

Heisenberg's "intuitive" interpretation turns out to be not intuitive at all. Jordan, just a few months earlier, contra Heisenberg, used the ab-

---

13. For every quantum theoretical quantity one can find a coordinate system in which the quantity has an exact value.

stract character of electromagnetic theory as an argument against the temptation to look for an intuitive interpretation. The fact that such features of the quantum formalism as, say, the terms of a position matrix can be connected through bridge principles with a position measurement is not an argument for *Anschaulichkeit* as Heisenberg's contemporaries understood it. Even less is it an argument for the "indispensability of classical concepts," Bohr's notorious doctrine. Heisenberg, let us note, did not argue in the uncertainty paper for the "indispensability of classical concepts." On the contrary, he claimed that in the case of relativity theory, "the possibility of employing usual space-time concepts at cosmological distances can be justified neither by logic nor by observation" (1927b, 62). Similarly, in the microdomain "the applicability of classical kinematics and mechanical concepts . . . can be justified neither from our laws of thought nor from experiment" (1927b, 82).[14] Yet because Heisenberg resurrected in the uncertainty paper the kinematic interpretation of matrix theory (as opposed to the earlier electromagnetic interpretation) and because he analyzed thought experiments in terms of classical space-time concepts, the impression can arise that he was aligning himself with the Bohrian doctrine of their indispensability.

The tension between *Anschaulichkeit* and *Unanschaulichkeit* is mirrored in the discussion of the applicability of space-time concepts with which Heisenberg opens the uncertainty paper. We clearly discern there many conflicting voices, including Heisenberg's own past voice. Thus Heisenberg repeated his conclusions from the earlier *Naturwissenschaften* paper (1926b) that no contradiction-free interpretation of quantum mechanics in terms of the usual kinematic concepts is possible because of the internal discrepancies that "show themselves in arguments about continuity versus discontinuity and particle versus wave" (1927b, 62).[15] This argument is consistent with the original philosophy of matrix theory, which arose "exactly out of the attempts to break with all ordinary kinematic concepts" (1927b, 62). Yet, rather than deducing the need to renounce space-time geometry, as he originally did, Heisenberg now announced that a revision of space-time geometry at small distances is "unnecessary," thus implicitly disputing with Campbell, Sentfleben, and his own past voice. The reason for this conclusion, according to Heisenberg, is that: "We can make the quantum-mechanical

---

14. Later, Heisenberg would at times align himself with Bohr's doctrine, introducing many inconsistencies into his arguments. See chapters 8 and 9.

15. No wonder Heisenberg was not eager to accept Bohr's insistence on the indispensability of both waves and particles. The appended postscript, where Heisenberg supported Bohr's position, stands in striking contrast to these opening lines of the uncertainty paper.

laws approximate the classical ones arbitrarily closely by choosing sufficiently great masses, even when arbitrarily small distances and times come into question" (1927b, 62). Heisenberg did not substantiate this claim, which again contrasts with his reasoning just a year earlier.

Still, the revision of kinematic and mechanical concepts is necessary, Heisenberg claimed, because such a need follows from the basic relation $pq - qp = -i\hbar$ (1927b, 63). These two voices are not easy to harmonize: one calling for the resurrection of regular space-time geometry (so that thought experiments can be discussed), the other demanding the revision of kinematic concepts, such as position and momentum (as dictated by the mathematical formulas). In Heisenberg's paper, they simply coexist and constitute yet another source of inconsistency in his writings (on this score the original philosophy of the matrix theorists was more consistent).

In the uncertainty paper Heisenberg argued that in a discontinuous world "contradiction between the concepts of 'position' and 'velocity' is quite plausible" (1927b, 63). He substantiated this claim by arguing that in a continuum theory the velocity of a particle is given by tangents to the position curve at every instant. In a theory based on discontinuity, we have instead of this curve a series of discrete points. In such a case, Heisenberg claimed, "it is clearly meaningless to speak about one velocity at one position (1) because one velocity can only be defined by two positions and (2), conversely, because one point is associated with two velocities" (1927b, 63).

This intuitively appealing argument is problematic—what meaning can one attach to a discontinuous world in which points are distributed with finite separation, unless one presupposes a continuous space-time container underlying the discontinuous one? Here Heisenberg's reasoning contrasts with the inspiration of the original philosophy of matrix mechanics, which arose precisely from attempts to eliminate the superfluous continuous substratum altogether.[16] Ingenious Heisenberg attempts the impossible: to show, by simple "intuitive" arguments and without appeal to the quantum mechanical formalism, why quantum mechanics is "inevitable" or, at the very least, "quite plausible" (1927b, 63).

16. Matrix theorists, claiming to have established a "truly discontinuous theory," objected to the quantization procedure that chose the quantized mechanical orbits of an atom from all possible mechanical ones; a satisfactory discontinuous theory, so the original argument went, should not contain the unrealizable motions at all.

CHAPTER 6

The Dialogical Birth of
Bohr's Complementarity

*His [Bohr's] turn of mind was essentially dialectical, rather than
reflective. . . . He needed the stimulus of some form of dialogue to start
his thinking.*

Leon Rosenfeld 1967, 117

*It is, of course, possible to simplify the medium in which a scientist works by
simplifying its main actors.*

Paul Feyerabend 1975, 19

Introduction

A dialogical analysis of a scientific paper differs from the usual conceptual analysis. The dialogical approach is opposed to static structures
and fixed meanings—it is inherently contextual and historicist.

Such an analysis, as we saw in the previous chapter, can change our
understanding of a scientific text: we uncover diversity of meaning,
complexity of argumentation, unresolved tensions. When we perceive
that the text is populated with invisible interlocutors, we realize what
the central issues of the paper are. This realization modifies our understanding of the content of the paper. Dialogical analysis is a potent
tool for deciphering opaque and obscure texts. Such is the case, I argue
in this chapter, with respect to the analysis of Bohr's Como lecture
(1927c)—Bohr's initial formulation of his complementarity principle.

The usual reading of the Como lecture identifies the announcement
of wave-particle duality as the main message of Bohr's initial presentation of complementarity. My dialogical reading challenges the accepted
reading: Bohr continues to reject pointlike light quanta, and his discussion is heavily asymmetrical in favor of waves. The usual reading of
Bohr's paper assumes the similarity of Bohr's and Heisenberg's positions, while a dialogical analysis reveals an incompatibility between

their positions at the time. The usual reading also assumes the centrality of measurements and of operational definitions of concepts, while
the dialogical reading of the Como lecture uncovers Bohr's more modest emphasis on the "harmony" between the wave theoretical definition of concepts and the possibilities of observation. Significantly, the
dialogical reading discloses that the central message of Bohr's paper
was not the resolution of wave-particle duality by the complementarity
principle but rather an extensive defense of his concept of stationary
states and discontinuous energy changes, or "quantum jumps."

As with Heisenberg's uncertainty paper, plurality of meaning permeates Bohr's initial elaboration of complementarity. An analysis of
Bohr's discussion of the complementarity between space-time and causality reveals different, even incompatible uses of this concept. Scholars have attempted to find an unequivocal connection between wave-
particle complementarity and space-time–causality complementarity
(Murdoch 1987). Some have argued that causal description is associated
with particles and space-time with wave propagation, while others conversely connect space-time with particles and causality with waves. In
fact, both of these contradictory readings are present in the text.[1] They
occur in different dialogues and address different issues. Complementarity between space-time and causality is an imprecise umbrella concept that allows Bohr to cope locally with interpretive issues while entrenching his initial conception of stationary states and discontinuous
energy jumps.

A dialogical reading of Bohr's Como lecture also brings back to life
"lesser" scientists. We will see that the work of Campbell triggered
Bohr's deliberations on interpretive issues as it did Heisenberg's (chapter 4). Campbell's name figures in a draft of Bohr's paper (1927b, 69), as
well as in the published text (1927c, 131).

A dialogical reading of Bohr's Como lecture also provides a new perspective on the famous clash between Heisenberg and Bohr over the
uncertainty paper. A dialogical analysis reevaluates Bohr's and Heisenberg's intellectual positions so that the confrontation between them becomes conceptually meaningful, no longer merely the result of "mis-
understandings" and "confusion." Such a presentation allows us to
merge the "conceptual" and "anecdotal" aspects of the history of quantum physics (Beller 1996b).

1. While discussing the Compton effect, Bohr associated conservation laws—causality—with particles and space-time descriptions with wave propagation. Yet while discussing the complementarity between space-time and causality in the case of stationary
states of an atom, Bohr associated causality with waves (a stationary state is described by
a proper vibration and has a definitive unchanging value of energy) and space-time descriptions within the atom with particles. See the analysis below.

Bohr's first presentation of the complementarity principle took place at the International Congress of Physics, held in the Italian city of Como. In the history of scientific thought it is hard to find another contribution about which opinions continued to differ sharply more than half a century after its appearance. Some physicists, such as Leon Rosenfeld, considered complementarity the most profound intellectual insight of the twentieth century, the pinnacle of the physical understanding of nature, no less inevitable than "the emergence of man himself as a product of organic evolution" (1961, 384). Others criticized Bohr's complementarity as an obscure "double-think" that impeded clear thinking and scientific progress, or as a crutch that helped initially but was eventually no longer needed (Landé 1967; interview with Alfred Landé, AHQP).

Attitudes toward Bohr's complementarity principle even in the camp of "believers" were more ambivalent than the published sources disclose. As Dirac expressed it: "I never liked complementarity. . . . It does not give us any new formula. . . . I believe the last word was not said yet about waves and particles" (interview with Dirac, AHQP). Similarly, Heisenberg disclosed, "I know that, besides Landé, many other physicists had been upset by this situation, and they felt it was a dualistic description of nature" (interview with Heisenberg, AHQP).

Despite the strong reactions it provoked, what exactly Bohr's complementarity means continues to be an enigma. Jammer, commenting on the difficulty of comprehending Bohr's complementarity principle, described Carl Friedrich von Weizsäcker's extensive effort to elucidate the original meaning of complementarity. However, when Weizsäcker asked Bohr "whether his interpretation accurately presents what Bohr had in mind, Bohr gave him a definitely negative answer" (Jammer 1974, 90). Recent students of Bohr's thought continue to face the "formidable difficulty," even incomprehensibility, of his writings dealing with complementarity. Thus Folse claimed, "Even with almost one full year's work on this paper, innumerable rewrites, two public deliveries, and three different sources of publication, the essay contains many obscurities and never makes very clear what this new 'viewpoint' is supposed to be" (1985, 108).[2] Don Howard recently proclaimed that "now there are signs of growing despair . . . about our ever being able to make good sense out of his [Bohr's] philosophical view" (1994, 201).

A dialogical perspective allows us to decipher Bohr's philosophy in general and the meaning of Bohr's original presentation of the comple-

2. Similarly pessimistic is the conclusion of Gibbins: "Niels Bohr went to great lengths to refine and then to clarify his thoughts on quantum mechanics, but, sad to say, his writings do not appear to have benefited from these efforts. It is very difficult to say . . . how his ideas hang together" (1987, 53).

mentarity principle in particular. Without identifying the interlocutors of each sentence of the Como lecture, it is impossible to understand the meaning of these sentences and the connections among them. Yet when we realize that the text is filled with implicit arguments with the leading physicists of the time—Einstein, Heisenberg, Schrödinger, Compton, Born, Dirac, Pauli, and the lesser known Campbell—the fog lifts and Bohr's presentation becomes clear.

A dialogical reading reveals that the central message of Bohr's Como lecture was the announcement of a compatibility, indeed, a "happy marriage," between Schrödinger's successful continuous wave mechanics and the quantum postulate, especially Bohr's original notion of discrete stationary states. In Bohr's presentation, the de Broglie–Schrödinger wave packet was sufficient to resolve the paradoxes of atomic structure, of the interaction of radiation with matter, and of the interaction of matter with matter—paradoxes that had confounded Bohr since his brilliant and troublesome debut into atomic theory in 1913. The existing accounts of complementarity center almost exclusively on the "wave-particle duality" of a free particle, while the significant physical problems—those that invoke physical interactions (bound electrons in atoms, collisions)—are neglected. Yet it was the paradoxical, nonclassical aspects of these problems that were central to the efforts of physicists at the time.

The Como lecture announced a resolution of these long-standing problems by harnessing the wave theoretical framework to the quantum postulate. In fact, each of the dialogues centers on Bohr's solution of one of these problems—atomic structure, the interaction of radiation and matter, and collisions. In each case, Bohr implied that the abstract idea of a point-mass particle is inadequate and must be replaced by a wave theoretical superposition of light waves or matter waves. The fertility of this idea for Bohr's skillful weaving of the argumentation in the Como lecture clearly demonstrates why, from the very beginning, he ascribed physical significance to Schrödinger's theory, unlike his younger colleagues Heisenberg and Jordan. Bohr's implicit argument for the superiority of wave concepts over particle concepts does not mean that waves represented for him a literal picture of reality. Rather the fruitfulness of the wave model is in the English tradition: the wave model is both heuristically useful and theoretically adequate. Bohr's Como lecture constituted a striking contrast to Heisenberg's efforts to develop an exclusively corpuscular ontology.

The fruitfulness of the concept of wave packets is especially evident in the case of interactions. Because we never observe either an isolated particle or a monochromatic wave (both are "abstractions"), but only cases of superposition of light or matter waves, the idea of a wave packet is particularly suitable for demonstrating the harmony between the possibilities of wave theoretical definition and those of observation (Bohr 1927d). The agreement claimed between the possibilities of definition and observation was initially much weaker than the operational assertion that measurement is the primary and indispensable part of the interpretation of concepts—the stand later taken by Bohr. Mere "harmony" between observation and definition of concepts is, of course, a necessary part of any interpretive attempt.[3]

My claim that Bohr's original announcement of complementarity was not a symmetrical solution of the wave-particle dilemma, and that in all physically interesting cases—those of interactions—one must necessarily deal with wave packets (light waves or matter waves), will be substantiated by discussing Bohr's dialogues with Einstein, Compton, and Campbell. It is generally assumed that the roots of Bohr's complementarity lie in the experimental refutation of the Bohr-Kramers-Slater theory. This theory used the wave theoretical framework of light exclusively rather than Einstein's theory of light quanta. After the Bothe-Geiger experiments, many authors assume, Bohr had no choice but to assimilate light quanta, which he had rejected vigorously beforehand. The foundations of Bohr's complementary framework, they hold, lay in his acceptance of light quanta in 1925, before all the subsequent breakthroughs—Heisenberg's matrix mechanics and Schrödinger's wave mechanics—occurred (Jammer 1966; MacKinnon 1982).

I argue for a substantially different perspective. Bohr did not adopt the idea of pointlike light quanta (as used in the explanation of the Compton effect), even after the Bothe-Geiger experiments. Bohr's complementarity principle implied further *rejection*, not acceptance, of the idea of pointlike material particles. It is incomprehensible that Bohr would have developed his interpretation of quantum physics without responding in a significant way to the pivotal theoretical developments that occurred during the years 1925–27. I will discuss the nature of Bohr's response in the Como lecture to Heisenberg's matrix mechanics and to Schrödinger's wave mechanics. In particular, I will describe Bohr's understanding of the space-time problem in the interior of the atom, and his defense and elaboration of his idea of a stationary state. This description further substantiates my claim that it is around Bohr's concepts of stationary states and quantum jumps—which Schrödinger aimed to eliminate—rather than around the wave-particle dilemma or indeterminism, that the new philosophy of physics was erected. Bohr's full-fledged defense of his own idea of stationary states is elaborated in

3. This was argued by both Born and Heisenberg in 1927 for their respective statistical and indeterminacy interpretations.

paragraph 5 of the Como lecture—a part either neglected or considered the most obscure (the opinion of the editors of Bohr's collected works; BCW, 6:30).

Bohr's defense and elaboration of the idea that an atomic system is adequately represented by a sequence of stationary states that are, in turn, adequately described by Schrödinger's wave function reveals a deep conceptual gap between Bohr's wave theoretical and Heisenberg's particle-kinematic interpretations of atomic systems—a gap that was circumvented rather than resolved by subsequent developments. This incompatibility between the positions of Bohr and Heisenberg is one of the historical roots of the inconsistencies that plague the Copenhagen interpretation of physics. My discussion of this gap also provides an insight into Einstein's and Schrödinger's early dissatisfaction with the Copenhagen interpretation. I argue that their initial criticism focused on the inconsistency of amalgamating the incompatible positions held by Bohr and Heisenberg. Part of the incomprehensibility of the Como lecture derives from Bohr's attempt to conceal this gap by uniting forces against the opposition.

Dialogue with Schrödinger: The Structure of Atoms

My discussion of Bohr's predisposition to de Broglie–Schrödinger concepts is based on the text of the Como lecture (Bohr 1927c), together with two manuscripts written before the lecture (Bohr 1927a, 1927b). I will also quote from a manuscript (1927d) written only three weeks after the Como conference. These three manuscripts are conveniently available in Bohr's *Collected Works*, volume 6.[4] As is clear from these writings, Bohr in his original struggle with the physical interpretation of quantum theory leaned heavily on the idea of a wave packet—a superposition of waves of different frequencies that results in a wave field limited in space and time. Bohr used the imagery of a wave packet whenever he described light quanta or electrons, both in cases of free individuals and in cases of interactions between them. The position of the light quantum is the position of such a limited wave field, rather than that of a mass-point.[5] Using the idea of wave packets, Bohr directly derived the uncertainty relations, a derivation that is opposed to Heisenberg's, based on the idea of pointlike electrons and photons.

4. Page references for Bohr (1927a, 1927b, 1927c, 1927d) are to BCW, vol. 6.
5. "Only by the superposition of harmonic waves of different wavelengths and directions is it possible at a given time to limit the extension in space of the wave-field. . . . If we ask about the position of a light quantum, we find that no more than in the case of its energy and momentum, we can define a position of a light quantum at a given time, without consideration of complementary waves" (1927b, 69–70).

In a manuscript written just a few days before the Como lecture (Bohr 1927a), Bohr introduced de Broglie wave packets to represent both light quanta and electrons. Bohr described de Broglie's ideas "of ascribing a frequency to any agency carrying energy" and consequently "a phase-wave to a material particle." According to Bohr, the matter wave theory, or "the representing of a particle by means of a wave-packet, represents a direct generalization of the light quantum theory." Such a point of view strongly suggests not only that the pointlike corpuscular light quantum is an abstraction but that a singular harmonic light wave is similarly merely an abstraction that is never actualized by itself in any physical situation: "As emphasized by de Broglie the abstract character of the phase-wave is indicated by the fact that its velocity of propagation . . . is always larger than the velocity of light c." All this suggests also that "a light wave may be considered as an abstraction, and that *reality can only be ascribed to a wave group*" (Bohr 1927a, 78, my italics).

Recognizing the physical significance that Bohr attached to a wave packet allows us to see the following often-quoted sentence in a different light: "Radiation in free space as well as isolated material particles are abstractions . . . their properties on quantum theory being observable and definable only through their interaction with other systems" (Bohr 1927c, 116).

This sentence does not imply, or at least did not originally imply, a strong instrumentalist approach, where no reality can be ascribed to atomic systems in themselves—the view Bohr gradually developed by countering later objections. Neither does it indicate a symmetrical solution of the wave-particle dilemma. Rather only the idea of a wave packet represents the "real thing" that allows the untangling of the mysteries of the atomic world.

My interpretation is confirmed by Bohr's analysis of the possibilities of observation in principle, as well as by his analysis of actual physical interactions—the interaction of radiation and matter and the Davisson-Germer experiments. According to Bohr, any observation necessarily involves superposition. In the case of light, "as stressed by de Broglie the only way of observing an elementary wave is by interference" (Bohr 1927a, 78). Here the possibilities of definition of a wave packet and the possibilities of observation are in close harmony. On the other hand, "the only way to define the presence of the waves is through analysis of the interaction between light and matter" (Bohr 1927b, 70). It was from this phenomenon that the corpuscular character of light was deduced. Here, according to Bohr, the discussion "was until recently most unsatisfactory" because "the behaviour of the so-called material particles rested so entirely on corpuscular ideas," that is, on the idea of localized pointlike particles. All the difficulties are removed "through the introduction of an *essential wave feature in the description of the behaviour of material particles* due to the work of de Broglie and Schrödinger" (Bohr 1927b, 70, my italics). The Davisson-Germer experiments, according to Bohr, directly confirm the matter wave theory in the spirit of de Broglie–Schrödinger: "The discovery of Davidson [*sic*] and Germer ... prove[s] the necessity of applying a wave-theoretical superposition principle in order to account for the behaviour of electrons. ... the wave character of the electrons is by these experiments shown just as clearly as is the wave character of light. ... As [is] well known the experiments are in *complete accordance* with the ideas of de Broglie" (1927a, 77, my italics). Though reference to these experiments is omitted in the text of the Como lecture, this discussion, written just a few days before Bohr's presentation in Como, forms a necessary background for understanding his argument in the lecture.

According to Bohr, an adequate visualizable interpretation of physics is possible only with the help of the idea of a wave packet: "The possibility of identifying the velocity of a particle with a group velocity indicates the applicability of space-time pictures in the quantum theory" (1927c, 118).

The idea of a wave packet allowed Bohr, following de Broglie, to rationalize the "irrationality" of the basic quantum relation $E\tau = I\lambda = h$, where "corpuscles of light" with energy E and momentum I had the characteristics of infinitely extended waves (period of vibration τ and wavelength λ). It was this irrationality that Bohr cited in his earlier work as the reason for his opposition to the idea of light quanta. The concept of a wave packet demonstrates that one is dealing not with two "rivalizing [competing] concepts" but rather with a description of two "complementary sides of phenomena (Bohr 1927b, 69). The relation $\Delta t\Delta E = \Delta x\Delta I_x = \Delta y\Delta I_y = \Delta z\Delta I_z = h$ indicates the reciprocal accuracy with which the space-time and energy-momentum vectors of such wave packets can be defined, suggesting the complementarity of space-time and causality descriptions. The relation indicates "the highest possible accuracy" in the definition of "individuals associated with wave fields." In general, the wave packet would "in the course of time be subject to such changes that [it would] become less and less suitable for representing individuals" (Bohr 1927c, 119). Here, according to Bohr, is the source of the "paradoxes" of quantum theory.

As we will see, Bohr did not accept the particle-kinematic interpretation of an atom in a given stationary state (as Heisenberg, Pauli, and Born would have it), nor did he entertain the idea of wave packets moving along visualizable Keplerian orbits (as Schrödinger at some point had hoped). In Bohr's interpretation of the interior of the atom, the possibilities of visualization were closely connected with the possibilities of experimentation. The wave theoretical model allowed Bohr finally to decipher the limits of the application of space-time to the atomic domain—a problem that had occupied him since 1913, when he himself had introduced the incomprehensible quantum leaps into an inexplicable space-time abyss. On the other hand, the wave model enabled Bohr to resurrect his idea of the stationary states of an atom—an idea that seemed to have been abandoned in the matrix approach and to be in danger of elimination by Schrödinger's attempts at interpretation.

The problem of stationary states and the problem of the application of space-time description in the atomic domain are closely connected. The idea of stationary states (when out of all classical mechanically realizable motions, only certain motions were assumed to be actualized), and especially the idea of instantaneous transitions between such states, implied from the very beginning a radical departure from classical space-time models. Yet how radical the departure would eventually become nobody initially grasped, not even Bohr. Both his 1913 and 1918 papers show that Bohr did not rule out the possibility that the mechanism of the transitions would eventually be understood. The concept of the stationary states of an atom was beset with conceptual inconsistencies from the start (Beller 1992b). Still, Bohr and other physicists used this idea, which successfully explained the stability of atoms and powerfully deciphered the structure of spectral lines (Heilbron and Kuhn 1969; Jammer 1966). As far as Bohr and his followers (but not Schrödinger) were concerned, the idea of stationary states and discontinuous transitions between them was fully and definitely corroborated by the Franck-Hertz experiments.

In retrospect, we can see how, step by step, the conceptual price of the utility of this idea rose. In the Bohr-Kramers-Slater theory the idea of stationary states and quantum jumps necessitated such major departures as nonconservation of energy and momentum, as well as the introduction of the strange "virtual field" that an atom in a definite stationary state emitted. The Bohr-Kramers-Slater theory also implied further departures from ordinary space-time visualization, for example, in the explanation of the Compton effect. Similarly, the analysis of the phenomenon of collisions, when a fast-moving particle collides with an atom in a certain stationary state, demanded resignation from strict conservation of energy and momentum (Bohr 1925). When Bohr learned of the results of the Bothe-Geiger experiments, confirming strict conservation, he did not question the formulas of the wave theory of light or the adequacy of the description of atoms in terms of stationary states. Rather it was a further departure from classical space-time models for the interior of the atom that Bohr found mandatory. Bohr was supported in this direction by the wave theoretical ideas of de Broglie and Einstein. The paper Bohr wrote on these matters was published in 1925, just before Heisenberg's reinterpretation paper—the paper that provided a panacea for all the ills of the old quantum theory by eliminating the classical space-time container from the interior of the atom.

Heisenberg's solution, however, was too radical. Matrix mechanics could not theoretically describe the states of atomic systems and the evolution of phenomena—the main reason for the lukewarm reception it received initially (chapter 2). Schrödinger indeed perceived the matrix mechanical formalism as eliminating Bohr's concept of separate stationary states. Schrödinger himself attempted a continuous wave description that would further eliminate the concept of stationary states with definite energies, and discontinuous transitions between these states, by suggesting instead a resonance model in terms of frequencies. As I have argued, the Göttingen and Copenhagen physicists joined forces in response to the perceived threat from Schrödinger—his attempt to reduce Bohr's concepts to the status of Ptolemaic epicycles. It was around the adequacy of Bohr's concepts of stationary states and quantum jumps that the crucial interpretative attempts revolved.

Bohr, who accepted the great usefulness of Schrödinger's formalism, could not see initially how solutions to Schrödinger's wave equation, in terms of the superposition of different proper vibrations, could be reconciled with the idea of separate stationary states. His heated argument with Schrödinger, during the latter's visit to Copenhagen in the fall of 1926, had centered on quantum jumping. After Schrödinger's visit, the direction of Bohr's efforts became clear: to achieve compatibility between wave theoretical ideas and the quantum postulate. As Heisenberg put it: "Bohr realized at once that it was here we would find the solution to those fundamental problems with which he had struggled incessantly since 1913, and in the light of the newly won knowledge he *concentrated all his thought on a critical test of those arguments which had led him to ideas such as stationary states and quantum transitions*" (1967, 101, my italics).

Bohr's Como lecture was the culmination and resolution of these efforts. Bohr criticized Schrödinger's attempts to replace "the discontinuous exchange of energy . . . by simple resonance phenomena" (1927c, 127; 1927d, 97). Schrödinger's theory must necessarily "be interpreted by an explicit use of the quantum postulate," and "in direct connection with the correspondence principle." Moreover, "in the conception of stationary states we are . . . concerned with a characteristic application of the quantum postulate" (1927c, 130). Countering Schrödinger, Bohr asserted: "The proper vibrations of the Schrödinger wave-equation have been found to furnish a representation of the stationary states meeting all requirements" (1927c, 126; 1927d, 97). If characteristic vibrations represent stationary states, then "a fundamental renunciation of the space-time description is unavoidable" (1927c, 129). The reason for this is the fact that "every space-time feature of the description of phenomena is based on consideration of interference taking place inside a group of such elementary waves" (1927c, 129; 1927d, 97). Stationary states, having definite energy, are adequately described by a single elementary wave (because "the definition of energy and momentum is attached to the idea of [a] harmonic elementary wave"; 1928, 113). The above sentences clearly indicate that no particle-kinematic description can be applied to a separate stationary state in principle, and that "a consistent application of the concept of stationary state excludes . . . any specification of the behavior of the separate particles in the atom" (1928, 112). Here lies a crucial retrospective insight into the past failures of the old quantum theory, which assumed visualization of these stationary states in terms of mechanical electron orbits. Since no time mechanism for the description of these states can be conceived, no prediction of the time of the transition is possible. This, in turn, discloses why, for describing these transitions, one must be content with probabilities—another crucial insight into the source of past struggles!

The initial difficulty of reconciling the idea of stationary states with the Schrödinger wave function was the fact that the general solution of Schrödinger's equation is a superposition of proper vibrations, and at first it seemed "difficult to attribute a meaning to such a superposition as long as we adhere to the quantum postulate" (Bohr 1927d, 97), according to which an atom is always in some definite stationary state. Schrödinger regarded this difficulty as conclusive, arguing that an atom exists in a superposition of several proper vibrations, and that the appropriate characteristic of a proper vibration is its frequency and not its energy. Bohr's ingenious resolution of this difficulty centered on an exploration of the compatibility of the possibilities of definition and observation. A superposition of proper vibrations is an adequate description in the case of interactions (observations). An atom in a definite stationary state with precise energy is, however, a closed system, not accessible to observation, and therefore it can be correctly described by a single proper vibration. As such, the idea of a stationary state becomes an abstraction, both because it is represented by a single wave and because a system "not accessible to observation . . . constitutes in a certain sense an abstraction, just as an idea of an isolated particle" (1927d, 97). Despite its being an abstraction, the concept of a stationary state is indispensable in the interpretation of phenomena (1927d, 97). Because "the conception of a stationary state involves, strictly speaking, the exclusion of all interactions" (1928, 115), the constant energy value associated with such states "may be considered as an immediate expression for the claim of causality contained in the theorem of conservation of energy" (1927c, 130). Thus the complementarity of space-time and causality in the quantum domain follows. The idea of an atom in a stationary state as a closed system with constant energy accords well with Bohr's initial introduction of this idea to explain the stability of matter: "This circumstance justifies the assumption of supramechanical stability of the stationary states, according to which the atom, before as well as after an external influence, always will be found in a stationary state and which forms the basis for the use of the quantum postulate" (1927c, 130). Schrödinger's interpretive aspirations are deficient precisely because they cannot deal with the stability postulate, for which the assumption of stationary states is essential.

Bohr devoted all of paragraph 5 of the Como lecture to arguing for the consistency of the concept of a stationary state. Bohr's strategy was to demonstrate that complementarity between stationary states and corpuscular space-time descriptions—or between causality and space-time, for the interior of an atom—accords fully with the possibilities of observation. For example, if one is to inquire about the behavior of separate particles in an atom, then one has to neglect their mutual interaction during the observation, thus regarding them as free. This necessitates a very short time of observation, shorter than the periods of revolution of electrons. This, in turn, implies a big uncertainty in the energy transferred during the observation, and thus the impossibility of ascertaining the energy values of stationary states.

The concept of complementarity between stationary states and corpuscular space-time descriptions, however, seems to present a serious difficulty when large quantum numbers are approached. According to Bohr's correspondence principle, in the limit of large quantum numbers the concept of stationary states must approach the classical space-time orbits along which intraatomic particles revolve. This problem occupied Bohr from the time matrix mechanics emerged with its total renunciation of space-time descriptions. Nor was the problem ignored by Heisenberg, who though not reluctant to abandon space-time description in the interior of the atom, was nevertheless eager to understand how the transition from the micro- to the macrodomain might be achieved. Because an electron moving along an orbit with a large quantum number should necessarily be represented by a wave packet (a superposition of many vibrations), the idea of a stationary state as a single proper vibration seemed to Schrödinger basically inadequate. How, then, can complementarity, or the mutual exclusion of the concepts of stationary states and individual particles, possibly be maintained in the case of large quantum numbers, where these ideas are not only no longer contradictory but simultaneously applicable?

Here again the harmony between the possibilities of observation and definition comes to the rescue, and Bohr demonstrates with characteristic ingenuity how even in this case his conception of stationary states and transitions between them can be preserved.[6]

The possibility of describing stationary states by means of Schrödinger's wave function, in contrast with the inadequacy of matrix mechanics on this score, indicated to Bohr the profound physical significance of wave mechanics. The fact that such a definition of a stationary state is theoretically adequate is demonstrated, according to Bohr, by Born's work, which provided "a complete description of the collision phenomena of Franck and Hertz, which may be said to exhibit the stability of st[ationary] states" (1927b, 71). Bohr's enthusiasm for Schrödinger's wave mechanics was manifest in his correspondence at the time. Shortly after Schrödinger's visit to Copenhagen, Bohr wrote to Ralph Fowler: "Just in the wave mechanics we possess now the means of picturing a single stationary state which suits all purposes consistent with the postulates of the quantum theory. In fact, this is the very reason for the advantage which the wave-mechanics in certain respects exhibits when compared with the matrix method." Two days later Bohr argued the same point in a letter to Kronig. Bohr made this point again in his first written attempt at interpretation leading to the Como lecture (Bohr to Fowler, 26 October 1926, AHQP; Bohr to Kronig, 26 October 1926, AHQP; Bohr 1927b, 70).

Closely connected with the idea of the quantum postulate and the concept of stationary states was the idea of quantization. The rules of quantization determined the choice of certain mechanical motions, to be associated with the stationary states, from a manifold of all possible classical motions. In this way a set of integers—quantum numbers—was associated with every stationary state. This choice by dictating integers seemed at best arbitrary, at worst a return to Pythagorean mysticism. Matrix mechanics, which dispensed with the description of stationary states, also eliminated the idea of quantization. Initially, as I have argued, Heisenberg and Born were quite happy to get rid of a postulate of quantization, deducing all the quantum effects, including the existence of the sequence of discrete stationary states, from the mathematical formulas of the matrix formalism. The concept of quantization reappeared, however, in Dirac's version of quantum mechanics,

6. When we identify the exact value of the energy of a stationary state by means of collisions or radiation reaction, we inevitably imply "a gap in the time description, which is at least of the order of magnitude of the periods associated with transitions between stationary states. In the limit of high quantum numbers these periods, however, may be interpreted as periods of revolution. Thus we see that no causal connection can be obtained between observations leading to the fixation of a stationary state and earlier observations of the behavior of the separate particles in the atom" (Bohr 1927c, 135).

indicating that perhaps this idea might have deeper significance than that conceived by matrix physicists.

Schrödinger's version of quantum mechanics allowed the resurrection not only of stationary states but also of the idea of quantization. According to Bohr, visualization in terms of the wave theoretical model points to the physical interpretation of quantization: "The number of nodes in the various characteristic vibrations gives a simple interpretation to the concept of quantum number which was already known from the older methods but at first did not seem to appear in the matrix formulation" (1927c, 126). While matrix mechanics defied any attempt at physical interpretation, Schrödinger's wave mechanics allowed an understanding of the electric and magnetic properties of atoms. The wave model also allowed an understanding of the failure of the concept of mechanical orbits: "From characteristic vibrations with only a few nodes no wave-packages can be built which would even approximately represent the motion of a particle" (Bohr 1928, 113).

Bohr's discussion of the problem of interaction, and especially of the problem of bound particles in the interior of an atom, is particularly instructive for seeing both the advantages of a wave ontology and its limitations. Because of the complementarity of the space-time and energy-momentum coordinates, for a free particle (and even more so for a bound one) precise knowledge of its momentum and energy excludes exact specification of its space-time coordinates. This implies the inadequacy of a particle-kinematic framework and the inapplicability of the classical concept of mutual forces and potential energy in Heisenberg's approach. This difficulty is "avoided by replacing the classical expression of the Hamiltonian by a suitable differential operator" (1928, 111). However, it is this advantage that shows the limits of visualization in the interior of the atom in terms of waves, because the Schrödinger equation contains imaginary numbers and is associated with a multidimensional space that is "in general greater than the number of dimensions of ordinary space." Therefore, intraatomic particles, though adequately represented by a wave theoretical framework, cannot be visualized in terms of ordinary space-time pictures. It is only a three-dimensional wave packet of a free particle that can represent a particle's space-time location. It is here that the most striking difference between classical and quantum mechanics lies: while in the former "particles are endowed with an immediate 'reality,' independently of their being free or bound" (1928, 114), in quantum mechanics intraatomic, bound particles are not visualizable. Consequently, Born's interpretation of the wave function as denoting the probability of position should not be applied uncritically to the interior of the atom.

This analysis was addressed simultaneously to Schrödinger and to

Heisenberg, who were engaged in a controversy over the relative merits of wave and matrix mechanics. Schrödinger claimed superiority for his version, due to its greater visualizability, or intuitiveness. To counter Schrödinger, Heisenberg, in his uncertainty paper, presented an exclusively corpuscular interpretation. Bohr found both approaches inadequate. Bohr agreed with Heisenberg that Schrödinger's waves in multidimensional space could hardly be considered immediately intuitive. But neither was Heisenberg's approach satisfactory: the intraatomic reality was not a particle-kinematic one. We begin to see the depth of the gulf between Heisenberg's and Bohr's positions. Only with extreme caution, argued Bohr, can one employ ordinary categories in the interior of the atom; for example, the particle concept may be used only during very short times of observation. A consistent interpretation in such situations is possible only when the compatibility of the possibilities of observation and definition is closely analyzed. In general, the idea of individual particles has "just as much or as little 'reality'" as the idea of stationary states.

Not surprisingly, what Schrödinger perceived as the focus of Bohr's argument in the Como lecture was Bohr's defense of the concept of a stationary state. Schrödinger struck back with Bohr's own weapon, arguing that the combination of uncertainty relations and the idea of stationary states destroys compatibility between definition and observation: "It seems to me that there is a very strange relation between Heisenberg's uncertainty relation and the claim of discrete quantum states. On account of the former the latter can really not be experimentally tested" (Schrödinger to Bohr, 5 May 1928, AHQP; Schrödinger also discussed this issue in a letter to Einstein, 30 May 1928, Przibram 1967). This controversy was not resolved. In the 1950s, Schrödinger (1952a) criticized Bohr's idea of stationary states and quantum jumps extensively (see chapter 10).

Dialogue with Einstein and Compton

Another central problem in atomic physics at the time was that of the interaction of radiation and matter. More than any other, this problem cried out for nonclassical treatment. Nonclassical departures had already been made in Max Planck's and Albert Einstein's early work, including Einstein's idea of light quanta (Jammer 1966; Kuhn 1978). For Bohr the problem of the interaction between radiation and matter led to the famous Bohr-Kramers-Slater proposal. When certain conclusions of this theory were refuted by the Bothe-Geiger experiments, many physicists considered the reality of quanta to have been proved unequivocally (Klein 1970; Stuewer 1974). According to the usual accounts, Bohr accepted the failure of the Bohr-Kramers-Slater theory, as well as the existence of light quanta, striving from this point on to incorporate quanta into his interpretive framework. The principle of complementarity, announced by Bohr in his Como lecture, was the culmination of these efforts, resulting in a symmetrical solution of the wave-particle dilemma.

In what follows I argue for a substantial revision of this reading. I see no evidence of Bohr's having accepted pointlike quanta. In fact, by using the de Broglie–Schrödinger idea of a wave packet, Bohr gained a dramatic insight into past difficulties, enabling him to rehabilitate the apparently discredited Bohr-Kramers-Slater theory.

The Bohr-Kramers-Slater paper, a programmatic rather than a technical work, had mapped out a comprehensive framework for the general problem of the interaction of radiation with atoms, eliminating the need for the idea of light quanta. In this theory, the kinematic model of the atom was replaced with a virtual oscillator model.[7] The virtual field produced by the atom determined its own probability for spontaneous emission, as well as the probabilities for the processes of emission and absorption induced in other atoms. Such a probabilistic description implied not only renunciation of exact conservation laws but—contrary to the light quanta point of view—independence of the processes of emission and absorption in atoms that are far apart.

The conflict between Bohr and Einstein on the nature of radiation caught the attention not only of the scientific community but of the general public as well (Klein 1970). The results of the Bothe-Geiger experiments, which were reported in April 1925, did not accord with the ideas of Bohr-Kramers-Slater. These experiments detected correlations between Compton recoil electrons and scattered X-rays, a result more in keeping with the light quanta concept of radiation that with the wave concept. Most physicists took the experiments as crucial evidence in favor of the corpuscular idea of light quanta. Pauli, for example, declared that from then on light quanta were "as real" as material electrons. There is no indication that Bohr ever considered the Bothe-Geiger and Compton-Simon experiments crucial with respect to the nature of light, even though it was clear to him that the program outlined in the Bohr-Kramers-Slater theory could not be retained without modification. Bohr explained his position in a letter to Hans Geiger, who had informed Bohr of the results of his experiments. Bohr further argued for the renunciation of ordinary space-time pictures—a necessity that

7. In each of its stationary states, an atom continuously emits a virtual wave field, which is "equivalent to the field of radiation which in the classical theory would originate from the virtual harmonic oscillators corresponding with the various possible transitions to other stationary states" (Bohr, Kramers, and Slater 1924, 164).

was already apparent, according to Bohr, from an analysis of collision phenomena.[8]

The need to depart from ordinary space-time pictures has a clear meaning in Bohr's discussion of the interaction of radiation with matter. Wherever Bohr used the notion of space-time pictures, he discussed models using material particles traversing continuous spatial trajectories. The Bohr-Kramers-Slater theory clearly indicated why such ideas could no longer be applied. In the case of the interaction of light with free electrons (the Compton effect), the wave picture of light indicated kinematic peculiarities in the description of the electron's motion. In the Bohr-Kramers-Slater theory, the notion of the atom as a kinematic system of orbiting electrons was replaced by a virtual oscillator model. The visualization of a stationary state of the atom as a state in which electrons move along well-defined orbits was abandoned (a step also indicated by Bohr's analysis of collision processes).[9] Bohr's reaction to the "crucial" Bothe-Geiger experiments was greater conviction that the ordinary space-time concepts of material particles were inadequate, and that a still more far-reaching revision of such concepts was necessary. Perhaps, he argued, de Broglie's modification of the idea of material particles might help to resolve this situation (Bohr 1925).

This thought, entertained as early as 1925, found its full expression in the Como lecture. The instantaneous interaction of light (wave phenomena extended in space and time) with matter (pointlike particles) is not reconcilable with exact conservation laws—the reason for introducing statistical conservation in the Bohr-Kramers-Slater theory. However, if the space-time and energy-momentum vectors of electrons are not sharply defined (as in the case of a wave packet, as opposed to a localized particle), the wave picture of light can be reconciled with the conservation laws. This idea is expressed in all the manuscripts preceding the Como lecture, as well as in the lecture itself. It is the complementarity of the space-time and energy-momentum vectors of a wave packet that makes it possible to unite space-time coordinates and

8. "Conclusions concerning a possible corpuscular nature of radiation lack a sufficient basis" (Bohr to Geiger, 21 April 1925, AHQP; *BCW*, 5:353). A similar position was expressed in other letters by Bohr at the time, as well as in the postscript to his "Über die Wirkung" (Bohr 1925), where he attempted to extend the nonconservation ideas of Bohr-Kramers-Slater to collisions between atoms and material particles.

9. Bohr 1925. The misinterpretation that Bohr accepted light quanta is related, in my opinion, to the fact that many authors assume Bohr's statement about "space-time pictures" applies to the wave propagation of light as well and consequently conclude that the failure of space-time pictures applies to classical wave theory. However, I am not aware of anything in Bohr's writings at the time supporting this point of view: wherever Bohr used the phrase "failure of space-time pictures," it referred to the failure of models using pointlike material particles.

conservation principles. What seemed a contradiction is now removed through the introduction of an essential wave feature in the description of the behavior of material particles: "The general character of this relation [the complementarity of the sharpness of definition of the space-time and energy-momentum vectors] makes it possible to a certain extent to reconcile the conservation laws with the space-time coordination of observations, the idea of a coincidence of well-defined events in space-time points being replaced by that of unsharply defined individuals within finite space-time regions" (Bohr 1927d, 93). As opposed to the earlier description of the Compton effect, where "in the change of the motion of the electron . . . one . . . is dealing with an instantaneous effect taking place at a definite point in space," the new wave theoretical view of matter indicated that "just as in the case of radiation . . . it is impossible to define momentum and energy for an electron without considering a finite space-time region" (Bohr 1928, 96).

All these ideas signaled nothing less than the resuscitation of what seemed a discredited (Bohr-Kramers-Slater) theory, and a further rejection of the idea of light quanta. Though Bohr did not argue this point explicitly in his Como lecture, the message is implicit in his analysis. What Bohr only hinted at, Hendrik Kramers, his faithful disciple, chose to announce unambiguously in the discussion following Bohr's lecture. Kramers opened his remarks by saying that he "shall not be able to add anything fundamental to Professor Bohr's exposition." Kramers intended only to call the attention of the audience to the examples, which illustrated the resolution of past paradoxes and difficulties with the help of the wave theory of matter: "I am thinking especially of the principle of conservation of energy and momentum, which seemed to contradict the wave-theory of light. . . . The difficulty, that the *results of these experiments* [Bothe-Geiger and Compton-Simon] *were at variance with the wave theory of light, disappears definitely,* if the de Broglie wave theory of matter is taken into account" (*BCW,* 6:139, my italics).[10]

Thus the victory of Einstein and his conception of light quanta, in the view of Bohr and Kramers, was only apparent. Bohr was more outspoken about this issue in private correspondence and discussions than in formal addresses. As Bohr informed Einstein: "In view of this new formulation . . . it becomes possible to reconcile the requirement of energy conservation with the implication of the wave-theory of light,

10. Kramers repeated Bohr's considerations, emphasizing that similar considerations apply to the question of the correlation of processes of emission and absorption in distant atoms: while an exact correlation (emission of a light quantum from the source and absorption of the "same" light quantum in the absorbing matter) is out of the question, an "approximate" correspondence is, in fact, maintained, in agreement with the basic ideas of wave mechanics (*BCW,* 6:139–40).

since according to the character of the description the different aspects never manifest themselves simultaneously" (Bohr to Einstein, 13 April 1927, AHQP; also *BCW*, 6:420, my translation).[11]

Bohr's complementarity viewpoint, formulated in the Como lecture, represented a "second phase" in his lifelong dialogue with Einstein. This dialogue was, in Pais's incisive phrase, "Bohr's inexhaustible source of identity" (1967, 219). Bohr never yielded; his entire epistemological edifice was constructed in dialogical response to Einstein's ceaseless challenge.

Dialogue with Campbell

We now turn to another central problem of the atomic domain—the interaction of matter with matter—and to yet another crucial dialogue with the lesser known Campbell. Bohr's first notes for the Como lecture were triggered by an exchange of letters between Campbell and Jordan on the interpretation of quantum theory (Campbell 1927; Jordan 1927c). Campbell took an early interest in Bohr's theory and its interpretation. In 1913, Campbell published a review of Bohr's new atomic theory. In the 1920s, Campbell suggested that the idea of orbiting electrons in stationary states was a purely formal assumption and that no reality could be ascribed to Bohr's planetary model. At the time, Bohr was not ready to consider such a suggestion—a stand he modified gradually when mechanical visualization led to a crisis of the old quantum theory. In 1921 Campbell suggested that the paradoxes of quantum theory could be avoided by modifying the concept of time and endowing it with statistical significance only. Again, Campbell's idea was not greeted with enthusiasm, probably because at the time Bohr's theory was scoring impressive successes. Encouraged by recent developments that acknowledged the need to modify space-time concepts and to renounce visualization (Bohr 1925; Heisenberg 1925), Campbell repeated and extended his earlier suggestion that time be treated statistically in

11. Similarly, in discussions following Compton's talk at the fifth Solvay conference, held a month after the Como meeting, Bohr continued to argue against the idea of point-like light quanta, using explicitly de Broglie–Schrödinger wave concepts. According to Bohr's description of the scattering process, "we must work with four wave-fields of finite extension" (two for electrons, before and after the phenomenon, and two for the incident and scattered light quanta), which are localized in the same space-time region. This modification of space-time concepts of particles preserves the ideas of the wave picture of light (and matter) together with conservation laws. Moreover, as Bohr pointed out, in the Compton experiment "the frequency shift produced is measured by means of instruments, the functioning of which is interpreted according to the wave theory" (*BCW*, 5: 211). Compton (1927) apparently did not find Bohr's arguments conclusive, arguing that the last word on this issue had not yet been said.

the programmatic paper "Time and Chance," published in 1926. This paper spurred Bohr's reasoning, as it did Heisenberg's (chapter 4), leading to the formulation of Bohr's complementarity principle.

Campbell (1926) reminded his readers both of his earlier proposal that the quantum and classical theories be reconciled by abandoning the concept of time in the interior of the atom and of Bohr's recent declarations (arrived at "of course quite independently") that the description of the quantum domain in terms of space and time might be impossible. Campbell referred to Bohr's famous postscript to his paper on the behavior of atoms in collisions, where Bohr took this stand (Bohr 1925). The paper itself was written before Bohr was familiar with the results of the Bothe-Geiger and Compton-Simon experiments, and it constituted Bohr's attempt to provide further guidelines for resolving the difficulties of quantum theory by extending the idea of nonconservation of energy from the realm of interactions between radiation and matter (Bohr-Kramers-Slater theory) into the realm of interactions between matter and matter (scattering, atomic collisions, ionization). Bohr's paper clearly displayed his thinking in terms of concrete models of atoms where stationary states were represented by mechanically orbiting electrons, and where processes were categorized as "reciprocal" or "irreciprocal" depending on whether the interaction time of particles passing by the atom is longer or shorter than the natural periods of revolution of the atomic electrons. When an atom collides with an electron of slow velocity, the time of collision is comparable to the periods of revolution of the intraatomic electrons; so during the time of collision, the passage of the atom into its final stationary state can be completed and energy is conserved (a "reciprocal process"). In the case of swiftly moving particles (such as α- and β-particles), "the collision [with the atom] must probably be regarded as finished long before one can speak of the completion of a possible transition of the atom from one stationary state to another" (Bohr 1925; BCW, 5:198).

Bohr suggested that, as in the case of radiation, so also in the case of collisions of swiftly moving particles with atoms, only statistical laws of energy conservation apply. This consideration was also connected with the fact that, while the reaction of an atom upon a particle is adequately expressed by continuous parameters, the change of state of the atom is described by discontinuous values. Bohr applied similar considerations to the case of collisions between atoms. The discussion and the conclusion of nonconservation were firmly based on space-time imagery in the interior of the atom. It was the limitation of this imagery that Bohr acknowledged in the postscript, after he became familiar with the results of the Bothe-Geiger experiments implying strict conservation. This point is also stated explicitly in Bohr's nontechnical paper, summarizing recent developments: "For impacts in which the time of collision is short compared to the natural periods of the atom . . . the postulate of stationary states would seem to be irreconcilable with any description of the collision in space and time based on the accepted ideas of atomic structure" (Bohr 1926; *BCW*, 5:851).

This conclusion formed the necessary background to the development of matrix mechanics, as well as a starting point for Campbell's discussion. Campbell's proposal that time be treated as statistical in nature was intended to provide a way out of this predicament, indicating how the space-time description of atomic phenomena must be modified. Campbell tackled Bohr's concern with collisions of fast-moving particles with atoms, suggesting that if time is statistical, the paradox resulting from the comparison of "short" collision times with "long" periods of electron revolution does not arise: "The conclusion that it [time] is short depends entirely on the assumption that the motion of the particle and the oscillation of the atom are uniform. This, of course, we deny. Particles moving with uniform velocity or oscillating in fixed orbits are undergoing fortuitous transitions between the points of their paths" (1926, 1111).

Campbell's discussion played an important role, as I have argued, in the emergence of Heisenberg's uncertainty paper. It also provided impetus to Bohr's further retreat from the reality of stationary states in his Como lecture. However, Bohr could not accept a statistical conception of time, for it would threaten his overall research program in the atomic domain, which was intimately connected with classical electromagnetic theory. In this theory periodicity in time is explicitly assumed. The new quantum theory, advocated by Campbell, clearly "cannot be deduced from Maxwell's equations in their present form" (Campbell 1926, 1113). Bohr, who vigorously defended Maxwell's electrodynamics against Einstein's idea of light quanta, and who even after the Bothe-Geiger results would only admit a need to modify space-time imagery rather than to revise electromagnetic theory itself, could not agree with Campbell's assertion.

However, Campbell's focus on the nature of time in dealing with the space-time problem in the interior of the atom turned out to be very suggestive: "The singular position of time in problems concerned with stationary states is . . . due to the special nature of such problems" (Bohr 1927c, 131). As for the conflict between space-time descriptions and conservation laws, discussed by Bohr in 1925 and subsequently treated by Campbell, the resolution of this conflict constituted another example of the complementarity between space-time and causality.[12]

12. "If the definition of the energy of the reacting individuals is to be accurate to such a degree as to entitle us to speak of conservation of energy during the reaction, it is necessary . . . to coordinate to the transition between two stationary states a time interval long compared to the period associated with this process. This is particularly to be re-

Clash with Heisenberg: Setting the Historical Record Straight

We are now better equipped to evaluate the clash between Heisenberg and Bohr over Heisenberg's uncertainty paper, and to determine the effect this clash had on the consolidation of Bohr's position. An understanding of the wide gap between Bohr's and Heisenberg's interpretive positions and professional interests illuminates the intense emotional strain that surrounded the Bohr-Heisenberg dialogue. The strain was so acute that Bohr and Heisenberg needed to spend some time away from each other, during which Heisenberg was able to write his uncertainty paper, and Bohr to elucidate his formulation of complementarity. Intense also was the confrontation between Bohr and Heisenberg over the uncertainty paper: Bohr found the paper mistaken and premature, forcefully urging Heisenberg not to rush into print. Heisenberg did not yield, though occasionally "bursting into tears" as a result of immense emotional pressure coming from Bohr (interview with Heisenberg, AHQP). Heisenberg (1927b) merely agreed to append a postscript that acknowledged Bohr's criticism without revising the content of the paper itself. One of the mistakes Bohr found occurred in Heisenberg's analysis of uncertainty during measurement in the γ-ray thought experiment. In his analysis Heisenberg treated both photons and electrons as regular point particles (!), arguing that during their collision (interaction!) a photon transfers a discrete and uncontrollable amount of energy to the electron (Compton recoil). Yet the Compton recoil of point-mass particles would not lead to indeterminacy, as I discussed in chapter 4, but to precisely calculable changes, as Bohr pointed out. The correct explanation, Bohr insisted, relies on the wave nature of light and matter in an essential way. I have argued earlier that in opposition to Schrödinger's accusations, Heisenberg's uncertainty paper was aimed at demonstrating that the "frightfully" abstract matrix mechanics is amenable to a visualizable, intuitive interpretation. As opposed to Schrödinger's program of wave ontology, aimed at eliminating the matrix approach, Heisenberg in his uncertainty paper used exclusively particle-kinematic concepts. In this effort, Heisenberg wanted to avoid Schrödinger's waves altogether (interview with Heisenberg, AHQP; interview with Klein, AHQP). Clearly, correcting the mistakes in the way that Bohr suggested, which involved essential reliance on wave concepts, would undermine Heisenberg's aims. Nor would the intensely

---

membered when considering the passage of swiftly moving particles through an atom. According to the ordinary kinematics the effective duration of such a passage would be very small as compared with the natural periods of the atom, and it seemed impossible to reconcile the principle of conservation of energy with the assumption of the stability of stationary states" (Bohr 1928, 116).

ambitious Heisenberg withdraw his paper and miss the opportunity to make another pivotal contribution by letting Bohr reap all the fruits of their interpretive struggle. As Heisenberg himself put it: "Perhaps it was also a struggle about who did the whole thing first" (interview with Heisenberg, AHQP).

In the postscript to the uncertainty paper, Heisenberg acknowledged several mistakes in his argument and included Bohr's objection to his analysis of the Compton effect—Compton recoil applies rigorously only to free and not to bound electrons. Given this mistake, Bohr must have thought, Heisenberg had completely failed to provide a physical interpretation for the interior of the atom. We have seen in this chapter what an ingenious conceptual web of arguments Bohr had spun in order to comprehend "intuitively" intraatomic structure. No wonder Bohr found Heisenberg's paper unsatisfactory. Even in the case of free electrons in the Compton effect, as we saw earlier, Bohr found particle-kinematic concepts inadequate. In any type of physical interaction, reliance on wave concepts (superposition of waves) was essential, according to Bohr. Electrons and photons treated as point-masses in Heisenberg's discussion not only led to technical mistakes in the discussion; they were "abstractions," insufficient for the description of real physical situations. This was the focus of the confrontation between Heisenberg and Bohr.[13] The issue was not metaphysical preferences (wave-particles as opposed to the mathematical structure of the formalism), as Heisenberg's recollections would lead us to believe, but Bohr's dissatisfaction with Heisenberg's exclusively particle-kinematic interpretation. The mistake Heisenberg made in the discussion of the γ-ray experiment, as well as his failure to correct this mistake in the particle framework, strengthened Bohr's position. Bohr's elucidation of the concept of the stationary state, and his resolution of past paradoxes of the Bohr-Kramers-Slater theory in the wave theoretical framework, made the gap between himself and Heisenberg unbridgeable. Bohr's Como lecture clearly displays his dissatisfaction with Heisenberg's position, though in a subtle way. When discussing the uncertainty principle, Bohr pointed to the mistake in Heisenberg's derivation of uncertainty in the γ-ray thought experiment: "Such a change [momentum change during a position measurement] could not prevent us from ascribing accurate values to the space-time coordinates, as well as to the momentum-energy components before and after the process. The reciprocal uncer-

---

13. "I argue with Bohr over the extent to which the relation $p_1q_1 \sim h$ has its origin in the wave—or the discontinuity aspect of quantum mechanics. Bohr emphasizes that in the gamma-ray microscope the diffraction of the waves is essential; I emphasize that the theory of light quanta and even the Geiger-Bothe experiments are essential" (Heisenberg to Pauli, 4 April 1927, *PC*).

tainty . . . is an outcome of the limited accuracy with which changes in energy and momentum can be defined, provided the wave-fields used for the determination of the space-time coordinates of the particle shall be sufficiently limited" (1927c, 120–21).

Some contemporary physicists understood well the point Bohr was making. Ehrenfest, in a remarkable letter to Goudsmit, Uhlenbeck, and Dieke, referred to Bohr's derivation of uncertainty, relying on "wave kinematics" and "amending the error running through the Heisenberg paper" (*der durchlaufenden Fehler von Heisenberg*; Ehrenfest to Goudsmit, Uhlenbeck, and Dieke, 3 November 1927, AHQP, translated in *BCW*, 6: 37–41, quote on 39). Born, in a discussion following the Como lecture, accepted Bohr's derivation of uncertainty as following, not from uncontrollable changes during measurement, but from the wave nature of matter (*BCW*, 6:137–38). If, for Heisenberg, uncertainty followed from discontinuity (discontinuous changes), for Bohr, uncertainty followed from a dialectical combination of continuity and discontinuity, or "individuality and superposition." For Bohr, at this point, not the operational definition of concepts, but agreement between the possibilities of definition and observation was essential.

The Como lecture contains other, more subtle hints of Bohr's past disagreements with Heisenberg. The clash over uncertainty was not the first time Bohr had reason to be dissatisfied with Heisenberg's position. As Bohr's secretary, Betty Schultz, recalls (interview with Schultz, AHQP), Heisenberg was not very helpful to Bohr, certainly not as useful as the ever loyal Kramers or, later, the agreeable Klein (about Kramers's crucial role in Bohr's research program, see Dresden 1987). Nor was Heisenberg always scrupulous about acknowledging his debt to Bohr. In his reinterpretation paper (1925), Heisenberg does not cite Bohr's work at all, despite the fact that the paper was built on Bohr's correspondence principle in a fundamental way. Nor does Heisenberg mention Bohr's conclusion from his 1925 paper about giving up space-time visualization inside the atom. Instead, Heisenberg presented his work as flowing from the positivist principle of elimination of unobservables. I have argued that this was not a guiding principle, but a justification after the fact, and that the conceptual package—correspondence principle, nonconservation, wave theory of light—seemed to be discredited by the Bothe-Geiger results. It is likely that Heisenberg chose not to mention the correspondence principle for strategic reasons. Bohr must have been unhappy about this, as well as about the fact that Heisenberg's matrix mechanics suppressed the idea of quantization and "swallowed" the idea of stationary states.

In his Como lecture Bohr set the historical record straight. After citing work by Kramers and Rudolf Ladenburg as a "characteristic example of the correspondence," Bohr presented matrix mechanics as the culmination of his own research program, based on the correspondence principle: "It is only through the quantum theoretical methods created in the last few years that the general endeavors laid down in the [correspondence] principle . . . have obtained an adequate formulation" (1927c, 124). Bohr also expressed his reservation about the initial emphasis on observability in papers on matrix mechanics, finding this emphasis not only historically misleading but physically inadequate. Only in "a certain sense" may matrix mechanics be described as a "calculus with directly observable quantities," because it is "limited just to problems, in which in applying the quantum postulate the space-time description may largely be disregarded, and the proper question of observation therefore placed in the background" (1927c, 125). The correspondence principle, according to Bohr, continued to serve as a necessary basis for the interpretive framework of physics, together with the indispensability of classical concepts. This is the reason for the amalgamation of "historical" and "philosophical" aspects in the Como lecture, which has puzzled some scholars of Bohr's work (Folse 1985; Gibbins 1987). Bohr's subtle message registered fully with Heisenberg, who from that time on, in his historical expositions, presented his reinterpretation paper as following from Bohr's correspondence principle.

Confrontation with Pauli

According to the usual historical accounts, the confrontation between Heisenberg and Bohr ended when each comprehended the other's point of view, realizing that there was no real contradiction between their positions. But the text of the Como lecture, as well as Bohr's and Heisenberg's scientific correspondence at the time, reveals the continuing gap between their interpretations of the atomic interior. For Bohr, the Heisenberg-Born-Pauli statistical kinematic interpretation of particles ("abstractions") was not sufficient for a consistent interpretation of quantum physics; the wave nature of free "individuals" (electrons, photons) had to be taken into account in an essential way. The uncertainty relations, according to Bohr, apply rigorously only to free particles (as a direct consequence of the limitation of the associated wave fields) and in general cases of interaction "must always be applied with caution" (Bohr to Schrödinger, 23 May 1928, AHQP, translated in *BCW*, 6:49). In particular, the uncertainty relation between position and momentum for discrete stationary states of an atom cannot hold true without modification. According to Bohr, the application of the concept of a stationary state and the tracking of the behavior of individual particles in the atom are mutually exclusive.

As one of the main architects of the particle-kinematic interpretation of the interior of the atom, Pauli did not agree with Bohr. For Pauli it was meaningful to talk about the uncertainty relation for particles in the case of individual discrete stationary states, as was discussed in Heisenberg's paper. Nor did Pauli agree with Bohr that a visualizable, intuitive interpretation is possible only in three dimensions, arguing that his own interpretation of the wave function could "in principle be empirically ascertained by statistical utilization of results of observation" (Pauli to Bohr, 17 October 1927, *PC*). The Bohr-Pauli dialogue had a crucial effect on further elaborations of complementarity. Pauli, an early critic of continuous field theories, was not as enthusiastic as Bohr about the indispensability of classical electrodynamics. He did accept the Bothe-Geiger results as conclusive evidence of the physical reality of Einsteinian light quanta. Bohr's complementarity of space-time and causality was not always clear to Pauli (Pauli to Bohr, 13 January 1928, *PC*). The first comprehensive discussion of wave-particle duality, as far as I know, is presented in Pauli's 1933 encyclopedia article. Pauli, then, was most likely the architect of the symmetrical complementarity of waves and particles, as distinct from Bohr's different notion of the complementarity of space-time and causal descriptions.

Bohr's and Pauli's versions of the interior of the atom were not, as far as I know, ever reconciled. Subsequent elaborations of complementarity (two-slit experiments, interaction of atomic objects with measuring devices) centered on individual particles. Bohr and Pauli did, however, agree entirely on the importance of elucidating the concept of measurement for the interpretation of quantum physics (Pauli to Bohr, 17 October 1927, *PC*). Subsequently, the *Nature* version of the Como talk (Bohr 1928), which was conceived in dialogue with Pauli, placed heavier emphasis on measurement than had the original version delivered at the conference, written in collaboration with Klein and Darwin. If the Como lecture stressed harmony between the possibilities of definition and observation, the later *Nature* version ushers in the key theme of the "uncontrollable element" introduced by observation. Bohr's and Heisenberg's positions seem to have become closer. In addition, Bohr's enthusiastic praise of Schrödinger's theory was subdued in the *Nature* version. Bohr's claim in the Como lecture that "just Schrödinger's formulation of the problem of interaction seems particularly well-suited for the illustration of the nature of the quantum theory" was eliminated from the *Nature* article. Similarly eliminated was the claim about the utility of Schrödinger's wave mechanics for a "demonstration of the consistency of symbolic [matrix] methods" (1927c, 128). With Pauli's skillful assistance, the Göttingen-Copenhagen front was uniting and consolidating its stand.

Conclusion

The united public front did not imply that the Copenhagen interpretation was coherent or consistent. The "interpretation" was actually an amalgamation of the different views of Bohr, Born, Heisenberg, Pauli, and Dirac. There is no indication that Heisenberg fully accepted Bohr's views. Unlike Bohr, the "mathematical physicist" Heisenberg preferred one coherent set of concepts, rather than two incompatible ones.[14] In this respect, Heisenberg was much closer to the "enemy" Schrödinger, who considered Bohr's solution merely a judicious escape. Heisenberg's position was that wave language and particle language, being equivalent descriptions of the same reality, were mutually convertible. One could therefore use either language at will, that of particles or that of waves, without needing to use them simultaneously, as Bohr claimed (interview with Heisenberg, AHQP).

According to the usual accounts, soon after their heated arguments over the uncertainty paper, Bohr and Heisenberg reached complete agreement. Yet genuine unanimity of opinion between the two men never occurred. Rather they realized that "all that mattered now was to present the facts in such a way that despite their novelty they could be grasped and accepted by all physicists" (Heisenberg 1971, 79). The need to offer a unified explanation, capable of countering the opposition, was one of the reasons for the obscurity of Bohr's Como lecture, where the differences between Bohr's and Heisenberg's positions were subdued. Such obscurity was necessary in order to conceal the conceptual gap between their interpretations of physics for the interior of the atom. The uncertainty principle for the position and momentum of a particle cannot easily be reconciled with the representation of separate stationary states by harmonic partial vibrations. This difficulty was one of the early targets for Schrödinger's criticism. Einstein agreed fully with Schrödinger that the particle-kinematic framework, even when supplemented by the uncertainty principle, was deficient, and that the "shaky" concepts p and q should be abandoned: "The whole thing was invented for *free particles* and suits only this case in the natural way." It was not because of their "classical nostalgia" that Einstein and Schrödinger were not persuaded by the "Bohr-Heisenberg tranquilizing philosophy" (Einstein to Schrödinger, 31 May 1928, Przibram 1967).

Years later, Heisenberg conceded that Bohr's matter wave interpretation for the interior of the atom was perhaps "much closer to the truth"

14. "If one immediately starts with the supposition there are both waves and particles, everything can be made contradictory free," Heisenberg wrote disapprovingly to Pauli two months after the uncertainty paper appeared (May 1927, *PC*). See chapter 11 for further discussion of this issue.

than his own (1958, 51). Yet the conflict between Bohr and Heisenberg on this issue had been circumvented rather than resolved. Subsequent interpretive developments shifted the emphasis to the measurement problem, the algebra of Hilbert spaces, quantum logic—areas of inquiry dissociated from the initial struggles over what happens inside the atom. As Heisenberg revealed, the physicist had no choice but "to withdraw into the mathematical scheme" when the vague use of classical concepts, encouraged by Bohr's complementarity, led to difficulties and inconsistencies (1958, 51). This sort of escape was not open to Bohr—a "natural philosopher" rather than a "mathematical physicist" (see chapter 12). To the end of his life Bohr struggled to clarify and extend his complementarity framework. In order to comprehend further developments, as well as the numerous contradictions in the orthodox interpretation of quantum physics, a historical perspective on the initial interpretive efforts is necessary.

Nor is the genesis of Bohr's complementarity principle fully comprehensible without taking into account such psychosocial factors as ambition, professional interest, group dynamics. It is remarkable, though not surprising, how much the cognitive positions of different contributors coincided with their professional and personal interests. Thus Bohr, in his Como lecture, emphasized the wave aspect of matter and light; the wave ontology allowed him to preserve and entrench his major contributions to science—the idea of stationary states, the approach of the Bohr-Kramers-Slater theory, the statistical description of atomic collisions. Heisenberg, who had no direct investment in these ideas, was ready to dispense with the wave aspect altogether, in order to argue the superiority of matrix mechanics, to which he had contributed so decisively. Emotional intensity in a scientific dialogue is not an aberration; it is vital fuel for the shaping of ideas, the formation of stands, the achievement of breakthroughs. In dialogical accounts that acknowledge the essential formative role of scientific controversies, the line between the "cognitive" and the "social" becomes blurred.

Equally problematic becomes the idea of a definable research program. A simplistic division between "orthodox" and "opposition" is not adequate for describing a living, creative stage in the formation of ideas. We see, rather, flexibility, simultaneous openness to different conceptual options, at times genuine dialogue, at times "infiltration" or selective appropriation of opponents' ideas. We saw that on certain points Bohr was closer to Schrödinger than to Heisenberg, while on other issues Heisenberg was closer to Schrödinger than to Bohr. At times, emotions become fierce, but no concessions are made; this is what happens at the crossroads of vital interests.

〜

CHAPTER 7

The Challenge of Einstein-Podolsky-Rosen

and the Two Voices of Bohr's Response

While imagining that I understand the position of Einstein, as regards the
EPR correlations, I have little understanding of the position of his principal
opponent, Bohr. Yet most contemporary theorists have the impression that
Bohr got the better of Einstein in the argument and are under the
impression that they themselves share Bohr's view.

J. S. Bell 1987, 155

Two Voices in Bohr's Response to Einstein-Podolsky-Rosen

The strongest challenge ever posed to the orthodox philosophy of quantum physics is the Einstein-Podolsky-Rosen argument (Einstein, Podolsky, and Rosen 1935, hereafter EPR).[1] The argument, if one accepts some intuitive and, at the time, widely accepted notions of physical description (reality), pointed out that something peculiar, and perhaps unacceptable, is implied by the quantum formalism: Quantum theory is either incomplete, or inconsistent, or both (Beller and Fine 1994). In particular, the conjunction of the completeness of quantum theory and the separability of states of distant systems cannot be maintained. The argument came as a "bolt from the blue," and its effect was "remarkable" not only on Bohr but also on other quantum physicists (Rosenfeld 1967, 128–29). Dirac initially considered the argument devastating: "Paul Dirac said: 'Now we have to start all over again, because Einstein proved that it does not work'" (interview with Bohr, 17 November 1962, AHQP).

Bohr's pre-1935 philosophy contained as an intrinsic part the concept of a robust physical disturbance. The challenge of EPR undercut Bohr's idea of a direct physical disturbance. Einstein, Podolsky, and Rosen discussed a system of two particles that interacted initially and then moved apart. If one directly measures the value of either of two conju-

---

1. My presentation of Bohr's response to EPR (Bohr 1935a) follows Beller and Fine (1994) closely. I am deeply grateful to Arthur Fine for most rewarding work together.

gate variables for one system, one can predict with certainty the (unmeasured) value of the same quantity in the other system.[2] The authors of EPR investigated the state function of a two-particle system and its "reduction" to the state functions of the component systems during measurement. They reached the puzzling conclusion that as a consequence of two different measurements, "without in any way disturbing" the second, distant system, two different functions can be assigned "to the same reality." The unmeasured particle has a reality that is simultaneously describable by an eigenfunction of position and an eigenfunction of momentum—using a standard eigenstate-eigenvalue rule, the authors concluded that both position and momentum can, with certainty, be ascribed to the second particle. As quantum mechanics forbids assignment of definite values to two conjugate variables simultaneously, the authors concluded that quantum mechanics was incomplete.[3]

During the course of their argumentation, the authors proposed a "criterion of reality": "If, without in any way disturbing a system, we can predict with certainty the value of a physical quantity, then there exists an element of physical reality corresponding to this physical quantity" (EPR, 778).[4] Bohr did not challenge their definition, finding only an "ambiguity" in the expression "without in any way disturbing a system." Bohr's answer to EPR centered on the idea of disturbance. Two different, even incompatible, answers are concurrently present in Bohr's response to EPR. One is tempted to assume that Bohr must have presented only one unequivocal answer and the difficulty is that of the reader. Yet such an assumption renders much of Bohr's text superfluous or incomprehensible. The two voices in the paper are real, with one voice rooted in the past (before 1935), the other emerging into the future (after 1935).

Bohr's response (or responses) to EPR was built on the "physical actualization" of EPR's mathematical reasoning, rather than dealing directly with the consequences of the formalism. Bohr proposed for the measurement arrangement a diaphragm with two parallel slits through which both particles pass simultaneously. If such a diaphragm is suspended by weak springs, we can know $Q_1 - Q_2$ and $P_1 + P_2$. Because

2. In the EPR mathematical presentation, the variables are a position coordinate and linear momentum.

3. The argument of EPR is a complex one—it was first analyzed in Fine (1986, chap. 3) and extended in Beller and Fine (1994). In particular, Beller and Fine discuss the issues of "incompleteness" and "inconsistency" in the EPR argument, and the crucial differences between the EPR argument and Bohr's summary of it.

4. The authors did not use this criterion in the paper. The only time they referred to it was to demonstrate that such a criterion is consistent with the eigenstate-eigenvalue rule, whose applications constitute accepted "quantum mechanical ideas of reality" (EPR, 778; see also Beller and Fine 1994).

$Q_1 - Q_2$ commutes with $P_1 + P_2$, we can by measuring Q_1 calculate Q_2, or by measuring P_1 calculate P_2. What is the reason that contra Einstein, we cannot do both simultaneously?

Let us consider Bohr's first answer—the operational one:[5] In addition to the first two-slit diaphragm, we employ a second diaphragm suspended by weak springs or rigidly bolted, depending on whether we intend to measure the position or momentum of the first particle (see diagram on next page). We can measure P_1 of the first particle by using the second diaphragm, and deduce P_2—the momentum of the second particle. For the measurement of P_1 we use the movable diaphragm, so we exclude in principle the possibility of measuring Q_1, and thus we exclude in principle the possibility of predicting Q_2—the position of the second particle. This answer accords well with Bohr's statement that measurements on the first particle imply "an influence on the very conditions which define the possible types of predictions regarding the future behavior of the system" (1935a, 700).[6] If, operationally, we equate measurement, and associated predictions, with definability, we see that there is no possibility of simultaneous prediction, and therefore definability, of the position and momentum of the second particle.

This answer is both too short and too long. Too short, because there are many sentences in Bohr's text (in addition to his repetition of some "simple, and in substance well-known," examples of measurement arrangements) that seem to be either superfluous or irrelevant (they find, as I will shortly demonstrate, a natural place from the perspective of Bohr's second answer). Yet the first answer is also too long, because the operational reply does not require the "physical actualization" of the EPR argument—the operational reply is a general one, independent of the specifics of the experimental arrangement. Indeed, when Bohr summarized his response to EPR in 1949, instead of two particles, he considered "for the two parts of the system . . . a particle and a diaphragm." The operational answer is thus compressed into a few lines: "After the particle has passed through the diaphragm, we have in principle the choice of measuring either the position of the diaphragm or its momentum and, in each case, to make predictions as to subsequent observations pertaining to the particle. As repeatedly stressed, the principal point is here that such measurements demand mutually exclusive experimental arrangements" (Bohr 1949, 233). A simultaneous reality, according to the operational approach, can be ascribed to two variables only if they are simultaneously measured.

5. It is this answer that Bohr himself later singled out (Bohr 1949); most commentators follow Bohr's later presentations, unaware of the second voice, discussed in Beller and Fine (1994).

6. It was here that Bohr found "an ambiguity as regards the meaning of the expression 'without in any way disturbing the system.'"

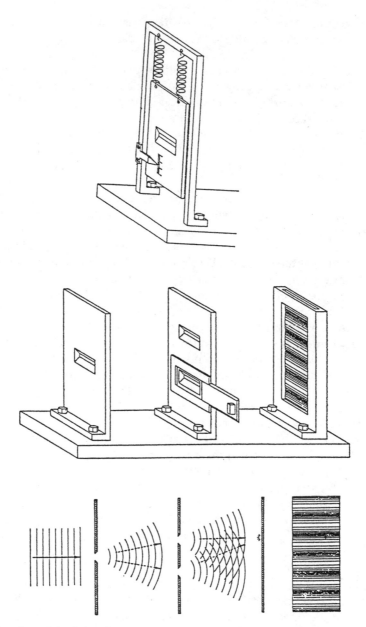

Bohr's analyses of simple thought experiments in terms of classical concepts were central to the rhetoric of the "inevitability" of the Copenhagen interpretation. These analyses created the illusion that no knowledge of the mathematical formalism of quantum mechanics is needed in order to comprehend its philosophical lessons.

The rigidly bolted apparatus (*center*) is for position measurement. The apparatus suspended by weak springs (*top*) is for momentum measurement. Diagrams adapted from Bohr (1949, 216, 219, 220).

Let us note that this simple answer was provided before Bohr by Arthur E. Ruark (1935).[7] His crisp reply does not leave much ground for the Bohr-Einstein dialogue—the opponents can only acknowledge a fundamental difference between their metaphysical presuppositions. Einstein envisaged such a simple (and for him, deeply unsatisfactory) countermove calling such a reply "unreasonable" (EPR, 780). Bohr's aim was not confined to amicably stating the metaphysical disagreements between Einstein and himself—Bohr wanted to "convince" Einstein, and to emerge "a victor" from this confrontation. If the operational reply is indeed what Bohr offered, it is not clear why he needed six weeks of "utmost concentration and unrelenting efforts" to fashion such a response (which was suggested, before Bohr, by Ruark and even Einstein himself!). Where in the operational reading do we find the "painstaking scrutiny of every detail" (Rosenfeld 1967, 131)? There clearly must be something more in the text than this simple operational countermove.

Another reading of Bohr's reply finds his second answer, which deals with the "ambiguity" in the concept of disturbance in a more direct way (Beller and Fine 1994).[8] It is this reading that makes good sense of Bohr's discussion of the details of the specific physical actualization that he proposes (details that do not make sense under the operational reading).

The measurement of momentum is similar to that outlined in the first answer—we measure $P_1 + P_2$ with the help of the two-slit diaphragm (through which both particles pass simultaneously). The two-slit diaphragm must be movable (suspended by weak springs or the like). We can subsequently measure the momentum of the first particle, using a second movable diaphragm, spatially separated from the first. Knowing $P_1 + P_2$, and measuring P_1, we can easily predict P_2 (because of the law of conservation of momentum, the two diaphragms can indeed be at any distance from each other).

The position measurement, however, presents a totally different conceptual situation. In fact, Bohr's physical actualization of the position measurement violates the EPR case. In Bohr's setup, as opposed to the EPR setup, $Q_1 - Q_2$ has a definite value at one instant only—when the two particles pass through the two slits of the first diaphragm. At any other point, the value $Q_1 - Q_2$ becomes indefinite, according to Schrödinger's equation. Yet this means that if we use a second diaphragm, spatially separated from the first (the two-slit diaphragm), to measure Q_1, we cannot predict Q_2! (We have lost our knowledge of $Q_1 - Q_2$.) In

7. "This conclusion [EPR's] can be attacked by anyone who prefers to say that P and Q could possess reality only if [they both] could be simultaneously measured" (Ruark 1935, 466).

8. The possibility of such a reading was first suggested to me by Alon Drory.

order to be able to predict Q_2 from the measurement of Q_1, we must measure Q_1 at the very moment the two particles pass through the two-slit diaphragm! This means either that the second diaphragm must be infinitely close to the first or that the two diaphragms (the two-slit one and the one for measurement of Q_1) actually merge into a single arrangement.

This reading accords well with certain passages in Bohr's reply to EPR, which seem to be strangely out of place under an operational reading. For example, if we measure the position Q_1 of the first particle, we lose the possibility of knowing $P_1 + P_2$. This loss of knowledge occurs only because the common diaphragm (which includes the two-slit one for the measurement of $P_1 + P_2$) must now be rigidly bolted and cannot serve for momentum measurement: "We have by this procedure cut ourselves off from any future possibility of applying the law of conservation of momentum to the system consisting of the diaphragm and the two particles" (1935a, 697). There are other sentences that seem to make sense only under the specific condition of merged diaphragms (Beller and Fine 1994, 14).

Clearly, the realization of position measurement introduces restrictions and physical effects not present in the EPR paper. In the EPR case both $Q_1 - Q_2$ and $P_1 + P_2$ are determinable simultaneously with either the position Q_1 or the momentum P_1 of the first particle, while Bohr's double-slit arrangement does not satisfy this requirement. In Bohr's example only one of $Q_1 - Q_2$ or $P_1 + P_2$ could be determined simultaneously with the variable one chooses to measure on the first particle. In addition, the EPR condition that "no real change can take place in the second system in consequence of anything that may be done to the first system" is violated in Bohr's arrangement. Bohr's physical realization contains an indirect mechanical disturbance, because changing the measurement from Q_1 to P_1 of the first particle demands a change in the mechanical arrangement of the two-slit diaphragm, through which both particles pass (and, consequently, in the second diaphragm, which must be attached to the first).

We can see now why Bohr does not challenge the EPR conception of reality and only finds an "ambiguity" in the expression "without in any way disturbing a system." A real physical disturbance does exist in Bohr's (incorrect) realization of the case proposed by EPR! Bohr's second reply is a veritable failure. No wonder Bohr never repeated this intricate, yet fundamentally flawed argument. Refusing (or unable) to explore quantum "wholeness" in terms of quantum ontology, Bohr's only choice was to land in positivism.

Two different voices, then, meet (or clash) in Bohr's response to EPR. The old voice, holding on to the notion of physical disturbance, brings this notion to a dead end. The new, emerging operational voice will culminate in unreserved verificationism and a future repudiation of the notion of disturbance. Such complex, polyphonic exploration is fitting for the interface between the old and the new—the transition to a substantially different position is not a Gestalt switch.

Other tensions, consequently, inhabit Bohr's paper, such as the tension between "consistency" and "inevitability" arguments: the first, operational answer introduces the inevitability motive, while the second answer is more compatible with Bohr's assertions of "soundness," "lack of contradictions," "rationality," and "consistency" of the quantum mechanical scheme (see chapter 9). Similarly, Bohr strikes two different chords on the theme of "the radical revision of our attitude towards the concept of physical reality" (1935a, 697). The first, operational answer forbids any talk about objective, observer-independent reality. This answer will culminate in Bohr's redefinition of physical phenomena—no elementary phenomenon is a phenomenon until it is an observed phenomenon. In relation to the revision of the concept of reality, the second answer does not contain anything new—Bohr merely repeats his pre-1935 arguments of "inseparability" of measuring and measured, of object and subject (such inseparability necessitates a change in the concept of reality, as I discuss below). This reply accords well with Bohr's assertion that his discussion of the EPR case does not contain any greater intricacies than those in his previous analyses of simple mechanical arrangements discussed in his publications before EPR.

Our description of the conflicting voices in Bohr's reply discloses why Bohr's readers have such difficulty following his arguments.[9] Unless we acknowledge the different voices in Bohr's paper, its tensions and contradictions, parts of the paper are incomprehensible. Nor does the great labor spent on it, and "the state of exaltation in which Bohr accomplished this work" (Rosenfeld 1967, 131) make sense.

Bohr's Victory?

There is a widespread myth that Bohr enjoyed a triumph over Einstein in their dialogue on EPR (see Rosenfeld 1967; Wheeler and Zurek 1983, 142–43). Yet none of Bohr's answers are satisfactory. I have already pointed out that the answer built on the concept of disturbance is fundamentally flawed. The other, operational answer can be more properly seen as an enforced, ad hoc legitimation move, rather than as an

---

9. Bohr himself wrote later that he was "deeply aware of the inefficiency of expression which must have made it very difficult to appreciate the trend of the argumentation aiming to bring out the essential ambiguity" (1949, 234).

adequate confrontation with the deep challenge of the EPR correlations. In particular, Bohr's operational answer conceals rather than reveals entanglement and nonlocality, which is the most powerful message of EPR.[10] Even if we accept Bohr's changing the rules of the game, and his refusal to enter into ontological inquiry, his operational reply falls apart under close scrutiny. Quite aside from the general weaknesses of the operational positivist stand,[11] operationalism is especially unsuited— in fact it undermines—Bohr's philosophy of complementarity.

In his discussion of a position measurement when the diaphragm is rigidly bolted, Bohr argues that an "uncontrollable" amount of momentum "passes into . . . support" (1935a, 697). The momentum transferred from the particle to the measuring device cannot be measured in principle. If it cannot be measured from an operational point of view, it has no meaning. Yet in this case, quantum uncertainty, contra Bohr, cannot arise (still less "inevitably" follow) from the physical interaction between an object and a measuring device (Beller and Fine 1994, 21).[12] The conclusion of Fine and myself was that

> from the positivist perspective that Bohr eventually adopted, the idea of an uncontrollable exchange of momentum, which is supposed to ground his physical picture of quantum uncertainty, is problematic. The only way around the problem seems to be to turn the ground upside down, and to make the measurable uncertainty the operational basis for the language of uncontrollable exchange. Thus despite the lively imagery, when Bohr talks of an exchange or transfer of momentum, there is literally nothing (and in particular, no momentum) that is transferred or exchanged. Bohr conjures up a robust physical picture: the feature of wholeness or "individuality" of the quantum phenomena connected to an uncontrollable interaction between object and apparatus—all giving rise to the quantum uncertainty. Upon scrutiny, however, this impression turns out to be the effect of a conjuring trick. Only the quantum uncertainty itself is independently meaningful. From the positivist point of view, the rest is a word-picture constructed around the experimentally verifiable uncertainty formulas, like a collage of printed words glued on to a radiant object. (1994, 22–23)

10. As argued in Beller and Fine (1994), Bohr's talk about the "wholeness of the experimental situation" reflected his positivist solution to EPR (where the operational presupposition implies the inclusion of all the aspects of conditions of measurement), and not his endorsement of a nonlocal or nonseparable conception of reality. In fact, Bohr, as did Einstein, considered the option of nonlocality unacceptable (Beller and Fine 1994). Bohr's notorious ambiguity creates room for later, charitable scholars to ascribe to him insights about the nonlocality of the EPR situation that he in fact did not express (see chapter 12).

11. I discussed these weaknesses in chapter 3. I will further analyze this issue in chapter 8 and argue that positivism is not a natural stand for the working scientist.

12. Beller and Fine discuss some potential responses to this difficulty and find them unsatisfactory (1994, 21).

We started with two superimposed answers to EPR by Bohr. We end up with no answer at all! It is bewildering that Bohr's response was ever considered, and is often still considered, an adequate (not to mention "triumphant"!) reply to EPR. I can suggest a few explanations for this strange state of affairs. The myth is in part connected with the general mythology of the Copenhagen interpretation, the hero worship of Bohr, the fabrication of the "winner's narrative"—issues I discuss in the chapters that follow. Yet we also have to take into account the specifics of the Bohr-Einstein confrontation over EPR. A few ingenious rhetorical moves characterize Bohr's response and create the illusion of victory. By giving a short, nonmathematical summary of the dense and complex EPR paper, Bohr ensured that few would bother to read the EPR paper itself. (Those who did, and immersed themselves in the details of the EPR argumentation without bias, obtained a rich harvest: Bell 1964, 1966; Bohm 1952.)

The overwhelming majority of presentations of the Bohr-Einstein debate use Bohr's nonmathematical summary of EPR. Yet this summary is misleading and introduces weaknesses not present in the original EPR paper. Those who lightly dismiss the EPR challenge (Pais 1991, 429–31) dismiss in fact Bohr's version of it (see Beller and Fine 1994, 2–6, for a full discussion of the difference between EPR and Bohr's EPR). A particularly obvious weakness is Bohr's assignment of simultaneous position and momentum values to the unmeasured particle. According to Bohr's summary, if we measure the Q_1, then, according to the criterion of reality, we can assign a definite value to Q_2; similarly, if we measure P_1, then, according to this criterion, we can assign a definite value to P_2. Yet, according to the criterion of reality, it is not clear why we can assign Q_2 and P_2 at the same time, unless we had measured Q_1 and P_1 simultaneously. The weakness is so obvious that it seems we can dismiss the EPR argument at once. Consequently, we might conclude, there is no reason to enter into the mathematical intricacies of EPR argumentation.

As Fine and I have argued, EPR's demonstration of simultaneous P and Q values depends not on the criterion of reality but on the state descriptions in accord with the eigenstate-eigenvalue rule (the criterion of reality is only introduced to show that it is consistent with this rule; Beller and Fine 1994). The main point, and the strength of the EPR argument, was to challenge the adequacy of the quantum mechanical characterization of a system's state by means of a wave function. A mathematical elucidation in the EPR paper demonstrated that such a description introduces nonclassical features of entanglement, or correlation, that are at odds with the deeply entrenched intuitions about the individuation of physical systems. Bohr's presentation of EPR conceals this crucial insight.

Thus the notion of "physical reality" is the focal point of Bohr's summary of EPR, but not of the EPR paper itself. Bohr, as we have noted, was not satisfied with stating his metaphysical disagreements with Einstein—he wanted to "win" the discussion.[13] He therefore defined the ground of the discussion as if Einstein and he shared some basic presuppositions about how physicists conceive of reality: "The extent to which an unambiguous meaning can be attributed to such an expression as 'physical reality' cannot of course be deduced *from a priori* philosophical conceptions, but—*as the authors of the article cited themselves emphasize*—must be founded on a direct appeal to experiments and measurements" (Bohr 1935a, 696, my italics).

Compare this with the EPR wording: "The elements of the physical reality cannot be determined by *a priori* philosophical considerations, but must be found by an appeal to results of experiment and measurements" (EPR, 777). Note the subtle change in wording and fundamental change in meaning: EPR talked about "elements of the physical reality," while Bohr talked about the "meaning" of the notion of reality. In the EPR discussion, the elements of physical reality are the physical variables that can be predicted with certainty—one can reformulate the EPR wording in terms of the adequacy of theoretical notions, without invoking the concept of reality. Bohr reformulated the passage from EPR into a metaphysical discussion of what physicists mean when they say "reality." This reformulation, together with Bohr's repetition of a few measurement procedures, has a strong rhetorical effect. Following Bohr's analysis of measurement procedures time and time again, the reader enters into Bohr's frame of mind and, without noticing, loses any critical perspective on the verificationist ground that Bohr gradually and carefully builds. By tinkering with the wording of EPR, Bohr creates an illusion that Einstein, Bohr, and the reader all share the same epistemological stand concerning the connection between theory and experiment. It is on this "common" ground that Bohr "defeats" Einstein.

The sloppy wording in the beginning of the EPR paper, where the authors write about "objective reality, which is independent of any theory," seems to support Bohr's ingenious rhetorical ploy. The opening lines of the EPR paper express a naive, simplistic notion of the "correspondence theory of truth," where theoretical "concepts are intended to correspond with the objective reality" (EPR, 777). Not much philosophical sophistication is needed in order to ask: What kind of access do the EPR authors have to this reality, which is "independent of any theory"? Is not their stand simply a metaphysical prejudice? Is Bohr not right to combat their unfounded position? In particular, is Bohr not right that a more adequate definition of reality is badly needed?

13. From the point of view of the second reading, Bohr probably imagined initially that he had won the discussion.

As I have mentioned, EPR did not use this unfortunate conception of physical reality in their paper. Yet their poor wording seems to call for a response to their formulation of the concept of physical reality. From the two notions of reality—the naive realist version of EPR and the verificationist version of Bohr—the latter is the more defensible and sophisticated. By fiat, the reader seems to have no choice but to follow the lesser of the two evils. Einstein appears to be a naive, outmoded scientific realist who "in regard to quantum physics is out to lunch" (Pais 1991, 434).

Quite apart from the fact that this poor definition of reality is never used in the paper, such a simplistic notion of reality is not Einstein's. Einstein himself ridiculed such a naive approach. Einstein wrote: " 'The world of bodies is real.' . . . The above statement appears to me to be, in itself, meaningless, as if one said: 'The world of bodies is cock-a-doodle-doo.' It seems to me that 'real' is *in itself* an empty, meaningless category" (quoted in Howard 1990, 368, Einstein's italics).[14] Einstein's realism has been described in Fine's (1986) pioneering analysis, and the discussion further extended and analyzed in Howard (1990, 1993) and Beller (forthcoming). For Einstein, the notion of scientific truth is Kantian and holistic: the truth of a scientific statement does not reside in its correspondence with reality but derives from the adequacy of the unified conceptual model to which it belongs (the empirical adequacy of such a model is one of the conditions for its truth; the logical simplicity of its foundation is another). Einstein's realism avoids the choice between naive realism and simple-minded positivism.

As Fine (1986) has pointed out, the final draft of the EPR paper was written by Boris Podolsky, so it seems that it is Podolsky's rather than Einstein's conception of reality that is reflected in the opening pages. Yet, due to the tendentious recollections of the Göttingen-Copenhagen physicists and to some careless wording in the EPR paper, Einstein's conception of reality is widely misrepresented and misunderstood. The myth of Einstein the naive realist and the myth of Bohr's triumphant defeat of Einstein in their debate over EPR go hand in hand.

Disturbance, Reality, and Acausality

The EPR challenge forced Bohr to make basic changes in his philosophy of complementarity, undermining the notion of disturbance on which his pre-1935 philosophy was built (Fine 1986; Beller and Fine 1994). Bohr's reaction to EPR was an opportunistic one, in the sense of

14. On other occasions Einstein also rejected the notion that the "truth" of a theory lay in its "correspondence with reality." "He [the scientist] will never be able to compare his picture with real mechanism, and he cannot even imagine the possibility of the meaning of such a comparison" (quoted in Fine 1986, 93).

"opportunism" that I discussed in chapter 3. As I will shortly argue, not only Bohr's concept of disturbance, but all the basic tenets of Bohr's philosophy—the problem of reality, acausality, and the nature of measurement—underwent a complete transformation, even a reversal, from his early (pre-1935) work to his later (post-1935) writings. Bohr's early and later philosophical writings cannot be unified into a coherent interpretive structure. No wonder scholars who assume that Bohr's philosophical framework underwent no substantial change over the years (at most, some refinement of terminology) experience formidable difficulties in understanding Bohr's thought.

After 1935, operationalism became a focal point of Bohr's philosophy.[15] Bohr (1935b) presented the operational definition of concepts not merely as plausible but as the only possible one, claiming that the account of measuring instruments constitutes the only basis for the definition of physical concepts. Despite the fact that Bohr criticized Heisenberg's fallacious reasoning in his description of the γ-ray experiment in the uncertainty paper, Heisenberg's idea of "uncontrollable disturbance" was the core of Bohr's pre-1935 writings. It inspired Bohr's understanding of the nature of measurement, his stand on the interrelation between observation and definition, and his elaboration of the philosophical problems of causality and reality in the quantum domain.

The concept of disturbance, inaugurated in Heisenberg's uncertainty paper, is an ill-fated and inconsistent one: it presupposes the existence of objective exact values that are changed by measurement, contrary to the desired conclusion of indeterminacy. Eventually, Bohr (1939) would repudiate the disturbance concept. Bohr's followers consequently minimized its significance in Bohr's writings (Pais 1991). Yet disturbance imagery is entrenched in Bohr's thought at the time. Disturbance is the reason for the inseparability of phenomena and the means of observing them, for complementarity (rather than a later identity between definition and observation), and for Bohr's initial conflation of the problems of objectivity and causality. According to Bohr, both objectivity and causality presuppose the notion of the exact definition of the state of a physical system, excluding in principle all disturbances. Bohr argued that if in the quantum domain every measurement implied intervention, or finite, nonnegligible interaction, then the conclusion of inevitability and the final overthrow of both reality and causality immediately follow. Later, when Bohr abandoned the imagery of disturbance, the

15. As I have argued, in 1927 (the Como lecture) Bohr objected to the operational emphasis in Heisenberg's approach. The definition of concepts, claimed Bohr at the time, is independent of, and prior to, any procedure of measurement. As I argued in chapter 6, Bohr held that the only way to connect the quantum formalism with observable space-time concepts was through wave theoretical imagery.

reality and acausality problems became dissociated, and each underwent an independent transformation.

Bohr's early description of the nature of measurement invoked the realistic imagery of existing phenomena (no operational definition of concepts yet!) and of disturbance, or finite changes in the phenomena, during measurement. This idea of disturbance is elaborated in an early manuscript by Bohr: "Our usual description of physical phenomena is based entirely on the idea that the phenomena concerned may be observed without disturbing them appreciably. . . . the quantum postulate implies that no observation of atomic phenomena is possible without their *essential disturbance*" (1927d; *BCW*, 6:91, my italics).

This is not an accidental, unhappy choice of terminology—the idea persists in all of Bohr's writings in the late 1920s to early 1930s:

> Unavoidable influence on atomic phenomena [is] caused by observing them. (1929b, 100; *BCW*, 6:216)
>
> Phenomena are influenced by observation. (1930, 134)
>
> The measurement of the positional coordinate of a particle is accompanied not only by finite change in the dynamical variables. (1928, 103)
>
> The action of the measuring instruments on the object under investigation cannot be disregarded. (1933, 7)
>
> Interaction between these instruments and the atom itself [is] an exchange of such magnitude that it erases all trace of the phenomena we try to observe. (1935b, 219)

The idea of disturbance is intimately connected with the immediately appealing—but in fact wrong—idea that the mere fact that measuring devices are themselves composed of atoms necessitates the inseparability of atomic phenomena and the means of observing them. Bohr often emphasized the atomic structure of measuring devices as the reason for the finitude, or wholeness, of the quantum interaction, which, in turn, implies the inseparability of phenomena and observation. The underlying—incorrect—intuition is that because in the quantum domain the interaction of measuring is of the same order of magnitude as the phenomena being measured, such finite interaction (in contrast to the classical case) cannot be neglected, or "accounted for."[16]

The connection between finite disturbance and the atomic constitution of measuring devices is prominent in Bohr's pre-1935 writings: "We cannot close our eyes to the fact that not only the bodies under

16. The idea of disturbance has, of course, a counterpart in the quantum formalism: the change of a quantum state, or "reduction of a wave packet." If an atomic system is not in the eigenstate of an observable, the measurement of this observable alters its state from a superposition to the eigenfunction corresponding to the measured value. It is not clear at all how the rules of the quantum calculus are implied by, or even connected to, the "finitude" of the quantum of action.

investigation, but also the measurement instruments are built up of atoms" (1931b, 152).[17] The atomic constitution of measuring devices implies not only difficulty in separating atomic phenomena and observation but a "difficulty of distinguishing between object and measuring instruments. With the latter problem we are at once presented when it is necessary to take the atomic constitution of the measuring instruments into account" (1931b, 155). The idea of disturbance implies a symmetry between measuring devices and atomic objects: "The idea of the means of observation independent of the phenomena or of phenomena independent of means of observation cannot be maintained" (1927d, 91)—a claim that stands in striking contrast to Bohr's later position that the nature of atomic objects and the nature measuring devices are fundamentally different.

The idea of disturbance, of "interference" with the course of phenomena, underlay Bohr's pre-1935 writings on both objectivity and causality. The discovery of the quantum of action throws "a new light upon the old philosophical problem of the objective existence of phenomena independently of our observations. Any observation necessitates an *interference with the course of the phenomena*. . . . the limit of speaking about phenomena as existing objectively finds its expression . . . *just* in the formulas of quantum mechanics" (1929a, 115; *BCW*, 6:249; my italics). Bohr's understanding of the breakdown of objectivity initially implied the introduction of a perceiving subject into physics[18]—an idea that Bohr would later deny vigorously (when he identified the act of measurement with the permanence of the recordings of measuring devices). The "close analogy" Bohr drew between the impossibility of strictly separating phenomena and observation and the "general limits of man's capacity to create concepts which have their roots in our differentiation

17. The atomic structure of measuring instruments, according to Bohr, does not merely have far-reaching epistemological consequences; a more thorough and explicit incorporation of this fact will eventually lead to further development of the physical theory itself: "The present formulation of quantum mechanics in spite of its great fruitfulness would yet seem to be no more than a first step in the necessary generalization of the classical mode of description, justified only by the possibility of disregarding in its domain of application the atomic structure of the measuring instruments themselves in the interpretation of the results of experiment" (1937c, 247). A similar idea was developed in Bohr's (1937d) Hitchcock lectures at Berkeley.

18. "The feature which characterizes the so-called exact sciences is, in general, the attempt to attain to uniqueness by avoiding all reference to [a] perceiving subject. This endeavor is found most consciously, perhaps, in mathematical symbolism which sets up for our contemplation an ideal of objectivity to the attainment of which scarcely any limits are set, so long as we remain within a self-contained field of applied logic. In the natural sciences there can be no question of a strictly self-contained field of application of the logical principles, since we must continually count on the appearance of new facts" (Bohr 1929b, 96–97, *BCW*, 6:212–13).

between subject and object," as well as Bohr's idea of introspection when no separation between object and subject exists, is inspired by his notion of disturbance. Bohr illustrated the "unavoidable" introduction of "subjectivity" by describing misleading analogies with relativity theory: "[The] theory of relativity . . . was destined to reveal the subjective character of all the concepts of classical physics" (1929b, 97; *BCW*, 6:213).

After 1935, as a response to the EPR challenge, Bohr abandoned the terminology and underlying imagery of disturbance. The problems of reality and causality became dissociated, and Bohr made an about-face in his opinions on the nature of measurement. In contrast to the earlier symmetry between the measuring and the measured, Bohr introduced the idea of a fundamental distinction between the nature of atomic objects and that of measuring devices. In Bohr's later writings an experimental device must be classical, "heavy," and its atomic constitution must be, "in principle," disregarded. The measurement interaction cannot be separated from phenomena, not because one cannot neglect the quantum, but precisely because one must neglect it: "The essentially new feature in the analysis of quantum phenomena is, however, the introduction of the measuring *apparatus and the objects under investigation*. This is a direct consequence of the necessity of accounting for the functions of the measuring instruments in purely classical terms, excluding, in principle, any regard to the quantum of action" (1958c, 3, Bohr's italics). The crux of Bohr's later arguments is precisely that the measurement interaction is nonformalizable in principle.[19]

As Bohr's ideas on the nature of measurement took an about-face, his accounts of objectivity and causality changed fundamentally. In Bohr's early writings the main psychological analogue of the inseparability of phenomena and observation was introspection, in which no sharp distinction can be made between object and subject. As in psychology, so in physics, the idea of objectivity, which avoids any reference to a perceiving subject, can no longer be maintained. Bohr's implicit definition of objectivity was initially metaphysical—the existence of objects and events having an independent reality regardless of being observed or not. In contrast, Bohr's later definition of objectivity was intersubjective: Bohr identified an objective description with a consistent method of recounting facts that can be understood clearly by others. After 1935, Bohr was eager to retain objectivity, not to dispense with it. It is not clear how the use of imprecise common language ensures unambiguity

19. "It is decisive to realize . . . that the description of the experimental arrangements . . . must be based on common language. . . . This circumstance . . . excludes any separate account of the interaction between the measuring device and the atomic objects under investigation" (Bohr 1960c, 773).

of communication—one can easily argue that esoteric technical language is more suitable, as Heisenberg, in fact, did (see chapter 8).

This change in the conception of objectivity, from observer independence to unambiguous communicability (a change forced by EPR), corresponded directly to Bohr's about-face on the nature of measurement: communication is only unambiguous if we put a separation line between subject and object. Consequently, Bohr's analogies between quantum physics and psychology change in character, from the example of introspection, in which no separation between subject and object exists, to the famous case of Møller's student, who becomes confused trying to untangle his different selves. Some students of Bohr's thought read this example incorrectly; Bohr advances it to argue that the line between object and subject must be drawn sharply, not that the two are inseparable. If we blur the line, as the poor philosophy student does, we become incapacitated, unable to function in daily life and to look for employment. Eventually, we may even fall into madness: "This situation can give rise to what we call splitting of personality." The chilling moral of Møller's humorous story is "how essential it is to pay attention to the separation lines as for example, in physics, separation between system and the observer" (1958b, 715).

For the later Bohr, separating the "content of our consciousness" and "the background loosely referred to as ourselves" is not only not impossible—it is mandatory for "unambiguous communication," even though the separation lines may be placed differently in various contexts due to the "richness of the reality of conscious life" (1960c, 13).

Bohr's Doctrine of the Indispensability of Classical Concepts and the Correspondence Principle

Bohr's doctrine of the indispensability of classical concepts underlies his philosophy of complementarity. Give up this doctrine, and the inevitability of complementarity in physics dissolves. Even today, the most sympathetic interpreters of Bohr's thought do not feel at ease with Bohr's categorical assertions of the impossibility of concepts other than the classical (Hooker 1994; Howard 1994). In the past, some of Bohr's closest collaborators, Heisenberg and Born, rejected this peculiar doctrine (see chapter 8).

Bohr presented his doctrine of the indispensability of classical concepts in a deceptively convincing way. Classical concepts, Bohr argued, being direct extensions of our intuition, are necessary to describe experiments and to communicate to others what we have done and learned. Bohr's claim that classical concepts constitute a necessary extension of our intuition is, however, both historically and philosophically inaccurate. This claim disregards the tremendous gap between our essentially Aristotelian intuition and the sophisticated, abstract framework of Newtonian physics (and thus ignores the huge intellectual efforts of the founders of modern science, who replaced the intuitive Aristotelian world with the counterintuitive Galilean-Newtonian one). Nor is it obvious that the most essential feature of a measuring device is the classically described space-time configuration of its components. Dirac, for example, at least initially, was at odds with Bohr on this point. As Oskar Klein recalled: "Who did not quite agree with Bohr at first was Dirac. . . . he made some kind of observation philosophy which had to do with permanent marks. Bohr objected a lot to that because it . . . is not the essential thing in the observation. But Dirac made that, so to say, as the essential thing . . . because one part of Bohr's view was that you had to . . . use the whole classical theory in describing observation. And I think that was not near to Dirac's mind at the time" (interview with Klein, AHQP).

The historical context of Bohr's initial idea of the indispensability of classical concepts was as follows: As I have argued, the biggest philosophical quandary of the new matrix mechanics involved its elimination of the space-time container from the atomic world (chapter 2). Heisenberg, Born, and Jordan accepted lack of visualization as the necessary price—and perhaps, in fact, a bonus—of an outstanding technical advance. Even Bohr became "more and more convinced of the need of a symbolization if one wants to express the latest results of physics" (Bohr to Harald Høffding, quoted in Jammer 1966, 347).

The great success of Schrödinger's competing version of quantum mechanics changed Heisenberg's stand on this issue. As Klein recalled, throughout fall 1926 and winter 1927, Bohr and Heisenberg searched feverishly for a way "to introduce space and time into these complex formulae" (interview with Klein, AHQP). Heisenberg finally came out with the answer in his uncertainty paper—all classical space-time concepts can be retained in the quantum domain if one gives up their precise simultaneous use; yet, in contrast to classical theory, variables in the quantum domain are subject to the uncertainty relations.

What for Heisenberg was an exercise in possibility (though at times he confused it with the rhetoric of inevitability; chapters 5 and 9) became, for Bohr, the argument for indispensability. The indispensability of classical concepts was discussed thoroughly in the correspondence between Bohr and Schrödinger at the time. Bohr's dismissal of the possibility of developing "new" concepts was expressed initially as an argument against Schrödinger's plan to search for a new conceptual theoretical scheme that would avoid quantum theory's peculiarities and "irrationalities."[20]

When Bohr wrote that the "interpretation of the experimental material rests extensively upon the classical ideas" (1927d, 91) or that "it lies in the nature of physical observation that all experience must ultimately be expressed in terms of classical concepts, neglecting the quantum of action" (1929b, 94–95; BCW, 6:210–11), he was not referring to the idea of treating measuring devices as classical in nature, as some commentators have understood it, reading Bohr's later ideas backward. Instead, Bohr was making a broad Kantian statement about the impossibility of describing physical experience in general by any concepts other than classical ones.

As in Heisenberg's uncertainty paper, so in Bohr's numerous discussions of thought experiments in his pre-1935 writings, classical concepts apply simultaneously to microscopic objects and macroscopic measuring devices. In Bohr's early writings, he spoke of the "preestablished harmony" between quantum theoretical concepts and the possibilities of measurement: a measuring device seems to be a perfect match to the wave theoretical definition of particles: "Uncertainty cannot be avoided. . . . the experimental devices . . . are seen to permit only conclusions regarding the space-time extension of the . . . wave fields [associated with particles]" (1928, 101)[21]

The measurement interaction in Bohr's early writings was quantum mechanical—quantum uncertainty applied even to the whole macroscopic γ-ray microscope as a measuring device: "A closer investigation shows, however, that such a measurement [of the momentum transmitted during measurement] is impossible, if at the same time one wants to know the position of the microscope with sufficient accuracy. In fact, *it follows from the experiences which have found expression in the wave theory of matter that the position of the center of gravity of a body and its total momentum can only be defined within the limits of reciprocal accuracy given by relation (2)* [the uncertainty relation]" (my italics). Bohr summarized: "The uncertainty equally affects the description of the agency of measurement and of the object" (1928, 101, 102).

20. "I am scarcely in complete agreement with your stress on the necessity of developing 'new' concepts. Not only, as far as I can see, we have up to now no cues for such a re-arrangement, but the 'old' experiential concepts seem to me to be inseparably connected with the foundation of man's power of visualizing" (Bohr to Schrödinger, 23 May 1927, AHQP; quoted from Murdoch 1987, 101).

21. A similar idea is expressed in the following passage: "Limitation on the possibilities of measurement is directly related to apparent contradictions in the discussion of the nature of light and of material particles" (contradictions that are removed, according to Bohr, by using the wave theoretical definition of particles; Bohr 1929b, 95; BCW, 6:211).

In striking contrast to his later analyses, before 1935 Bohr claimed that only by incorporating the atomic nature of measuring devices more thoroughly into the analysis of the microdomain could the interpretative program be advanced. After 1935 Bohr moved gradually to an intermediate position, according to which one can "to a *very high degree of approximation* disregard the molecular constitution of the measuring instruments" (1948, 451, my italics). This "approximation" later became a matter of principle—it was now a "logical" necessity to ignore the atomic structure of measuring devices and describe them in "purely classical terms, excluding in principle any regard to the quantum of action" (1958c, 3–4). This "essentially new feature" in the analysis of quantum phenomena is tied directly to the "heaviness" of the measuring devices as opposed to the microobjects: apparatus are "sufficiently large and heavy to allow an account of their shape and relative positions and displacements without any regard to any quantum features inherently involved in their atomic constitutions" (1962, 24).

Similar passages are scattered throughout Bohr's later writings (Bohr 1958b, 712; 1958c, 2–4). This "heaviness" is now crucial for unambiguous communication, and, indeed, in order to talk about well-defined experimental conditions at all.[22] After 1935 Bohr held that the "heaviness" of the measuring apparatus, and the fundamental distinction between measuring and measured that follows from it, is *the reason* phenomena and observation are inseparable (1958b, 712, my italics).[23]

This fundamental distinction between the measuring and the measured undermined Bohr's earlier arguments about the complementarity of space-time and causality. These arguments relied on the "uncontrollability" of the measuring interaction, which in turn was demonstrated by applying the uncertainty relations to the measuring devices regarded as objects of measurement.[24] But, once Bohr introduced the

22. "In the description of the experimental arrangements we must certainly use our ordinary language. . . . We can by experiments only understand something about which we are able to tell others what we've done and what we've learnt. . . . The reason that one can describe experimental conditions in this matter is that one uses as apparatus heavy bodies, bodies which are so heavy, immensely heavy, compared with the single atomic particles that we can, in the description entirely neglect all implications of the quantum" (Bohr 1958d, 727).

23. It is this unbridgeable gap between the classical and the quantum that implies the unsurveyability, or unformalizability, of the interaction, implying in turn a "wholeness," or the necessity of specifying the experimental conditions for any definition of quantum phenomena (Bohr 1961, 77–78). This gap is made even larger as Bohr talks of the necessity of using not even "classical" but only "plain" or "ordinary" language (1958c, 2–3; 1958e, 695).

24. Thus, in EPR, one cannot use the exchange of momentum between a suspended diaphragm and a particle to predict the exact value of the particle's momentum, because,

idea that the fundamental distinction between the measuring and the measured is tied directly to physical constitution ("heaviness"), the same object simply cannot be used as a measuring device and as a measured object even in different contexts.[25] In this case, not only Bohr's arguments about the complementarity of space-time and causality but the whole early enterprise of demonstrating compatibility between the quantitative conclusions of the quantum formalism (uncertainty relations) and the experimental situation (analysis of measurements) is undermined. It is only in some vague sense that one can claim, as Bohr did after 1935, that the unsurveyability of the measurement interaction and the consequent "wholeness" of quantum phenomena dictate a departure from classical description (see chapter 9). For example, it is not clear in Bohr's later writings why this "wholeness" indicates the necessity of statistical description—in fact, it is not clear how one assesses the necessary features of an appropriate description at all (see chapter 12). Bohr simply retained an intuitive connection between the necessity of classical concepts, the inseparability of phenomena and observation, and acausality—ignoring their separate sources in now incompatible arguments. This maneuver has resulted in the misconception that Bohr's philosophical framework remained unchanged.

While the meaning of Bohr's doctrine of the indispensability of classical concepts varied over the years, there is a sense in which it remained unchanged. The claim that classical concepts are necessary is connected with Bohr's principle of correspondence—a principle that guided Bohr's early work in quantum theory, and to which Bohr subscribed all his life. Bohr had introduced the correspondence principle as a heuristic guide, as a methodological principle of research—a "formal analogy between the quantum theory and the classical theory." The correspondence principle stated, not only that for large quantum numbers classical and quantum calculations should coincide, but also that for small quantum numbers there is a correspondence between various harmonic components of motion computed classically and the characteristics of various types of quantum transition from one stationary state to another (Jammer 1966; Hendry 1984; Darrigol 1992a).

Bohr's correspondence principle, applied with ever increasing rigor by Kramers, Born, and finally, in a fundamental way, by Heisenberg, led to the formulation of the new quantum formalism. Even before the

---

in order to measure a diaphragm's momentum, "this body can no longer be used as a measuring instrument . . . but must . . . be treated, like the particle traversing the slit, as an object of investigation, in the sense that the quantum-mechanical uncertainty relations regarding its position and momentum must be explicitly taken into account" (Bohr 1935a, 698).

25. This fundamental distinction is not merely a semantic one (the "cut" as discussed in chapter 9) of the context in which, for example, the diaphragm is used.

discovery of the new quantum theory, mathematical physicists (Sommerfeld, for example) perceived correspondence arguments as a temporary heuristic tool, criticizing Bohr's elevation of the principle into a "law of the quantum theory" (Mehra and Rechenberg 1982). After the discovery of matrix mechanics the natural conclusion was that the consistent mathematical formalism should supersede Bohr's correspondence principle. Yet Bohr never abandoned it, both as a heuristic and as an interpretive principle. Bohr had no doubt that correspondence considerations must guide the development of quantum electrodynamics.[26] And it served as a touchstone in his later writings: "It is of course important in each case to remember how far from the point of view of correspondence we can acquaint ourselves with the new situation" (1957c, 666).

Bohr's insistence on the continuing importance of the correspondence principle seems puzzling. Dirac, who had himself initially used correspondence arguments in a most ingenious way (Darrigol 1992a), did not share Bohr's attitude: "He [Bohr] still referred to the correspondence principle for some years, I think, after quantum mechanics really made definite equations which would replace the correspondence principle. . . . When one gets so absorbed with one idea, one does stick to it always" (interview with Dirac, 14 May 1963, AHQP). Why did Bohr need correspondence arguments to acquaint himself "with the new situation"? Why did he not refer in his writings to features of the quantum formalism, and to such associated interpretive schemes as the projection postulate? Why did he never consider the quantum formalism with articulated bridge principles a viable interpretive option?

I can suggest part of the answer: For Bohr, the necessity of the correspondence principle and the indispensability of classical concepts were not merely the subjects of an abstract philosophical inquiry. The purely philosophical considerations were invoked to "dress up" a methodological question—the question of the success of a particular program of investigation. His heuristics, which he internalized because of long and successful use by himself and others, eventually became the most essential feature of all possible research programs for Bohr.[27] It is

26. "That for the moment the paradoxes connected with the use of the idealization of point charge for the electron are preventing the development on correspondence lines of a comprehensive relativistic quantum electrodynamics, must indeed rather be imputed to our failure . . . to grasp some deeper feature of the stability of the individual particles themselves than to any lack of soundness of the general lines on which the incorporation of the quantum of action in atomic theory has been achieved" (Bohr 1939, 389).

27. In this sense, Bohr is similar to Einstein. An enlightening analysis of Einstein's philosophical makeup was given by Arthur Fine (1986), who argued that for Einstein philosophical questions about "realism" and "determinism" were not abstract philosophical questions but questions of the success of research programs assuming these notions.

precisely because Bohr did not (and possibly could not—see chapter 12) participate in the mathematical elaboration and consolidation of quantum theory that classical theories with their underlying mathematical structures appeared to him the only tools one could use to grapple with the quantum world. In this way, personal idiosyncrasy was transformed into an overarching principle of knowledge. Since the claim that classical concepts are necessary is the least defensible of Bohr's assertions, its rhetorical underpinnings are especially strong. It is usually presented as "obvious," as a "simple logical demand" (1948; 1950, 512–13; 1958d, 727).

I have argued that it is impossible to combine Bohr's pre-1935 and post-1935 writings into a unified, coherent structure, and that a basic change occurred in Bohr's notions of disturbance, reality, acausality, and the indispensability of classical concepts. It is tempting to assume that perhaps Bohr's early and later writings, taken separately, are systematic and free of contradictions. Yet textual evidence does not support this assumption either. Bohr's discussions of causality are a case in point.

Despite Bohr's lifelong preoccupation with the issue of causality, his use of this concept was unsystematic and contradictory. His notion of causality was very "thick": Sometimes it was a cause-effect relationship. Sometimes it was "determinism." Sometimes it was an epistemological, other times an ontological, definition. Sometimes causality was equated with the applicability of the conservation laws of energy and momentum, other times with the simultaneous applicability of space-time and energy-momentum concepts. Sometimes Bohr's understanding of causality was probabilistic, applied to an individual system; and sometimes it was a statistical interpretation, applied to an ensemble of similar systems. Bohr often conflated determinism and predictability, which are in fact different notions.

In Bohr's later writings, he often employed a definition of causal description as predictability: "In physics, causal description . . . rests on the assumption that the knowledge of the state of a material system at a given time permits the prediction of its state at any subsequent time" (1948, 445). Note that such a definition of "causality," and correspondingly of "acausality," is compatible both with the idea that the laws of nature themselves are statistical (Born's view) and with the position that the laws of nature are deterministic and statistics is introduced only because of an inability to determine simultaneously and exactly all the physical variables of a state (Heisenberg's view). Bohr employed both of these mutually contradictory concepts of acausality at roughly the same time. He spoke in an ontological vein, about a "free choice of nature": "The specification of the state of a physical system evidently cannot determine the choice between different individual processes" (1948, 446). In an earlier paper Bohr argued that "quantum postulates . . . imply an explicit renunciation of any causal description. . . . as regards its possible transitions from a given stationary state to another stationary state . . . the *atom may be said to be confronted with a choice* for which, according to the whole character of the description, *there is no determining circumstance*" (1939, 385, my italics). Yet, in the same paper, a few pages later, Bohr used Heisenberg's argument for acausality, incompatible with the position just cited: "The essentially statistical nature of this account [is] a direct consequence of the fact that the commutation rules prevent us to identify at any instant more than a half of the symbols representing the canonical variables with definite values of the corresponding classical quantities" (1939, 387).

I will argue in the next chapter that similar changes and contradictions characterize the interpretive pronouncements of other quantum physicists. The Copenhagen interpretation was erected, not as a consistent philosophical framework, but as a collection of local responses to changing challenges from the opposition.

PART TWO

Rhetorical Consolidation

CHAPTER 8

The Polyphony of the Copenhagen Interpretation and the Rhetoric of Antirealism

All that mattered now was to present the facts in such a way that, despite their novelty, they could be grasped and accepted by all physicists.

Werner Heisenberg 1971, 79

Introduction

If intellectual activity is by nature addressive and communicative, if thoughts are suffused with the voices of visible and invisible interlocutors, a dialogical analysis is bound to reveal polyphony and incoherence in a scientist's philosophical accounts. Solid structures and conceptual frameworks, presumably built on stable, paradigmatic presuppositions, turn out to be ahistorical abstractions or misleading illusions.

The Copenhagen paradigm is supposedly built on two incontestable pillars—indeterminism and a revision of the classical notion of reality. We have seen that the concept of indeterminism had no fixed meaning for the Göttingen-Copenhagen physicists: Bohr, Heisenberg, and Born each used different notions of causality in different theoretical and social circumstances. If no coherence can be found among all the utterances on a "given issue" made by a single person, it is unreasonable to expect that the writings of a group of thinkers will cohere into a unified consistent framework. I have argued that Bohr's notion of complementarity in his Como lecture was not compatible with Heisenberg's particle ontology. Similarly, wave-particle complementarity was not compatible with Born's probabilistic interpretation of quantum mechanics.

Certain positions seem to be more stable, or perhaps more recurrent, than others—Bohr's wave-particle complementarity, which Heisenberg and Born often supported publicly, is a case in point. Yet such positions only appear to be stable because they are repeatedly presented to lay audiences, from whom a serious scientific challenge can hardly be expected. The variety of audiences to which orthodox quantum physicists addressed their statements was a major source of contradictory elements in their writings. Heisenberg adopted Bohr's positivist approach for mathematically unsophisticated audiences, yet he employed elements of a realistic ontological interpretation when addressing his mathematically skilled colleagues. Sometimes, as I argued in chapter 5, two contradictory approaches were present simultaneously. Contradictions and inconsistencies in the writings of the founders of the quantum revolution are not an anomaly, but the order of the day. In this chapter I will articulate this point with respect to two central pillars of the orthodox interpretation: antirealism and the necessity of classical concepts.

Strong realistic and positivist strands are present in the writings of the founders of the quantum revolution—Bohr, Heisenberg, Pauli, and Born. Militant positivist declarations are often followed by fervent denials of adherence to positivism. No wonder different scholars, with good textual evidence, have arrived at conflicting interpretations of these writings. While Popper (1963b) presented Bohr as a "subjectivist," Feyerabend (1981a) found him an "objectivist." More recently, Murdoch (1987) has concluded that Bohr was a realist, while Faye (1991) has argued with equal competence that Bohr was an antirealist. Whether Bohr's philosophy was "realist" or "antirealist" is a topic of great interest, as is demonstrated by a recent collection of essays (Faye and Folse 1994). Although scholars have invested much serious thought and ingenuity in rendering Bohr's position consistent, I hold that conflicting opinions of realism and positivism (in Bohr's instrumentalist or Heisenberg's operationalist version) are both undeniably present. My aim is not to cure this "schizophrenia" (the characterization in Fine 1986) by eliminating the inconsistencies but to analyze the sources, uses, and aims of such shifting philosophical positions, on both the conceptual and the sociological level.

This chapter further elaborates the stand, now commonplace in the historiography of science, that ideas can properly be understood only by an analysis of their local theoretical and sociopolitical emergence and use.[1] It also supports an approach critical of what Skinner (1969) christened a "mythology of doctrines" and a "mythology of coher-

---

1. This stand, from Collingwood (1939) and Wittgenstein (1953), through Rorty (1979) and Fine (1986), now has an overwhelming following among scholars dealing with social studies of science and with the rhetoric of science. Some prominent examples include Collins (1992), Shapin and Schaffer (1985), Knorr-Cetina (1981), Galison (1987, 1997), Latour (1987), and Pickering (1992). However, these intellectual currents hardly touch the historiography and philosophy of quantum mechanics. One of the best recent histories of early quantum physics—Darrigol (1992a)—exemplifies this point.

ence."[2] The first mythology assumes that philosophical writings necessarily deal with "eternal" epistemological or ontological issues, such as the "realism" of atoms. Bohr, for example, is often classified as a "realist" because he "never doubts" the reality of atoms. The second mythology assumes that philosophical writings are in principle intended to produce a systematic contribution to their subject, and that to "understand" such writings means to produce, at whatever cost, the most coherent and contradiction-free presentation of their content. I suggest that such "mythologies" are especially inapplicable to interpretive explorations by working scientists, who espouse their philosophical positions in dialogical responses occurring in changing theoretical and sociopolitical circumstances. I argue that different pronouncements by quantum physicists on the realism-antirealism issue can be understood by differentiating between what scientists "need not" and "must not" do. "Need not" is a notion linked to addressivity, which is apt to head off potential and existing objections to a scientific innovation; "must not" is an overarching binding dictum that physicists abstract illegitimately from the local "need not." It is the local "need not," rather than the universal "must not," that turns out to be most relevant to understanding the philosophical deliberations of scientists.

I also argue that Bohr's antirealism served as a potent cohesive force inside the community of physicists, and as a way to prevent its alienation from wider cultural circles. For this reason, I suggest, Heisenberg and Born often supported Bohr's position, one incompatible with their own. This public espousal of antirealism is one of the major sources of contradictions in Heisenberg's and Born's writings. Under closer scrutiny the Copenhagen paradigm has neither coherence nor stability, despite the mass of rhetoric by Bohr and his followers asserting its "inevitability."[3] The Copenhagen philosophy can thus be seen as a contingent composite of different philosophical strands, the public face on a hidden web of constantly shifting differences among its founders. If philosophical pronouncements by quantum physicists are most adequately understood as local, shifting, and opportunistic, the question arises of how the appearance of consensus is achieved. The concluding pages of this chapter address that issue.

What Scientists "Need Not" and "Must Not" Do

One of the most useful hints to how to deal with the realism-positivism problem can be found in Feynman's distinction between what scientists

2. I am grateful to Menachem Fisch, who made me aware of the similarity between the ideas expressed in the first draft of this chapter and the writings of Skinner.

3. For an analysis of Bohr's rhetoric of inevitability, see Beller (1993, 1996a) and chapter 9.

"must not" and "need not" do (Feynman, Leighton, and Sands 1969). It is wrong to say that scientists "must not" presuppose reality, aiming at the elimination of all unobservable theoretical entities. Feynman stated that no creative scientific theorizing can proceed in such an imaginative vacuum. Yet, when certain realistic and seemingly indispensable notions lead to contradictions, one can eliminate them, provided a consistent interpretation of experiments is possible without their use. In the two-slit experiment, one "need not" assume that a particle traverses a well-defined path between the two-slit diaphragm and the detector (Feynman, Leighton, and Sands 1969, 38-8, 38-9). "Need not" is an integral part of scientific practice, without which such breakthroughs as the rejection of absolute simultaneity in relativity theory and the rejection of a strict deterministic framework in quantum theory would not have been possible. "Must not" is a positivist excess, at odds with the practice of science, which relies on realistic inspirations as a source of motivation for research, and on realistic models as a heuristic guide to discovery.[4]

Much confusion on the realism-antirealism issue in quantum physics is caused by the failure to discriminate between the local "need not" and the overarching "must not." When Heisenberg dispensed with electron orbits and the space-time container in his reinterpretation paper, he did so to obtain a novel solution, incompatible with the existence of such orbits. He realized that he "need not" keep the electron orbits, because they "cannot be observed anyway" (Heisenberg to Kronig, 5 June 1925, AHQP). As I have argued, Heisenberg adopted the principle of elimination of unobservables a posteriori, in order to legitimate his unconventional, formal solution. Yet Heisenberg soon realized that he had been too hasty in elevating the local "need not" to the status of universal "must not" and changed his stand accordingly (chapters 2, 3, and 4).[5] Born's positivist statements are also best understood on the level of the local "need not." When Born provided a probabilistic solution of the collision problem and a probabilistic interpretation for the wave function, he did not hold that a more detailed deterministic description in terms of microscopic coordinates is forbidden in

---

4. On "motivational realism," see Fine 1986. MacKinnon (1977) argued that the virtual oscillator heuristic model was crucial in determining Heisenberg's road to the new quantum theory. On the strong presence of realist notions in what have usually been taken as developments characterized by positivist and instrumentalist philosophy, see Beller (1992b).

5. Heisenberg admitted in his Chicago lectures: "To avoid these contradictions [between quantum theory and experiment], it seems necessary to demand that no concept enter a theory which has not been experimentally verified. . . . Unfortunately, it is quite impossible to fulfill this requirement" (1930, 1–2).

principle. Theoretical physicists "need not" seek such a description because Born's own probabilistic theory is sufficient, being in perfect accord with the practice of experimentalists: "Earlier, it was supposed . . . that there is meaning in the question: 'When and where an election is ejected.' . . . Suppose that we decide to renounce this question, an act which is the easier inasmuch as no experimenter would think of asking it" (Born 1928, 32).[6]

In order to protect quantum theory from the charge of incompleteness, or perhaps of inconsistency, Bohr countered the EPR challenge by arguing that one "need not" assume that certain atomic attibutes exist independent of the process of measurement (chapter 7). Reality becomes, by definition, "what quantum mechanics is capable of describing" (Rosen's remark, quoted in Heilbron 1987, 219). Pauli and Heisenberg (who often supported Bohr publicly) did not feel at ease with an explicit positivist approach. We should not confuse their reluctant concession to a positivist legitimation with the voluntary embrace of a positivist stand.

According to Pauli, one does not have to be an empiricist in order to realize that certain realistic forms of thought (Denkformen) are not necessary for the successful construction of a physical theory (1954, 123). Heisenberg's notion of "practical" versus "dogmatic" realism was similarly aimed to provide enough leeway for a positivist protection of quantum theory from possible objections, without thereby losing the realistic ground altogether (1958, 81–83). According to Heisenberg, practical realism and dogmatic realism differ in the way they objectify experience (the English translation coins the word "objectivate"). A statement is objectified if its content does not depend on the conditions of its verification. Dogmatic realism assumes that there are no statements concerning the material world that cannot be objectified. Practical realism assumes that many, but not all, statements can be objectified. Practical realism is indispensable for scientific work: "Every scientist who does research work feels that he is looking for something that is objectively true." Practical realism, according to Heisenberg, "has always been and will always be an essential part of natural science." What quantum mechanics demonstrated is that good science is possible without a basis of dogmatic realism (Heisenberg 1958, 82). While Heisenberg's pragmatic notion of practical realism is not compatible with Bohr's operational definition of physical phenomena, both are directed toward the same aim—to unburden physics from those troublesome realist notions that "need not" be presupposed in quantum description and to ease the acceptance of a new and controversial knowledge.

6. Page references for Born's papers are to the collection Born (1956).

The Appeal of Antirealism: Some General Considerations

Antirealist philosophical stands have wide sources, both philosophically and historically. The appeal of antirealism varies substantially in different theoretical and sociopolitical contexts. In the past, antirealist sentiments were often invoked because it had been recognized that equivalent accounts of the same domain of phenomena were possible. Demonstrations of the possibility of equivalent descriptions, by Hipparchus in antiquity, Poincaré in the nineteenth century, Schrödinger and Pauli in the twentieth century, strengthened the notion of science as merely a device "to save the phenomena."[7] Thus P. W. Bridgman argued that the "quest for underlying reality" is a meaningless one, when one realizes that two explanations of the same phenomena can be equally true: "However much one might have been inclined fifty years ago to see some warrant for ascribing physical reality to the internal processes of a theory . . . certainly no one of the present generation will be capable of so naive an attitude after our illuminating experience of the physical equivalence of the matrix calculus and the wave mechanics" (1930, 21).

The attraction of antirealist attitudes becomes strong when one is eager to avoid a clash between the findings of science and religious teachings and aspirations, be it Osiander's preface to Copernicus's treatise or the defense mounted by scientists against the accusations of theologians in the nineteenth century. Antirealism is especially valuable when one wants to protect science (and society) from undesirable extensions of scientific metaphysics into the political realm. Philipp Frank, in letters to Bohr in 1936, tried to convince Bohr to join the camp of outspoken positivists. The duty of every physicist, pleaded Frank, is to counter mystical interpretations of quantum physics, which were exploited by national socialist elements to support a reactionary political philosophy and a barbaric regime. Only a consistent positivist stand can insulate the words of physicists from such misuse (Heilbron 1987; Beller and Fine 1994, 20).

An antirealist approach can also serve as a potent cohesive force in the scientific community. At the beginning of the century, Pierre Duhem argued that a positivist view of science, avoiding the minefields of metaphysics, gives the greatest chances of achieving consensus in the

---

7. On the demonstration by Hipparchus of equivalent descriptions of celestial motions using either epicycles or eccentrics, see Duhem (1991). Poincaré's leadership in the antirealist "descriptionistic" movement is explored in Heilbron (1982). Convincing modern arguments in favor of this view of science are made by van Fraassen (1980). On the conflict between scientific realism and the possibility of equivalent theories, see Ben-Menahem (1990).

scientific community. The discord that can result from competition between different models, each claiming a privileged road to truth, was eloquently displayed in the fierce and bitter confrontation between Heisenberg and Schrödinger over the superiority of their equivalent theories. Bohr's Como lecture, and the subsequent antirealist philosophy of complementarity, was intended not only to "harmonize" the discordant theoretical notions but to pacify the antagonized factions in the community of physicists: "I shall try . . . to describe to you a certain general point of view . . . which I hope will be helpful in order to harmonize the apparently conflicting views taken by different scientists" (1928, 87). Pauli, the (sometimes) hidden architect behind the major interpretative breakthroughs in quantum physics, understood the crucial importance of Bohr's unifying mission. "Niels Bohr," he announced, "integrated, in lectures and international congresses, and at those carefully planned conferences in Copenhagen, the diverse scientific standpoints and epistemological attitudes, and thereby imparted [to the physicists] . . . the feeling of belonging, in spite of all their dissensions, to one large family" (1945, 97).

As the head of this large family, Bohr did not tolerate quarrels. He considered Einstein's criticism of the orthodox interpretation of quantum mechanics "high treason." Being the official spokesman for this community, he worried that too much dissent might harm its image and power. In an interview with Kuhn, Bohr complained about Einstein's objections to the Copenhagen philosophy: "That, of course, is really the difficulty . . . that the philosopher says: When the atomic physicists do not agree, why should we trouble?" (17 December 1962, AHQP).

Bohr's antirealist philosophy was uniquely designed, not only to prevent discord inside the collective of quantum physicists, but also to prevent the alienation of this collective from the rest of the scientific community. The appeal of Bohr's teachings was considerably enhanced by his immense authority, personal charisma, and powerful institutional position. Bohr's status as a father figure goes a long way toward explaining the wide diffusion of the Copenhagen spirit (see chapter 12; Heilbron 1987). Yet it is the special role of Bohr's antirealism as a source of cohesion that was largely responsible for the widespread appeal of the Copenhagen interpretation.

The new quantum physics was a theory of unprecedented abstraction, especially in its matrix and operator formulation. By renouncing visualization in space-time, quantum theory challenged intuition *(Anschaulichkeit)* and common sense even more severely than general relativity had just a few years earlier. The ascendancy of the profession of theoretical physics at the beginning of the twentieth century threatened to upset the established hierarchy in the German physics community (which placed experimentation and collection of data at the top). Theoretical physicists seemed to alienate themselves even further from the rest of the scientific community with the rise of the new quantum theory. The nonintuitiveness of the abstract matrix mechanics was anything but appealing: Heisenberg faced the painful sight of Schrödinger's intuitive theory immediately enchanting most theorists. Schrödinger objected openly to the formidable complexity of matrix mechanics, warning that its lack of *Anschaulichkeit* would be detrimental to scientific progress. I have argued that Heisenberg's response—the restoration of space-time and *Anschaulichkeit* by an operational analysis of scientific notions in thought experiments in his uncertainty paper—was a clever defense of this highly abstract theory (chapters 2 and 4). This operational analysis, despite Bohr's initial objections in his Como lecture, became a paradigmatic example in Bohr's later legitimation and elaboration of the quantum philosophy.

Bohr's numerous illustrations of experimental arrangements with bolts, springs, diaphragms, and rods made experimentalists feel at ease when venturing into unfamiliar quantum territory. These simple considerations, supplemented with Bohr's potent persuasive technique, which disguised demonstrations of consistency as compelling arguments of inevitability, made the far-reaching and controversial conclusions appear to be inescapable (see chapter 9). No mathematical training and only the most elementary physics was needed to follow Bohr's arguments. The development of the quantum theory, declared Bohr in one of his numerous talks, "has given rise to much anxiety, to much doubt both among physicists as well as philosophers, and one [could] overcome these problems in a very simple way" (1958d, 726–27). Bohr's simple way alleviated anxiety, transforming every reader into a "virtual witness," to use Shapin and Schaffer's (1985) term. It removed the threat of esoteric and nonintuitive theory, making the quantum mystery widely accessible. English experimental physicists felt especially at home with Bohr's drawings, so reminiscent of Duhem's description of an English factory. The English experimentalist Frederick Donnan was enthusiastic: "As you point out, physical reality can only be defined by means of experiments and measurement. . . . Your logic—based on the complementary possibilities of different experimental arrangements in assigning definite and unambiguous meaning to the values of any pair of canonically conjugate variables—seems unassailable. . . . It is not a question of abstruse math but of an entirely new standpoint" (Donnan to Bohr, 26 July 1935, AHQP).

In the early 1930s, Bridgman described the anxiety and resentment caused in the physics community by the rapid shifts in the foundations of physics, especially those due to changes implied by the new quantum theory. Physicists felt beleaguered by new "mathematical theories, which are being continually formulated at an ever increasing tempo, and in complexity and abstractness increasingly formidable." Yet the "average physicist . . . who flounders in bewilderment" need not feel lost: "Since the new theories are formulated [so] as to be consistent with the cardinal principle that the properties of the thing have no meaning which is not contained in some describable experience, our intuition should be able to tell us what to expect in various experimental situations." An analysis of thought experiments, using simple intuition, is open to every physicist and "would be found by many . . . to give a more illuminating insight than a painful acquisition of the details of the present mathematical picture" (Bridgman 1930, 21). Operational analyses, of the sort advocated by Heisenberg and Bohr, were the anchor and the chart that Bridgman offered to boost confidence in an age of uncertainty.

Yet Bridgman's work shows eloquently the dangers of operationalism applied too diligently. Operationalism can undermine any well-established scientific theory, be it statistical thermodynamics or cosmology, by attacking its foundations as going beyond the possibility of experimental verification. Cosmologists, for example, resented Bridgman's attacks on the integrity of their discipline (Beller 1988). It is this kind of offense that Bohr carefully and ingeniously avoided. The unique appeal of Bohr's philosophy resided precisely in the way it protected the macrorealm and well-established classical theories from the excesses of operationalism, while fully enjoying the fruits of antirealism in the microdomain. This situation was achieved by Bohr's division of the physical world into two radically different parts—classical and quantum—and by Bohr's doctrine of the indispensability of classical concepts. Before elaborating this claim, I will describe Bohr's position as it contrasted with that of many mathematical physicists, including Born and Heisenberg.

Reality, Classical Concepts, and Symbols

One of the greatest philosophical differences between Bohr and his more mathematically minded colleagues concerns the questions of quantum reality and the necessity of classical concepts. Bohr transformed the fact that one "can" describe a macroscopic measuring device by classical concepts (as opposed to a microobject, which cannot be so described) into the idea that one "has to"—into the doctrine of the indispensability of classical concepts. Supplemented by the operational stand, this doctrine resulted in Bohr's peculiar denial of the possibility of quantum concepts and quantum reality. According to Bohr, a symbolic description, removed from common sense and familiar classical visualization, can be nothing but a tool for the correlation of experiments. Bohr's doctrine of the indispensability of classical concepts rests on several key ideas that were not supported even by his closest collaborators—Heisenberg and Born. Notorious among them are the alleged proximity between classical and commonsense notions and the direct accessibility of classical reality to sense perception.

Bohr's belief in this direct accessibility is rooted in the Kantian heritage of space-time concepts as forms of intuition (*Anschauung*) and in his lifelong reliance on visualizable classical space-time models of the atom as a basis for correspondence arguments (on the former, see Faye 1991 and Chevalley 1994; on the latter, Darrigol 1992a). The idea that the classical world is revealed directly to our senses is a recurrent theme in Bohr's writings. This direct and necessary connection between classical concepts and sense impressions ensures the absolute indispensability of classical concepts: "There can, of course, be no question . . . of abandoning the conceptual frame of our conscious recording of sense impressions" (Bohr 1938a, 379). Bohr's intuitions about what is "real" and what is not are connected with the possibility of visualizing conceptual notions, *not* with the possibility of observing them. Atoms are therefore "real"—the calculation of their masses and charges is based on the visualizable laws of classical electrodynamics.

Yet when we deal with the unvisualizable symbolic description of the same atom, we should leave all questions of its "nature" behind. Because classical concepts are inapplicable, or limited, in the quantum domain, and only classical concepts are *anschaulich* and therefore "real," there can be no question of "quantum reality." There is no compelling argument for the reality of classical description as opposed to quantum description, except this alleged Kantian kinship between the visualizability of classical physics and our sense perceptions.

According to Bohr, commonsense conceptions and ordinary language are the foundation of all communication, and therefore ultimately of all science. According to Born and Heisenberg, common sense is profoundly mistaken. In Born's opinion, the guiding factor in the development of science is "a belief in a real external world and an ability to distrust sensation" (1928, 20). Born held that commonsense and scientific concepts are greatly removed from each other. The concepts of classical mechanics are not extensions of common sense. The fact that classical physics uses such everyday terms as "mass" and "force" is misleading: "Their sound is the same as words of ordinary speech, yet their meaning can be found only from specially formulated definitions," and they can be apprehended "only with the aid of mathematical tools." The word "gravitation" is an "artificial concept" that has little in common with the simple idea—the feeling of force. Far from being an extension of commonsense regularities, the laws of classical mechanics "cannot be enunciated without ideas which lie far outside the natural limits of thought." Justification of classical concepts is based on nothing else but "their place in the system of objective natural science" (Born 1928, 21). Because the justification of classical concepts is due to their role in the abstract conceptual framework of classical theory, their ontological status is no different from that of the concepts of quantum theory. There is no essential difference between classical and quantum concepts.

If reality is ascribed to the best entrenched theoretical concepts, they become "real" no matter how abstract. The exposure to general relativity weaned physicists from their familiar classical intuitions and developed such realistic symbolic sensitivities. Einstein and his followers ascribed genuine reality to four-dimensional space-time and tensor fields, not to familiar classical space and time notions. Despite the demanding symbolization, these concepts became real for physicists by being included into a new comprehensive and consistent theory. And conversely the regular notions of space and time became less real: "The familiar geometry of Euclid and the corresponding time are now reduced to mere approximation to reality" (Born 1928, 25). Similarly, Heisenberg, redefining *Anschaulichkeit* in his uncertainty paper, argued that the simplest interpretation of space coordinates in quantum theory is obtained by identifying them with the diagonal terms of the appropriate matrices.

Heisenberg's published works, and the reminiscences of his student Weizsäcker, testify that Heisenberg did not embrace Bohr's radical splitting of the world into two unbridgeable realms: "Heisenberg never seriously considered the idea that quantum theory might be limited to microscopic objects; . . . he considered quantum theory as universally valid and classical physics as macroscopic approximation" (Weizsäcker 1987, 283). This willingness to reeducate their intuitions and to redefine reality characterized mathematical physicists working in quantum theory. Far from believing that no new conceptual schemes of reality are possible, Born held that the development of such schemes is the very essence of scientific advance. In developing new invariants, physicists learn to handle them intuitively. As a result of such a process, "new conceptions sink down into the unconscious mind, they find adequate names, and are absorbed into the general knowledge of mankind" (Born 1936, 52).

Bohr, being a "natural philosopher, and not a mathematical physicist" (Heisenberg's characterization), did not possess the experience or dexterity needed to handle the abstract mathematical formalism of general relativity. In fact, his familiarity with general relativity was very limited (Pais 1991). His objections to the possibility of new conceptual schemes are peculiarly reminiscent of the objections of the neo-Kantian opponents to general relativity. Bohr did not internalize the new quantum symbolism as others did who made the necessary investment in everyday mathematical labor.[8] For him, only the classical realm remained intuitively accessible and therefore real.

In his later years, Bohr replaced his initial, Kantian emphasis on the impossibility of new conceptual forms with a different argument for the indispensability of classical concepts. This argument was based on Bohr's post-EPR use of operationalism and on his doctrine of the unambiguous nature of communication in terms of common language. Bohr held that only classical language, as a natural extension of common language, is capable of communicating physical experience. According to Bohr, the unambiguous communication of measurement results is the only possible way to ensure the objectivity of the scientific enterprise (a penetrating analysis of this position can be found in Hooker 1972, 1991). Complementing his instrumentalist insistence that quantum formalism is merely a tool for coordinating measurement results, Bohr declared the classical realm the only one where realistic talk can be meaningful. Any creative efforts to construct the quantum world become pointless and are bound to fail. Yet the doctrine of the indispensability of classical concepts for the unambiguous description of measurement does not by itself prohibit the construction of a quantum reality. An emphasis on experiment is not at odds with realist strivings, as is eminently displayed in writings by Born, Heisenberg, and Pauli. As Born noted: "In physics, all 'experience' consists of the activity of constructing apparatus and of reading instruments. Yet the results thereby obtained suffice to re-create the cosmos by thought" (1928, 36).

Nor did Bohr's mathematical colleagues share his opinion that unambiguous communication demands the use of common language. Quite the opposite stand is expressed by Heisenberg: The concepts of ordinary language are "inaccurate and only vaguely defined. . . . the concepts of the general laws must in natural science be defined with complete precision, and this can be achieved only by means of mathematical abstraction." Far from immutable, scientific language evolves as scientific knowledge grows: "New terms are introduced and the old ones are applied in a wider field or differently from ordinary language. . . . in the theory of general relativity the language by which we

---

8. On Bohr's attitude toward mathematics, see chapter 12.

describe the general laws actually now follows the scientific language of the mathematicians" (Heisenberg 1958, 172).

Although few working physicists abided by Bohr's strictures on quantum concepts and quantum reality, their disagreement with Bohr rarely reached the printed page. Yet the disagreement was often unequivocal and forceful. Bohr's correspondence with his lifelong colleague Born and with the Russian physicist Vladimir Fock provides characteristic examples. Fock attacked the central points of Bohr's philosophy in his letter to Bohr of 23 February 1957 (AHQP), and in a paper published the next year (Fock 1958). Both Fock and Born were amazed by Bohr's rejection of quantum concepts and frustrated with his resignation from the attempt to comprehend the underlying quantum reality. Fock argued: "It is not only the limitations characteristic of a description of phenomena . . . that have philosophical significance, but also the constructive part of quantum mechanics and the new fundamental concepts connected with it. . . . The brilliant demonstration given by Bohr of the limitation of classical concepts is not accompanied by even the slightest indication of new concepts by which to replace them" (1958; quoted from Fock's translation, AHQP). Similarly, Born claimed that science aims beyond mere coordination of measurement readings, that there is something objective "behind the phenomena" (Born to Bohr, 28 January 1953, AHQP). Unless one takes a realistic stand, one "resigns oneself to answering any question about why one is investigating the [world] at all" (Born to Bohr, 10 March 1953, AHQP).

Fock disagreed vehemently with Bohr's opinion that the quantum formalism was merely a tool for the coordination of measurement readings. According to Fock, such a point of view is erroneous: "True enough, quantum mechanics (like any other physical theory) allows us, among other things, to coordinate the readings of the instruments used for the measurement, but that is not its fundamental importance. The purpose of a physical theory is always to describe the properties of physical objects in their relation to the outer world" (Fock to Bohr, 23 February 1957, AHQP).

Fock rejected Bohr's idiosyncratic claim that the "mathematical symbols of quantum mechanics, in contradistinction to the mathematical symbols of classical physics, have no physical meaning in themselves." He did not see any difference in the role played by mathematics in classical and in quantum mechanics. The notions of quantum state and quantum probability are in fact fundamental quantum concepts. "They are not symbolic, but quite physical. There is no reason to avoid them in the description of nature" (Fock to Bohr, 23 February 1957, AHQP).[9]

9. Similar criticism has been provided by Krips (1987).

The Appeal of Antirealism: Bohr's Version

Antirealism in Bohr's philosophy was associated with the unfamiliar and (according to Schrödinger) "frightfully abstract" quantum mathematical formalism, not with the classical realm known to every physicist. The proximity between commonsense notions and classical concepts, the direct accessibility of classical reality to sense perception, the need to use common and classical language for unambiguous communication, the need to describe measuring apparatus using classical concepts—Bohr presented them all as self-evident truths rather than as the controversial doctrines they, in fact, were. And their appeal to experimental physicists was very strong. Born understood it clearly: "Every experimental physicist treats his instrumentation as if he were a naive realist. He takes its reality as given a priori, and doesn't rack his brains about it. Niels Bohr has made this attitude the basis of his whole philosophy of physics" (1962, 21).

Born did not share Bohr's views. Yet Born understood the immense potential of Bohr's stand as a unifying force. So Born would sometimes endorse Bohr's radical split of the physical world in order to avoid a split in the world of physicists: "There has already developed a gap between pure and applied science and between the group of men devoted to the one or the other activity, a separation which may lead to a dangerous estrangement. Physics needs a unifying philosophy, expressible in ordinary language, to bridge this gulf between 'reality' as thought of in practice and in the theory" (Born 1953c, 151). "Plain," or "ordinary," language is the only kind the expert can use to tell the wider audience about new abstract technical advances: "The physicist may be satisfied when he has the mathematical scheme and knows how to use it for the interpretation of the experiment. But he has to speak about his results also to non-physicists who will not be satisfied unless some explanation is given in plain language, understandable to anybody" (Heisenberg 1958, 168). Heisenberg's writings are full of contradictions precisely because he simultaneously addressed two "complementary" audiences—the laity and the mathematically initiated.

Bohr's doctrine of the necessity of classical concepts worked wonders to alleviate the anxiety many physicists felt about symbolic quantum theory. His operational analyses of thought experiments were both a therapeutic compensation for the loss of *Anschaulichkeit* and a crutch (London's term; Landé 1967), though a brittle one, on which scientists could lean while familiarizing themselves with the quantum mystery. How brittle the crutch was soon became apparent because of inconsistencies connected with Bohr's and Heisenberg's initial ideas about the disturbance of atomic objects by observation—the notion Bohr eventually had to renounce. But even after his renunciation, insoluble difficulties remained, as I argued in the previous chapter. Heisenberg grasped that the operational analysis of quantum phenomena through the use of classical terms exclusively cannot be contradiction free (Beller 1993). Other physicists expressed similar reservations in unpublished correspondence with Bohr (for example, Fritz Bopp's letters to Bohr, AHQP). Yet in print not many dared, or cared, to enter the territory that only Bohr, so the legend goes (Rosenfeld 1961; Blaedel 1988), had mastered fully.

Antirealism and Opposition

I have argued that scientific philosophizing in general, and that dealing with the realism-positivism issue in particular, is most profitably understood by juxtaposing the universal "must not" with the local "need not." By its very nature, "need not" brings us to a specific historical situation, to local argument designed to dispose of deeply entrenched but no longer useful habits of thought, to an appeal addressed to a specific audience. While philosophically questionable, it is rhetorically wise to endow a local "need not" with universal validity. Attempts to mesh different conceptual strands into a systematic position lead to such philosophical hybrids as Murdoch's characterization of Bohr as an "instrumentalistic realist" (1987, 222). It is not surprising therefore that what is called the Copenhagen interpretation is so riddled with vacillations, about-faces, and inconsistencies. For quantum philosophy—which was, more often than not, elaborated as a series of local responses to challenges from the opposition, rather than as an attempt to construe a systematic framework—by the very nature of its development cannot be found to be free of contradictions.

Many of the antirealist statements of the Copenhagen camp came in response to Schrödinger's and Einstein's critiques of quantum theory. As the nature and forcefulness of the opposition's threat changed, different, often contradictory claims on the same issues appeared in writings by Bohr, Heisenberg, and Born. Opinions about the "reality" of the ψ-function constitute a prominent case in point. At the end of the 1920s, when Schrödinger, hoping to eliminate quantum jumps, attempted to substitute a wave model for the Göttingen-Copenhagen particle model, his efforts were discredited by reference to the multidimensionality of the wave function. Because the ψ-function was multidimensional, the matrix physicists argued, it was not something "real," but rather an abstract notion, having great mathematical utility but no physical significance. This argument was clearly targeted at an audience who identified *anschaulich* with "real," in the neo-Kantian spirit. Such an identification was at odds with the more sophisticated notions of Heisenberg and Born, who viewed reality as flowing from features of the mathematical model. Heisenberg and Jordan's pronouncements on the unreality of the ψ-function exasperated Schrödinger. In vain, Schrödinger appealed to Born, who backed his younger colleagues fully on this issue (chapter 2). The possible criterion of reality by which Born considered particles real, yet the wave function abstract, was a mystery to Schrödinger. He explained to Wien: "I believe that Born thereby overlooks that . . . it would depend on the taste of the observer which he now wishes to regard as real, the particle or the guiding field. There is certainly no criterion for reality if one does not want to say: the real is only the complex of sense impressions, all the rest are only pictures" (quoted in Moore 1989, 225).

When the original challenge posed by Schrödinger's interpretive efforts faded, Born changed his presentation of the significance of the ψ-function. In a lecture given in 1950 Born asked: "What about the waves? Are they real and in what sense? . . . Though the wave-functions are representing, by their square, probabilities, they have a character of reality. That probability has some kind of reality cannot be denied. How could, otherwise, a prediction based on probability calculus have any application to the real world?" Born concluded that "the scientist must be a realist" and that the abstract notions of a mathematical formalism can—and must—be assigned a realistic significance: "He [the scientist] uses ideas of very abstract kind, group theory in spaces of many or even infinitely many dimensions and things like that, but finally he has his observational invariants representing real things with which he learns to operate like any craftsman with his wood or metal" (Born 1950, 106).

Three years later Born made another about-face in his evaluation of the reality of the ψ-function. Schrödinger (1952a) had renewed his attack on the Copenhagen interpretation and revived his old proposal to eliminate quantum discontinuities and irrationalities by using an intuitive wave model. Born responded immediately with a counterattack that was a repetition of the argument against Schrödinger made thirty-five years before: "A multi-dimensional wave function is nothing but a name for the abstract quantity ψ of the formalism, which by some of the modern theorists also goes under the more learned title of 'state vector in the Hilbert space'" (1953b, 143). Born repeated his strong commitment to the reality of particles as opposed to waves. Thus the assignment of reality or lack thereof can be understood properly only in the local context of a controversy; realism or antirealism regarding the same theoretical construction is often a rhetorical weapon, whose significance can only be grasped within the definite sociohistorical circumstances in which it is used.

Attempting to present an "objective" definition of what is real, Born (1928, 1964) equated real things with invariants of observation and mathematical manipulation. He contrasted his scientific approach with that of a layman, for whom "the concept of reality is too much connected with emotions to allow a generally acceptable definition" (1964, 103). Yet perhaps the difference between a layman and a quantum physicist in this respect is not as great as Born claimed. One could say that for scientists, as for laymen, "the real things are those things which are important for them" (1964, 103).

The Appearance of Consensus and Conclusion

In reading philosophical accounts written by scientists, it is important to explore fully the contradictions therein, rather than selecting and constructing an artificial, consistent system of belief. For it is precisely those claims that refuse to be neatly systematized that are the most illuminating. Idiosyncratic contentions, puzzling stands, seeming reversals—these are the most useful guides to the fascinating local context, so different from the rational reconstruction that ultimately serves the interests of the orthodox.

Full acknowledgment of its contradictions casts great doubt on the existence of a common basis for the Copenhagen interpretation, and perhaps on the notion of a conceptual framework in general. The central pillars of the Copenhagen philosophy are indeterminism, wave-particle duality, and the indispensability of classical concepts. Yet none of them commands unwavering commitment even from Bohr's closest collaborators. Nor did these concepts have unchanging and definite meanings for Bohr himself.[10] Born and Pauli, as their correspondence discloses, did not think that the bankruptcy of determinism could be compellingly deduced; neither did Born and Heisenberg fully share Bohr's understanding of wave-particle complementarity;[11] nor did Bohr's mathematically skilled colleagues adhere to his dictum of the indispensability of classical concepts, as I have argued. It is difficult to find any common denominator, any idea, that enjoyed apparent "commitment" that did not also see that "commitment" broken at one time or another. This is not to say there was no consensus that the advent of

10. The opposition also lacked a coherent response. Fine (1993) argues that Einstein subscribed to three distinct kinds of interpretations of the quantum theory—subjective, instrumental, and hidden variables—giving himself considerable interpretive leeway.

11. "It would be silly and arrogant to deny any possibility of a return to determinism" (Born 1964, 108). Heisenberg, as he revealed in an interview with Kuhn, never liked the "dual" complementarity approach. "We have one mathematical scheme that allows many 'transformations': The fact that we can use two kinds of words ["wave" and "particle"] is just an indication of the inadequacy of words" (interview with Heisenberg, 25 February 1963, AHQP). See chapter 11 for a full discussion.

quantum theory necessitated a radical departure from classical physics. It is exact and consistent agreement about what would constitute such a departure that was lacking. The bankruptcy of the classical idea of motion was as fully grasped by the opposition—Schrödinger and Einstein—as by the orthodox quantum physicists. It is true that Einstein had serious reservations about quantum theory, and it is also true that Bohr and his followers were ready to accept the finality of quantum theory, no matter what far-reaching philosophical revisions it might imply. But finality is an ideological, not a conceptual, position, and this is perhaps the reason scientific controversy often looks more like a political campaign, with one side discrediting and caricaturing the other, than an open-minded dialogue about fundamentals.

Now we have to face the question: How does the impression (illusion) of the integrity, of the solidity, of a conceptual framework arise? By what rhetorical means are we led to believe that there is more stability in the views of a certain scientist, and more consensus among different participants, than is actually the case? The question about the views of an individual scientist is perhaps the easier of the two: In order to persuade others or to appear credible, scientists often present changes in their views as natural elaborations of previous positions, rather than as major revisions or reversals. Substantial changes, as we have seen in previous chapters, unacknowledged by the authors, are characteristic of the philosophical thought of Heisenberg, Bohr, Pauli, and Born. And there is nothing distinctive here about quantum physicists—to give just one example, Newtonian scholars have found that Newton, in the controversy over the nature of white light, similarly presented what was a forced departure from a previous stand as merely an elaboration of it (Shapiro 1989).

The more difficult question is how the appearance of consensus among different participants is achieved. Here again the distinction between "need not" and "must not" is helpful. Closely related to the issue of "need not" versus "must not" is the question of what one "can" and what one "has to" do. Failure to discriminate between the two leads to the impression of a firmer consensus between quantum physicists than actually existed. What Heisenberg and Born often promoted as a possible epistemological stand, Bohr advanced as an inevitable one. In the fact that one "can" describe a macroscopic measuring device by classical concepts, Bohr transformed "can" into "has to," elevating the fact into the doctrine of the indispensability of classical concepts. In contrast, Heisenberg, more often than not, did not say that one "has to" use classical concepts—rather his claim is that we are "satisfied" to use classical concepts, or that we "feel entitled" to use them (Heisenberg 1958).

Wave-particle complementarity is another prominent case in point (for further discussion, see chapter 11). Bohr argued that one "has to" use both waves and particles in quantum mechanical description, presenting wave-particle complementarity as an epistemological necessity. Born and Heisenberg held that one "can" describe quantum mechanical phenomena using both notions, and they often did so. Neither believed that one "has to" employ both wave and particles for a comprehensive description of every quantum phenomenon, nor did Born and Heisenberg hold that these concepts are necessarily mutually exclusive. Neither was eager, of course, to publicize his disagreement with Bohr.

In the early 1930s Born was lukewarm to Bohrian elaborations of complementarity. In 1929 he claimed that the resolution of the wave-particle dilemma resulted first and foremost from his own statistical interpretation: "If we disregard all philosophical aspects, the contradiction between the corpuscular and wave properties of radiation would be insoluble without this statistical viewpoint" (Born 1928, 34). Born's philosophical discussions in the 1930s were concerned with indeterminism and hardly mentioned complementarity. In the 1950s, the public image of physics and the status of physicists underwent a drastic change, due to the use of the atomic bomb in Japan—a disaster that affected Born deeply (Born 1951). Born subsequently embraced the complementarity philosophy because of what he perceived as its conciliatory message: "The world which is so ready to learn the means of mass-destruction from physics would do better to accept the message of reconciliation contained in the philosophy of complementarity" (1950, 108). Considering complementarity a "healthy doctrine," suitable to "remove many violent disputes in all ways of life," Born insisted that the "real enrichment of our thinking is the idea of complementarity" (1950, 107).

My presentation of antirealism in the interpretation of quantum physics distances itself from such notions as "belief," "commitment," and "metaphysical influence" and denies the very possibility of presenting the Copenhagen interpretation as a coherent philosophical framework. Whatever cohesion scientific paradigms or conceptual frameworks exhibit has to do more with the integrity of the tools (mathematical, experimental) than with "shared" metaphysical presuppositions.[12] We should beware of conflating "exposure to" (a certain viewpoint) with "influence," of hastily translating "familiarity" into "predisposition"

12. One interesting attempt to find some stability, objectivity, and cohesion, despite the impressive impact of social studies of science that deny those characteristics to science, is that of Ian Hacking (1992b). Hacking sees stability and objectivity in developed "styles of reasoning" that, proving their value and eventually shaking off their social origins, become universal.

and "predisposition" into "commitment." The point is not merely that it is exceedingly difficult to formulate a consistent framework even when one intends to do so.[13] It seems to be impossible to reconstruct a coherent philosophical framework from a multitude of utterances and deliberations that were aimed at meeting challenges in shifting scientific and sociopolitical circumstances.

13. "We characteristically spill over the limits of our intelligence . . . and get confused, and . . . attempts to synthesize our views may in consequence reveal conceptual disorder at least as much as coherent doctrines" (Skinner 1969, 49).

Above, According to the received history, Werner Heisenberg (*left*) and Niels Bohr (*right*) came to complete agreement after their heated debates in 1926 over the content of Heisenberg's uncertainty paper. Yet an analysis of their dialogue reveals that despite a common public stand, the disagreements between them were never resolved (see chapters 6, 8, and 11).

Below, The dialogue between Bohr (*left*) and Albert Einstein (*right*) lasted from the early 1920s until the end of Bohr's life. While it is often asserted that their debate over EPR was a milestone in Bohr's "victory" over Einstein concerning interpretive issues of quantum theory, a critical analysis undermines such a reading (see chapter 7).

Above, Heisenberg (*right*) and Wolfgang Pauli (*center*) corresponded frequently during the crucial years of the creation of the new quantum mechanics and its interpretation. Heisenberg wrote his uncertainty paper in an intense dialogue with Pauli (see chapters 2 through 5). Otto Stern appears at the far left.

Right, Paul Dirac (*left*) and Heisenberg (*right*) in the 1930s. The dialogue between Heisenberg and Dirac was crucial for the development of quantum mechanics and its physical interpretation in general, and for the emergence of Heisenberg's uncertainty principle in particular (see chapter 4).

Above, The experimental work of James Franck (*center*) provided crucial support for the concepts of stationary states and quantum jumps proposed by Bohr (*left*). The issue of the "reality" of quantum jumps was at the center of interpretive attempts between 1925 and 1927 (see chapters 2, 4, and 6). Hans Hansen appears at the right.

Left, Heisenberg (*right*) and Erwin Schrödinger (*left*) shared the Nobel prize in physics for their matrix and wave versions of quantum mechanics, respectively. Heisenberg's inner dialogue with Schrödinger was indispensable for the formulation of the uncertainty principle (chapters 2 and 4). The king of Sweden appears in the center.

Left, Experimental work on atomic collisions by Franck (*right*), Max Born's (*center*) colleague in Göttingen, was instrumental in Born's statistical interpretation (see chapter 2). Walther Meissner appears at the left.

Below, Pauli (*right*) was one of the main architects of the Copenhagen interpretation. Bohr's (*left*) dialogue with Pauli was instrumental in the initial formulation and later dissemination of complementarity (see chapters 6 and 12).

As Bohr's dialogue with Einstein (*left*) lasted a lifetime, so did Pauli's (*right*). Pauli's predilection for the philosophy of complementarity was reinforced in his later years by his spiritual and metaphysical longings (see chapter 12).

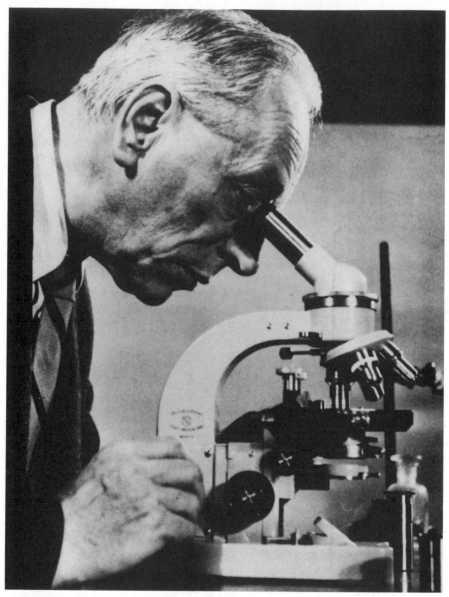

A dialogical analysis unveils the importance of "lesser" scientists, whose names rarely appear in the received history of quantum physics. Frits Zernike's work on the limits of precision in measurements spurred Heisenberg's deliberations on the theoretical uncertainty in quantum mechanics (see chapter 4).

The work of the American physicist William Duane, who using corpuscular light quanta explained the diffraction of light by a grating, stimulated the initial treatment of photons and electrons purely as particles in Heisenberg's uncertainty paper, as well as Alfred Landé's later particle interpretation (see chapters 4 and 11).

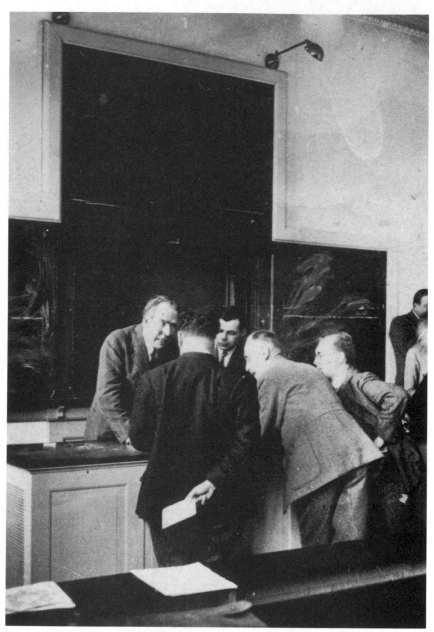

Bohr (*left, facing camera*) in a discussion with Pauli (*left, back to camera*), Lothar Wolfgang Nordheim (*next to Pauli*), Leon Rosenfeld (*next to Bohr*), and an unidentified scientist. During Bohr's professional life, the Copenhagen institute attracted physicists from around the world. Heisenberg and Dirac spent the winter of 1927 in Copenhagen—both Heisenberg's uncertainty paper and Dirac's paper on transformation theory were published from Copenhagen (see chapter 4).

CHAPTER 9

The Copenhagen Dogma: The Rhetoric of Finality and Inevitability

You cannot revolt: this is two times two equals four! Nature does not ask you, she cares nothing about your wishes and whether you like or dislike her laws. You must accept the way Nature is. . . . A wall, this means, is a wall . . . and so on . . . and so on . . .

Fyodor Dostoyevsky, *Notes from the Underground*, my translation

Make it look inevitable.

Louis Pasteur's advice on the best strategy of persuasion, quoted in Holton 1991, 174

"If it is so, it must be necessary," was a remark which he [Bohr] would often make in a discussion, and he would not be satisfied until the necessity was completely elucidated.

Jørgen Kalckar 1967, 233

Introduction

The founders and followers of the Copenhagen interpretation advocated their philosophy of physics not as a possible interpretation but as the only feasible one. Attempts at basically different approaches, albeit by such prominent scientists as Einstein, Schrödinger, Landé, and Bohm, were dismissed and ridiculed (see chapter 13). Niels Bohr advanced the Copenhagen interpretation relentlessly. He published extensively, lectured widely, and spread the Copenhagen spirit in direct contact with a generation of physicists who visited the Institute of Theoretical Physics in Copenhagen (Heilbron 1987). Bohr was a skillful rhetorician of the quantum revolution, no less ingenious than Galileo in weaving "natural interpretations"—arguments designed to introduce controversial knowledge in the guise of intuitively appealing propositions.[1]

1. The term "natural interpretation" is due to Paul Feyerabend (1975), who analyzed Galileo's rhetoric using this notion.

There is, no doubt, a striking contrast between the founder of complementarity and chief spokesman of the quantum revolution and that of the Galilean. Galileo—passionate, impatient, and breathtakingly eloquent; Bohr—desperately struggling for words and talking in an "inaudible manner" (Heilbron 1987, 201), or more reverently, a "divinely bad speaker" (Pais 1991, 11). Galileo was superb at antagonizing his audience, Bohr skillful in creating a favorable mood. Bohr rarely failed to praise the achievements of his listeners or to mention the contributions of the host institution. The kind of audience determined the way Bohr presented his epistemological efforts—as "philosophical" in a receptive Danish setting, or as those of a pure physicist, anxious not to be identified with philosophers, in a more pragmatic American environment: "I am very much afraid I shall appear to you as a philosopher. . . . I am just a physicist who has been forced to enter on such problems" (Bohr 1937d, 264). The word "forced" contains a subtle yet powerful rhetorical message. Not merely does it legitimate Bohr's continuing preoccupation with philosophical matters, it endows that preoccupation with inescapability and Bohr's conclusions with an aura of inevitability.

Einstein characterized the young Bohr as a thinker who "utters his opinions like one perpetually groping and never like one who believes he is in possession of definite truth." Pais (1991) chose these words as an epigraph to his book on Bohr. While this portrayal might be fitting to describe the young Bohr's attempts to cope with the quantum mystery, it is hardly true of the atmosphere of Bohr's later diffusion of the Copenhagen spirit. The later Bohr spoke as a prophet of truth, with the intimidating privilege of ultimate authority, in a categorical manner that excludes, "in principle," any questioning. Bohr's writings are thick with such expressions as "we see that it cannot be otherwise," "this is something there is no way around," "something we believe is going to stay forever," "the situation is an unavoidable one" (1935b, 210, 215). Bohr presents complementarity as "the most direct expression of a fact . . . as the only rational interpretation of quantum mechanics" (quoted in Landé 1965, 126), founded on an "obvious" analysis of the "epistemological situation." In view of such a compelling "logical" analysis, complementarity is an inevitable, irrefutable, final principle of physics, from which "we cannot deviate, whatever progress the future may bring" (Bohr to Born, 2 March 1953, AHQP). Bohr's later writings are especially dense, sentence after sentence studded with such expressions as "out of the question," "in no way," "of course," "surely," "leaves us no choice" (1937a, 1938a). Bohr's arguments are often prefaced with such phrases as "we must recognize," "we must realize," "the main point [to] realize," "it is important to realize," "it is imperative to realize" (1930, 1937a, 1938b, 1948). This strategy of persuasion, which infects the reader with the sheer intensity of Bohr's conviction, is skill-fully interwoven with the intimidating rhetoric of Bohr's authority. Surely, an intelligent reader cannot fail to realize, intimates Bohr, some-thing that is "obvious," that is "evident," that was "clear from the out-set," and that is, after all, "a simple logical demand" (1937a, 1948).

In addition to these forceful strategies of persuasion are others, more subtle. One of the most potent consists of Copenhagen's arguments of "inevitability," which are, as I have mentioned, merely disguised ar-guments of consistency. The opposition's resistance to this technique of argumentation, and its subsequent rejection of the Copenhagen phi-losophy, is therefore reasonable, and not "conservative" as is often claimed by adherents to the Copenhagen orthodoxy. The Copenhagen physicists used their arguments of inevitability to show the impos-sibility of observer-independent, deterministic description in the quan-tum domain. The rhetoric of inevitability was especially disturbing, as J. S. Bell mentioned, after the 1950s, when David Bohm provided pre-cisely the sort of description that the Copenhagen physicists had sup-posedly ruled out "in principle" (see Cushing 1994b for an excellent study of Bohm versus Copenhagen; see also the collection of papers in Cushing, Fine, and Goldstein 1996).

Referring to the orthodox denial of the possibility of a quantum on-tology, Bell invoked a vivid image: "It is," he wrote, "as if our friends could not find words to tell us about the very strange places where they went on holiday. We could see for ourselves whether they came back browner or fatter" (1987, 171). Why were quantum physicists mute for so long on the issues of quantum reality? Why did they voluntarily give up an exciting, though admittedly inconclusive, quest to unveil the mystery of the quantum world? Why did they hold fast to the notions of observer dependence and inevitable acausality, despite the existence of Bohm's alternative version? Bohm attributed this puzzling situation to the inability of the orthodox physicists to conceive of any concepts other than classical ones (waves and particles), and to their commit-ment to the philosophy of positivism: "There appears to have existed, especially among physicists such as Heisenberg, and others, who first discovered the quantum theory, a rather widespread impression that the human brain is, broadly speaking, able to conceive of only two kinds of things, namely fields and particles" (1957, 96). Another reason for their refusal to consider a "subquantum mechanical level," accord-ing to Bohm, was the adoption of the positivist thesis of the elimination of unobservables from physical theorizing, "a general philosophical point of view containing various branches such as 'positivism,' 'op-erationalism,' 'empiricism,' and others, which began to attain a wide-spread popularity among physicists during the twentieth century"

(1957, 97). I will reiterate Bohm's insight about the intimate interrelation among the finality of acausality, the indispensability of classical concepts, and positivism but provide a different account of how they are connected, along more sociological lines.

From a certain perspective, the stand that quantum theory is fundamentally acausal can seem natural and legitimate. If our best scientific theory is ipso facto acausal, what reason, beyond an unwarranted a priori belief in classical causality, would urge physicists to search for a causal version? After all, many physicists, including Bohm himself, found determinism neither indispensable nor appealing. Heisenberg's argument—that if physicists are convinced of the correctness of quantum theory, which is built on probabilistic foundations, then they should accept acausality (1958, 89)—is a characteristic example of a legitimate kind of reasoning that Arthur Fine called "entheorizing" (1986, 87). The metaphysical question of whether a certain property is a property of our world is translated into the question of whether a theory that entails this property is fruitful, or correct. Yet such sober entheorizing is rare. Copenhagen writings, Heisenberg's included, are laden with the heavy rhetoric of finality, of the inevitability of acausality, of the impossibility of a causal alternative. An analysis of the Copenhagen arguments for the inevitability of acausality and the indispensability of classical concepts is the main focus of this chapter.

Acausality and the Indispensability of Classical Concepts

As I have argued in previous chapters, the stand against causality, especially in Heisenberg's able hands, crystallized as a response to the threat posed by Schrödinger's competing theory. Schrödinger's declared ambition was to eliminate the "monstrous"—as he called them—quantum jumps, replacing the Göttingen-Copenhagen research program with a continuous, causal alternative. It was in response and in opposition to the threat from Schrödinger that Heisenberg resurrected the foundational status of discontinuity and swiftly developed arguments against causality.

As I have argued, the probabilistic formalism of quantum mechanics was developed by Dirac and Jordan as a direct extension of Heisenberg's fluctuation paper (1926a), in which, contra Schrödinger and his supporters, Heisenberg claimed that for two atoms in resonance, exchange of energy must occur via discrete jumps. One cannot obtain the energy of an atom as a function of time, argued Heisenberg, but only the time average of energy, or probability that the energy takes a particular value. Similarly, for any dynamic variable, both Dirac and Jordan postulated, one can only calculate the probability that it lies between two specified values (chapter 4). In this way discontinuity and acausality were entrenched in the very axiomatic basis of quantum mechanics to such a degree that they appeared to be "hard facts" of nature.

This reading dispels two widespread misconceptions concerning the history of quantum mechanics. The first misconception is that an abstract quantum formalism developed before any interpretive concerns were raised, simply as a result of disinterested mathematical activity. According to the second misconception, the Göttingen-Copenhagen interpretation came first, and only later did Schrödinger begin his competing attempts at interpretation. Both claims, we have seen, must be qualified substantially. From the beginning, the orthodox interpretation and the formalism itself were developed to defend the past achievements that Schrödinger threatened to undermine and, through its philosophical legitimation, to protect the Göttingen-Copenhagen research program against any challenge or alternative. The ambitious Heisenberg, who laid the foundation of matrix mechanics, was one of the central players in this enterprise. Strong sentiments against causality quickly arose precisely because causality was an intrinsic part of the threat from the opposition.

Perhaps another few words about why it was Heisenberg who initially led the campaign against causality are in place here. The interpretive attempts of quantum theorists took place against a background of philosophical controversies about the changed status of space and time concepts implied by Einstein's relativity. Young Heisenberg witnessed the emotional and politically charged confrontation between Einstein and the neo-Kantians (Cassidy 1992, 96–98). To ask whether the verdict passed on the a priori status of Newtonian-Euclidean space and time did not hang over the a priori status of causality as well was natural. An intensely ambitious man, who aspired, following Einstein's example, to leave his mark on the philosophy of nature and the public imagination, and not only to solve scientific "puzzles," must have kept an eye on this issue. None other than Heisenberg's teacher, Arnold Sommerfeld, posed the high-stakes challenge. Discussing "the present state of physics" and referring to the causality issue just a few months before Heisenberg's uncertainty paper, Sommerfeld called for a "new Kant" (1927, 235). Heisenberg wanted to be the new Kant—in his initial presentations of the uncertainty principle to academic audiences, he always described the abandonment of the "Kantian category of causality" as a natural continuation of Einstein's overthrow of Kantian space and time as forms of intuition (see, for example, Heisenberg 1928, 1931b).

Thus the philosophical framework of the orthodox interpretation arose initially as a forced response to the opposition and developed subsequently into a huge project for the legitimation of the Göttingen-Copenhagen program.[2] Because an argument of "finality," or inevitability, is perhaps the strongest way to advance any sort of claim, the rhetoric of finality and inevitability permeates many philosophical arguments in the Copenhagen framework. Following Bohr, Heisenberg presented the search for a causal quantum theory, not merely as hopeless, but as downright meaningless. Heisenberg argued that "there seems to be the strongest evidence" for the "final" renunciation of causality and objectivity. The concept of "causality," after the discovery of quantum mechanics became as devoid of meaning as the concept of the "end of the world" after the discoveries of Columbus and Magellan (Heisenberg 1934, 17). Bohr, claiming that acausality was "unavoidable," ridiculed the "odd wish" for a causal description of quantum mechanics: "It appears to me to be exactly that, that we would say that we are not satisfied with descriptions in English, and we might like, because we can't get the word 'causality' . . . to translate it into Sanskrit, to get everything different" (1937d, 338). And Heisenberg, quoting Bohr, asserted that the structure of Bohm's "strange" arguments for hidden variables is identical to the hope that "$2 \times 2 = 5$, for this would be of great advantage for our finances" (1958, 132).

One would expect that the proponents of the Copenhagen interpretation were in possession of some very strong arguments, if not for inevitability, at least for high plausibility. Yet a critical reading reveals that all the far-reaching, or one might say far-fetched, claims of inevitability are built on shaky circular arguments, on intuitively appealing but incorrect statements, on metaphorical allusions to quantum "inseparability" and "indivisibility," which have nothing to do with quantum entanglement and nonlocality. Acausality, in the writings of Bohr, Heisenberg, Pauli, and Born, is not a sharply defined concept.[3] Because acausality is to be used as a tool of legitimation, as a sword to wield against opponents, its meaning changes from text to text, from context to context.

Among Heisenberg's many arguments for the inevitability of acausality, one of the more frequent and careless ones is the argument based on the notion of a "cut" between quantum object and classical measuring device. This cut is mobile. Heisenberg asserted that the statistical aspect of quantum mechanics enters only *on* the cut, or dividing line (1979, 15, 49; 1958, 48–49) because on either side of the line determin-

2. On the "Copenhagen hegemony," see Cushing (1994b), and on the "diffusion of the Copenhagen spirit," Heilbron (1987).

3. In this chapter, I will follow Bohr, Heisenberg, and Bohm in using "determinism" and "causality" as interchangeable notions, despite the differences between them. These differences are not important for the issues discussed here.

istic lawfulness prevails (on one side, because of the laws of classical physics, and on the other side, because of the differential equations of quantum mechanics). If we supplement the variables at the dividing line by some "hidden variables" and then "move" the cut to a new place, we will come into conflict with the causal laws that prevailed at this place to begin with (the situation becomes overdetermined).

This argument is, of course, unconvincing. If we move the cut into what was the classical domain, it does not necessarily mean that the situation becomes overdetermined—the hidden variables for the cut can turn out to be the existing variables in the macrodomain, as when Bohm supplemented the quantum description with position variables. It is not too hard to find serious flaws in other arguments advanced by Heisenberg along similar lines. The more interesting question is: when, for what purposes, and in what contexts did Heisenberg use such arguments?

The crucial part of all "cut" arguments is the need to describe the measuring apparatus using classical concepts, which is connected with the more far-reaching Bohrian claim that to speak of quantum concepts and quantum ontology is meaningless. In many of his writings, Heisenberg distanced himself from this position. Thus he often said that measuring apparatus "can" (as opposed to the Bohrian "have to") be described by classical concepts. Nor did Heisenberg agree with Bohr that no new concepts and no new language beyond the classical can be conceived—in fact, while discussing the lessons of the theory of relativity, Heisenberg emphasized that new concepts arise and a new language develops "which adjusts itself after some time to the new situation" (1958, 175). For example, in "the theory of general relativity, the language by which we describe the general laws actually now follows the scientific language of the mathematicians" (1958, 177).

I argued in chapter 8 that mathematical physicists, especially those who had firsthand familiarity with relativity theory, such as Born, Fock, and Heisenberg, did not accept the special status Bohr accorded to classical concepts. Instead, they held that concepts are justified by their place in a coherent mathematical formalism of the physical domain—be it classical or quantum. It is unwarranted, therefore, to give special status to classical notions. David Bohm was of a similar opinion: "Mathematics in general and the differential calculus in particular have played a key role in guiding the development of a clear concept of accelerating motion. . . . it is practically impossible to gain such a clear concept on the basis of everyday experience alone, or even on the basis of the laboratory experience alone." Similarly, our concepts of waves come not from watching water waves, but from the mathematical treatment of interference and wave propagation (Bohm 1957, 97–98). Bohr was unique among quantum physicists in his denial of the possibility of new concepts. One cannot avoid the question, then: why and when did Heisenberg, very much a mathematical physicist, support this peculiar doctrine of Bohr's?

In the cut argument, analyzed above, Heisenberg described the mobility of the cut as "particularly important." The next sentence reveals why it is so important: "An appreciation of this fact also helps to dispose of an objection frequently made against the *finality* of quantum mechanics" (1934, 16, my italics). Heisenberg immediately proceeded to the "more general question of the finality" of changes promoted by Copenhagen proponents and concluded that the renunciation of objectivity and causality would be final (1934, 17). Similarly, Heisenberg presented the flawed cut argument in another lecture a year later, again using it as an argument for "the finality of the statistical character of quantum mechanics," and against "any hope" of "completing quantum mechanics on a determinist basis" (1935, 48). Again, in another article, Heisenberg claimed that the "root of the statistical character of quantum theory" is the "tension" between the need to use the concepts of classical physics and "the knowledge that these concepts do not fit accurately." Heisenberg mentioned that "it has been suggested that one should depart from classical concepts" and search for radically new ones that "might possibly lead back to a non-statistical, completely objective description of nature." He then immediately denied such a possibility, repeating the Bohrian argument that "the concepts of classical physics are just a refinement of the concepts of daily life" (1958, 56). Heisenberg's denial was a forceful one: "We cannot and should not replace these [classical] concepts by any others. . . . we cannot and should not try to improve them" (1958, 44).[4]

I argued in chapter 8 that the Copenhagen interpretation cannot be reconstructed as a coherent philosophical framework, for it is more adequately described as a collection of local, often contradictory, arguments embedded in changing theoretical and sociopolitical circumstances. Heisenberg's contradictory statements about the necessity of classical concepts in different contexts is more evidence for my claim. From this discussion it becomes clear that Heisenberg insisted on the indispensability of classical concepts in those contexts, and, as far as I am able to determine, only those contexts, where he attempted to argue for the finality of the indeterministic Copenhagen version of quantum mechanics.

Returning to Bohm's contention that it was Heisenberg's belief in the

---

4. David Finkelstein's pointed comment on these lines strikes the heart of the matter: "This seems like too much protest; if we cannot then the should-not is surely otiose, and if we should not then we probably can" (1987, 291).

indispensability of classical concepts (and Heisenberg's inability to conceive of categories other than waves and particles) that lead him to a belief in finality, my claim is rather the opposite. It was his vested interest in finality that led Heisenberg to insist on the indispensability of classical concepts in a local and opportunistic fashion. There was neither belief nor commitment on Heisenberg's part—only a selective and opportunistic use of Bohrian doctrine in those circumstances where Heisenberg's aim was to argue against the opposition and for the finality of the Copenhagen orthodoxy. This is a characteristic example of a powerful social strategy of legitimation disguised as an abstract theoretical argument.

It is quite instructive to see how Heisenberg argued in those articles where he discussed the lessons both of relativity theory and of quantum physics. For the main lesson of relativity, as Heisenberg presented it, was the invalidity of the a priori status of classical concepts, while the defense of the Copenhagen interpretation forced him to claim that classical concepts are indispensable and final in an a priori Kantian sense. Ingenious Heisenberg had no problem avoiding the impasse: he invented the notions of "practical *a priori*" and "relative synthetic *a priori*," which are, in fact, contradictions in terms. The only function of these self-contradictory notions, as they appear in Heisenberg's works, is to argue for the a priori status of classical concepts in the context of a defense of the Copenhagen framework, against Heisenberg's own better judgment when discussing relativity (1958, 90–92). Bohm, after his "conservasion" had taken place, understood that arguments for the inevitability of acausality cannot be compelling. They are, in fact, circular: "The conclusion that there is no deeper level of causally determined motion is just a piece of circular reasoning, since it will follow only if we assume beforehand that no such level exists" (Bohm 1957, 95).

Operationalism: From Consistency to Inevitability Arguments

Bohr's arguments for the inevitability of acausality and complementarity similarly cannot withstand close scrutiny. I have shown that Bohr's arguments for inevitability are in fact arguments of consistency in disguise (Beller 1993). "The uncontrollable disturbance," or uncontrollable interaction between the measuring and the measured, is the crucial component of all of these arguments and the basic support on which Bohr's web of arguments for complementarity rests. Bohr employed different strategies to make this uncontrollable interaction, founded on the "indivisibility of the quantum of action," a fundamental concept in quantum philosophy.

Yet how exactly does the "uncontrollability of the interaction" follow from the "indivisibility of the quantum of action," and what do these terms mean? Their use and their interconnection differed, in fact, significantly before and after 1935 (the EPR challenge). Before 1935, Bohr's arguments were incorrect and were based on the later discredited idea of disturbance. After 1935, the arguments were tautological or circular: textual analysis reveals a web of terms, where "indivisibility of the quantum of action," "individuality of quantum phenomena," "unsurveyability of interaction," and "uncontrollability of interaction" have the same meaning and are interchangeable.

The following lines are typical of Bohr's pre-1935 writings: "Quantum of action means that it is not possible to distinguish between the phenomena and observation. . . . we cannot make the interaction between the measuring instrument and the object as small as we like. Now, that means that phenomena are influenced by observation" (1930, 134). Before 1935 the "indivisibility of the quantum of action" simply meant "finitude" of the quantum of action. Measurement "disturbs" phenomena precisely because of the atomic constitution of measuring devices. The underlying intuition was that because in the quantum domain phenomena and measurement interaction are "of the same order of magnitude," such interactions, unlike in the classical case, cannot be neglected or accounted for (chapter 7). This intuitively appealing argument is, of course, incorrect. The concept of disturbance is an inconsistent one: it presupposes the existence of objective exact values that are changed by measurement, contrary to the desired conclusion of their indeterminacy. Moreover, the finitude of interaction does not assure uncontrollability—many classical measurements involve finite interactions that cannot be neglected yet can be calculated.

After 1935, when the EPR challenge undermined the notion of disturbance, Bohr's reasoning did an about-face. The uncontrollability of interaction was then tied inseparably to the necessity of describing measuring devices using classical terms only, neglecting "in principle" their atomic constitution.

In Bohr's later writings the uncontrollability of the interaction follows from its "unsurveyability," which in turn follows from the dogmatically postulated essential difference between quantum object and classical measuring apparatus. If, in Bohr's pre-1935 writings, uncontrollability necessitated acausality, in his later writings it is the statistical character of quantum theory that leads to the inseparability of the measuring and the measured and to the uncontrollability and unsurveyability of the measuring interaction.

This reversal of what is to be explained and what is foundational, of *explanans* and *explanandum,* is frequent in Bohr's and Heisenberg's writings. This reversal is an eloquent illustration of the circularity and ad hocness of many similar arguments. One striking example is Heisenberg's reversal of the idea of disturbance—if the original idea of disturbance was based on the indivisibility, or finitude, of microscopic particles, now the "unavoidable disturbance" is seen as the reason for their indivisibility: "Nature thus escapes accurate determination . . . by an unavoidable disturbance which is part of every observation. . . . In atomic physics it is impossible to neglect the changes produced on the observed object by observation. . . . The supposition that electrons, protons and neutrons, according to modern physics the basic properties of matter, are really the final, indivisible particles of matter, is only justified by this fact" (1979, 73).[5]

Heisenberg's and Bohr's arguments for the foundational, a priori status of uncontrollable interaction are illegitimate. The uncontrollability of interaction has one, and only one source—it follows from the mathematical framework of the quantum theory. It is necessary to ensure the consistency of quantum theory when the possibilities of measurement are taken into account. Bohm, after abandoning the orthodox Copenhagen stand, understood this clearly. Thus, discussing the uncontrollability of the interaction in Heisenberg's γ-ray experiment, Bohm pointed out: "The disturbance is unpredictable and uncontrollable because *from the existing quantum theory* there is no way of knowing or controlling beforehand what will be the precise angle, with which the light quantum will be scattered into the lens" (my italics). Bohm realized that the indeterminacy relations are "primarily . . . a deduction from quantum theory in its current form," and that attempts to prove their "absolute and final validity" are based on nothing but an a priori, illegitimate insistence on the finality of quantum theory (1957, 82 n).

The legitimate function of uncontrollability arguments is to argue the consistency, rather than the inevitability, of uncertainty and complementarity. Bohr (1935a) did discuss certain canonical thought experiments along these more legitimate lines. He argued, for example, the impossibility of measuring the position of a particle in a setup measuring its momentum. Using a movable diaphragm, we can measure a particle's momentum (by using an appropriate test body) before and after the passage of a particle through the slit. By applying the law of conservation of momentum to a system consisting of a particle and a moving diaphragm we calculate the momentum of a particle passing through the slit. However, because we have exact knowledge of the momentum of the diaphragm during, or immediately after, the particle's passage, we block the possibility of also knowing the position of the diaphragm, in accord with the uncertainty relations. Consequently, we are denied knowledge of the particle's position, which is identical with that of the diaphragm (chapter 7; see also Beller and Fine 1994). Why does the diaphragm obey the uncertainty relations? Either we just state it (it applies to quantum objects), and then we deduce uncertainty for the particle (this is, of course, an argument of consistency), or we say that while measuring the momentum of the diaphragm with a test body, the diaphragm undergoes uncontrollable interaction. We deduce consequently the uncontrollable interaction between the particle and the diaphragm—again, these are elementary considerations of consistency.

The crucial point of these thought experiments is that momentum and position cannot be *measured* simultaneously. This is, however, not satisfactory to Bohr. Why is it not satisfactory? The answer is clear from Bohr's own words. He warns against "misunderstanding by expressing the content of uncertainty relations . . . by a statement as 'the position and momenta cannot simultaneously be measured with arbitrary accuracy.' . . . such a formulation . . . would not preclude the possibility of a future theory taking both attributes into account" (1937c, 245). What is the strategy of precluding a theory, such as Bohm's, where particles have well-defined position and momentum, which however cannot be measured simultaneously according to the uncertainty relations? How do we argue that Bohm's theory is impossible in principle? How do we transform consistency arguments into those for inevitability?

With the benefit of hindsight the answer is obvious. Simply postulate that what cannot be measured—does not exist. By defining concepts operationally through a procedure for their measurement and then applying the quantum formalism to an analysis of the measurement procedure, we will obtain nothing but deductions from the quantum formalism (such as, for example, the uncertainty relations). In this way an illusion is created that features of the theory (such as uncertainty) belong to the very definition of the concepts used and that they follow inevitably from a logical analysis of the conditions of experience.

Thought experiments were indeed initially used by Heisenberg and Bohr to argue the consistency of the new quantum mechanical scheme. This was the original thrust of Heisenberg's analysis in the γ-ray thought experiment. Bohr (1929a), following Heisenberg, argued that the uncertainty relations are the basis of the logical consistency of quantum mechanics. Bohr's considerations of mutually exclusive experimental arrangements were initially aimed at proving the consistency of quantum theory: the wave and particle aspects "can never be brought into direct contradiction with one another, since their closer analysis in mechanical terms demand mutually exclusive arrangements" (1933, 5).

This mutual exclusivity, aided by operationalism, would soon develop into arguments for the inevitability of uncertainty and complementarity. The connection between operationalism and inevitability was substantially strengthened in Bohr's response to EPR. As Arthur Fine and I have argued (Beller and Fine 1994), Bohr was only able to counter the challenge of EPR by using an operational and relational definition of concepts. After EPR, operationalism became the strongest tool in legitimation arguments for the inevitability of quantum theory.[6] It is in this context that Bohr often used the word "logical"—the inevitability of complementarity and uncertainty follows, in his words, from "simple logical analysis," and it leads to "the situation where the very concept of determinism loses its possibility of logical application" (1957b, 601). With the "dictatorial help" of operationalism, to use Schrödinger's (1935, 157) apt expression, the central Copenhagen dogmas became unassailable and final.

Now we can return to Bohm's claim that it was the positivist commitments of the founding fathers of the Copenhagen interpretation that led them to believe in the finality of quantum theory and the impossibility of revising it along more objective or deterministic lines. As in the case of the inevitability of classical concepts, the situation is the opposite: it was the need, or the desire, to argue for finality against threats from the opposition that led Heisenberg and Bohr to take a forceful operational stance. I argued in chapter 8 that the proponents of the Copenhagen interpretation did not have, collectively or even individually, a consistent position on the realism-antirealism issue and that they employed different, often contradictory arguments, depending on the local context. We have now identified the context in which positivist operational arguments were selectively employed. Positivism in quantum philosophy has two central strands—operationalism and instrumentalism. Instrumentalism implies that the quantum formalism is merely a tool, an algorithm for the description and prediction of measurement results. The instrumentalist strand, denying the possibility of a quantum ontology, while emphasizing the Kantian inevitability of the classical ontology, is, of course, directly connected with the doctrine of the indispensability of classical concepts. Both varieties of positivism are aimed at arguing the finality of quantum mechanics.

Were physicists aware that unless they took an extreme and untenable positivist stand, what the Copenhagen interpretation offered were, at best, considerations of consistency, and not of inevitability? After

---

6. It is instructive to mention here that Bridgman's original aim of introducing operationalism was to prevent the "embarrassing" need for basic revision of physical theories, ensuring their certainty by a strict operational definition of concepts (Beller 1988).

abandoning the orthodox stance, Bohm was aware of this. According to Bohm, Bohr's essential contribution was to show that one can "*consistently* . . . renounce the notion of unique and precisely defined conceptual models in favor of that of complementary pairs of imprecisely defined models" (my italics). What Bohr was able to demonstrate was that such a use of complementary concepts provides "a *possible* way of discussing the behavior of matter in the quantum-mechanical domains" (Bohm 1957, 93, my italics). Not only such dissidents as Bohm, but orthodox quantum physicists as well, were aware that many of the philosophical pronouncements on complementarity and uncertainty were merely consistency arguments. Pauli announced that the central function of the notion of complementarity was "to symbolize the fundamental limitation of our . . . idea of phenomena as existing independently of the means by which they are observed. . . . It is indeed this limitation that makes the theory *logically consistent*" (1945, 99, my italics). Dirac, who was fond neither of pictures nor of complementarity, conceded that one may "extend the meaning of the word 'picture' to include any way of looking at the fundamental laws which make their *self-consistency* obvious" (quoted in Jammer 1974, 13, my italics).

Both Feynman and Weizsäcker realized that Heisenberg's analysis of the γ-ray experiment was merely a consistency argument, despite the fact that it is often presented as an argument for inevitability. Feynman announced in his *Lectures on Physics:* "It was suggested by Heisenberg that the new laws of nature could only be *consistent* if there were some basic limitation on our experimental capabilities not previously recognized" (Feynman, Leighton, and Sands 1969, 1–9, my italics). And, more eloquently, Weizsäcker: "The impossibility of simultaneously observing precise values of position and momentum does not mean the pseudo-positivistic nonsense 'what cannot be observed does not exist,' but the test of consistency: 'what does not exist in the theory cannot be observed in an experiment describable by the same theory'" (1987, 278). In this argument Weizsäcker was following his teacher Heisenberg, who revealed that he regarded all of Bohr's complementarity elaborations as demonstrating the consistency of the quantum formalism by an analysis of the possibilities of measurement. Heisenberg considered all these arguments "trivial": "Many experiments were discussed and Bohr again successfully used the two pictures, wave and particle, in the analysis. The results confirmed the validity of the relations of uncertainty, but in a way this outcome could be considered trivial, because if the process of observation is itself subject to the laws of quantum theory, it must be possible to represent its results in the mathematical scheme of this theory" (1977, 5–6). This position undermines, of course, all of Heisenberg's own arguments for inevitability, or finality.

The project of arriving at true, final, certain, indisputable knowledge was, of course, not a new one. Classical physics, despite the empiricist revolt against scholasticism, inherited the Aristotelian ideal of science as demonstration, of arriving at certain conclusions from certain premises. The "father of empiricism" himself, Galileo Galilei, used such scholastic demonstrative arguments: two bodies of unequal weight must fall at the same rate, for if they do not (if heavy bodies fall faster), we would arrive at a logical contradiction.[7] Heisenberg's cut arguments are peculiarly reminiscent of such pseudological proofs.

It was Kant, of course, who provided the most grandiose philosophical demonstration of indubitable knowledge—by his analysis of Newtonian space and time as forms of intuition, and causality as a necessary category. Newtonian mechanics received the status of synthetic a priori knowledge—knowledge that, though empirically arrived at, had the same certainty as the analytic knowledge of Euclidean geometry. Bohr's attempts to arrive at true, certain, final knowledge by the mere analysis of the conditions of experience is thoroughly Kantian in spirit. What is striking, however, is that such Kantian arguments were advanced after a full realization of the bankruptcy of Kantian arguments for a priori knowledge (the theory of relativity being the final stroke), and by those physicists who, excepting Bohr himself, had a thorough knowledge of relativity theory. Only "extrascientific" reasons can plausibly explain this fact. Not surprisingly, then, what the orthodox physicists advanced as arguments for inevitability were, in the best case, arguments of consistency. It is enlightening to mention here the views of the Göttingen mathematician David Hilbert, with whom the mathematical quantum physicists were in close contact. Pitowsky outlines Hilbert's stand succinctly: "Consistency, or rather the avoidance of a paradox, is ultimately the only safeguard in the pursuit of knowledge. We would have liked to do better for sure, but we have very little choice in the matter" (1994, 119).

The Copenhagen interpretation, and its rhetoric of inevitability, rests on two central pillars—positivism and the doctrine of the necessity of classical concepts. Both are unnatural for working mathematical physicists. No wonder the Göttingen-Copenhagen physicists, Heisenberg included, often denied their adherence to positivism. No wonder contradictory statements abound concerning the doctrine of the necessity of classical concepts.

7. If we combine the two bodies of unequal weight, the composite body should fall slower than the heavier of the two would have alone, because the lighter of the two bodies impedes the fall of the composite body. It also should fall faster than the heavier body, because its weight is greater. Hence all bodies, regardless of weight, fall at the same rate.

Bohm on Classical versus Quantum Concepts and on Indeterminism

Bohr advanced his doctrine of the indispensability of classical concepts, on which the whole framework of complementarity stands or falls, as an obvious fact. But, as I argued in chapter 8, it was not a doctrine that Bohr's closest collaborators, Heisenberg and Born, could subscribe to, public support of Bohr notwithstanding. Even those physicists eager to follow Bohr found it difficult not to question it. Bohm's discussion of classical concepts and causality, composed before his defection from the orthodoxy, is instructive.

Bohm's *Quantum Theory* is rightly considered the best Bohrian textbook. Indeterminism, indivisibility, and wave-particle duality—all are elaborated with rare clarity and force. The arguments are perfected as far as they can be. Because Bohm labored so hard to make them as convincing as possible, he could not avoid seeing their weak points.[8] It is perhaps not incidental that both Bohm and Alfred Landé, another prominent dissident, defected from the orthodoxy after completing textbooks on quantum theory in the Copenhagen spirit. While the central pillars of the Copenhagen interpretation are elaborated faithfully in Bohm's textbook, a careful reading finds a powerful undercurrent of tensions with Bohr—tensions that, when taken to their logical conclusions, should have, and historically could have, been the reason for Bohm's final break with the orthodoxy. Not surprisingly, these discrepancies concern Bohm's discussion of classical and quantum concepts. Bohm treats these issues in chapter 8, revealingly titled "An Attempt to Build a Physical Picture of the Quantum Nature of Matter": the wording holds a hint to—and possibly a seed of—his future quantum ontology.

Unlike Bohr, Bohm did not argue that classical concepts are either indispensable or natural. In fact, the Bohmian analysis undermines Bohr's arguments that classical concepts are extensions of common sense. Nor did Bohm seem to agree with Bohr that there can be in principle "no quantum concepts," only a quantum mechanical abstract algorithm for

---

8. Many physicists who did not trouble to go into the Bohrian arguments in depth had an easier time following them uncritically. Heisenberg complained as early as 1930 that physicists had more faith in the Copenhagen interpretation than clear understanding of it (1930, preface).

According to Bohm, "The whole development [of a search for a hidden variables alternative] started in Princeton around 1950, when I had just finished my book *Quantum Theory*. I had written it from what I regarded [as] Niels Bohr's point of view, based on the principle of complementarity. Indeed I had taught a course on the quantum theory . . . and had written the book primarily in order to try to obtain a better understanding of the whole subject. However, after the work was finished, I looked back over what I had done and still felt somewhat dissatisfied" (1987, 33).

the prediction and description of experimental results. Bohm called for the development of a new language, which would depart from classical intuitions and would be adequate for the new quantum conceptual framework. Bohm also undertook a thorough critical analysis of the classical idea of the continuity of motion and of classical determinism. His aim was to demonstrate that "there is no *a priori* logical reason for their adoption." The quantum concepts of indivisible transitions and incomplete determinism are "much more analogous to certain naive concepts that arise . . . in common experience" (1951, 144). Not surprisingly, Greek philosophers were unable to grasp the idea of continuous motion, as Zeno's paradoxes eloquently remind us (1951, 147). In fact, our intuitions revolt against thinking of a motion in which the position and velocity are both precisely determined and are thus in close harmony with the uncertainty principle. Using the example of photography, this point becomes especially clear. Our simplest intuition of an object with a definite position is an object that is not moving. To obtain a picture of movement, we must allow the position to blur slightly. It is the blurred picture of the speeding car that suggests some space was covered during some time that the car was moving. If we use a very fast camera and get a sharp picture of the moving car, we will not deduce motion. Our intuitive presupposition is that "a continuously moving object has a somewhat indefinite range of positions"—we simply "cannot visualize simultaneously a particle having a definite momentum and position." Bohm concluded that "quantum theory . . . gives a picture of the process of motion that is considerably closer to our simplest concepts than does classical theory" (1951, 146).

But does not our concept of an object at rest, with fixed position and zero velocity, contradict the uncertainty principle, according to which an object with a precise position has a completely indefinite momentum? Bohm argued that in fact there is no contradiction. When we think of an object at a definite position, we can imagine that a short time later it will move to a different position. Which position it will move to we cannot infer from the original position—all subsequent positions are equally consistent with it. This means that any velocity, or any value of momentum, is equally consistent with a picture of a particle in a definite position. Put differently, "if we think of an object in a given position, we simply cannot think of its velocity at the same time" (1951, 146). We see, according to Bohm, that the classical idea of the continuity of motion is an unnatural, non-commonsensical idea. On a purely logical basis, there is no reason to choose the concept of a continuous trajectory in preference to that of a discontinuous trajectory. The idea of continuous motion developed through our experience with planetary orbits, projectile trajectories, and the theory of differential calculus was tailored to deal with such motions. It is only by studying "such things for a while" and getting used to them that the "succeeding generation began to take the basic ideas for granted" (1951, 148). This analysis is of course a far cry from Bohr's assertion of the necessity of classical concepts because of the alleged proximity of classical to commonsense notions and the direct accessibility of classical reality to sense perception.

According to Bohm, there are quantum concepts that are inherently different from classical concepts. Quantum energy and momentum, despite being given the same *names* as concepts in classical physics, are examples. In classical theory, we regard energy and momentum as fundamental properties of matter because they obey conservation laws. But from a logical point of view they need not be considered fundamental, because energy and momentum can be expressed as functions of positions and velocities, so they are in a certain sense redundant. All the classical laws of motion can be expressed in terms of space-time relationships alone. This is not the case in quantum theory. In quantum theory, velocity cannot be defined as a limit of distance divided by time. In quantum theory, momentum, defined by the de Broglie relation, is to be considered an independent rather than derived property of matter—"the statement that an electron was observed to have a given momentum . . . stands on the same footing as the statement that it had a given position. Neither statement is subject to further analysis" (1951, 154).

Bohm departed from Bohr also in his position on the issue of classical versus quantum mechanical reality. He uses the word "convenient," rather than the Copenhagen "necessary," to differentiate between classical and quantum processes: "It is convenient to make a distinction between classically describable processes and those essentially quantum-mechanical in nature." Such a distinction is convenient when we deal with large objects, whose features of interest "do not critically depend on the transfer of [a] few quanta more or less." In direct, though unacknowledged, opposition to Bohr, Bohm asserted: "In the last analysis all processes are, of course, quantum-mechanical in nature" (1951, 165).

It is very enlightening to see how Bohm argued against classical Laplacian determinism. By presenting classical determinism as a historically contingent rather than necessary notion, Bohm aimed to ease the way for the acceptance of indeterminism. Yet precisely because of its historical contingency, Bohm implicitly undermined the Copenhagen stand of the inevitability of indeterminism. Bohm argued for replacing the notion of classical determinism with a more limited notion of causality as determining general tendencies (but not determining the behavior of systems completely). This notion is much closer to the way we relate to forces or causes in everyday life: "The idea most likely to have been used in connection with common experience is that a particular force or cause produces a tendency toward an effect, but that it does not guarantee the effect" (1951, 152). Quantum lawfulness, which determines only statistical trends, is much closer to our common sense than complete determinism (1951, 152).

Bohm offered a brief critical analysis of the history of the idea of classical determinism. The Greek idea of "fate"—the course of events that is beyond the power of humans to change—belonged to one of the earliest conceptions of complete determinism. Yet it was only with the invention of machinery, which allowed motion to be controlled with precision, that the idea of a mechanistic and deterministic worldview developed in the sixteenth through nineteenth centuries. When work was done mostly by hand or with the aid of animals, motions could not be controlled exactly—one could push in the right general direction, and push backward if one had gone too far. "It is very likely that the modern form of the idea of complete determinism was suggested . . . by its resemblance to complex and precisely constructed machines, such as clocks" (1951, 150). Bohm's analysis of machines—material means of production—as determining the basic conceptual framework of Newtonian mechanics resembles the analysis of the Soviet historian of science Boris Hessen, who in 1931 at a congress in London, launched the Marxist historiography of science with an analysis of Newtonian mechanics along similar lines. It would be rewarding to know whether Bohm was acquainted with the works of Marxist historians of science in general, and with Hessen's work in particular. Yet the crucial point here is this: Bohm's analysis of historical rootedness, of the sociohistorical contingency of our most cherished, most uncritically accepted concepts, does not agree with the rhetoric of finality of the Copenhagen interpretation. What Bohm's analysis rather discloses is that nothing more than continuous habit endows concepts with the aura of naturalness, necessity, inevitability. We need only take a small step to ask whether the same kind of critical analysis can be extended, or will be extended in the future, to the Copenhagen quantum philosophy itself. Even the idea of hidden variables surfaces favorably in this discussion (1951, 171). Bohm, in his later quantum ontology, was open on the issue of determinism versus indeterminism. Bohm's nonlocal ontology can be either causal or stochastic. No wonder Bohm was soon led to his idea of understanding nature in terms of an inexhaustible diversity and multiplicity of things, and of understanding scientific theory as approximate, conditional, and relative truth (1957, 164–65).

How strikingly different from Bohr's ideas this trend of thought was! Bohr, who considered classical concepts natural and necessary and who considered quantum theory "a rational generalization" of classical theory, could believe (as opposed, perhaps, to the more opportunistic Heisenberg) that his analysis of quantum theory was really the only possible one. Bohr was impatient with those who questioned him: "How could philosophers not understand, that this was an objective description, and the only possible objective description?" (interview with Bohr, AHQP; Bohr to Born, 1953, AHQP). As the official spokes-man of the orthodox interpretation, who had to fight against opposition such as Bohm's, Bohr preached closure, finality, inevitability, and the impossibility of alternatives. How fitting for Bohm, a victim of this ap-proach, to develop a diametrically opposed view of science—pleading for tolerance, for creative plurality, for peaceful theoretical coexistence, for a free play of imagination, for friendly, open-minded, and joyful scientific cooperation and communication (Bohm and Peat 1987).

CHAPTER 10

⤺

Constructing the Orthodox Narrative

The tendency in many historians [is] to write on the side of Protestants and Whigs, to praise revolutions provided they have been successful, to emphasize certain principles of progress in the past and to produce a story which is the ratification if not glorification of the present.

Herbert Butterfield 1931

Introduction: "Whiggish" History and "Winner's" Strategies

How does one construct the orthodox narrative, in which the development of "disembodied" ideas seems to follow so persuasively, if not inevitably? How does one retrospectively construct coherent conceptual frameworks from a multitude of contradictory attempts, predispositions, vacillations? How does one suppress the polyphony and temporality of the creative process by timeless, static paradigms? How does one obtain what professional historians, not without disdain, call a "Whiggish" history?[1]

It is often awkward to deduce the past from the present, in life as in science. The following conclusion of Schrödinger's biographer sounds strange: "Considering what a good swimmer he later became, he probably could swim well enough at ten" (Moore 1989, 19). Yet there is also something very natural, almost inescapable, about Whiggishness. We often assign meaning retrospectively, after some accumulation of information that allows a perspective to take shape. We all, so it seems, project our present knowledge of the outcome of our acts onto an appraisal of our own decisions in the past (Fischhoff 1975, 288). Writing history backward need not be a cynical, self-serving act, though it sometimes is. We have a whole spectrum of possibilities—from innocent attempts to understand the present by looking for coherence and meaning in the past, to conscious manipulations of history. The past is often manipu-

---

1. The passage from Butterfield quoted in the epigraph describes narratives of British political history. Historians of science have adopted the expression "Whig interpretation of history of science," or "Whiggish history of science," to characterize those works that interpret the past in terms of the present, ignoring historical material that cannot be interpreted as having contributed to the "triumph" of the present.

lated in Whiggish narratives, which make the "winners" look "right" so naturally.

Not all scientists are aware, as Feynman was, that what they take as the matter-of-fact history of their discipline is merely "a sort of conventionalized myth-story."[2] The history of quantum mechanics is usually written from the "winner's" perspective, with heavy reliance on the recollections of Bohr, Heisenberg, and such ardent followers as Rosenfeld. This Whiggish, winner-oriented perspective is clearly displayed, for example, in the widely used collection *Quantum Theory and Measurement* (Wheeler and Zurek 1983). The volume begins with Bohr's account of the past (Bohr 1949), and only later the actual historical papers, including those of the Bohr-Einstein debate, are introduced. Bohr's paper is prefaced by six pages of "commentaries" by Bohr and Heisenberg, with less than half a page left to a "commentary" by Einstein. Only a deliberate, concentrated effort allows a reader to avoid approaching the papers in this volume from the Copenhagen perspective.

In this chapter and the next I will describe the construction of historical narratives by which naturalness, or even inevitability, is ascribed by the orthodox to such central notions of the Copenhagen interpretation as quantum jumps, the impossibility of causal space-time models, wave-particle duality, Born's probabilistic interpretation, the indispensability of classical concepts, and complementarity. Orthodox physicists constructed the past as a major reinforcement of their rhetoric of inevitability. Different strategies—from describing later insights as the outcomes of earlier efforts, to outright deception—are employed. The complex, multidimensional theoretical activity of the past is conflated to a single perspective—the orthodox one.

Bohr proclaimed his research program—the rational generalization of classical physics incorporating the quantum of action—to be an unquestionable task shared by all physicists.[3] An extreme, yet characteristic case is Bohr's discussion of "Planck's discovery of the universal quantum of action" leading directly to the realization of "regularities of a novel kind," whose elucidation brings us directly to the inseparability between the "objects under investigation and our tools of observation" and to complementarity. Bohr subsequently presented the

2. "What I have just outlined is what I call a 'physicist's history of physics,' which is never correct. What I am telling you is a sort of conventionalized myth-story that the physicists tell to their students, and those students tell to their students, and is not necessarily related to the actual historical development which I do not really know" (Feynman 1985, 6).

3. "The task with which Planck's discovery confronted physicists was nothing less than . . . to provide room for the quantum of action in a rational generalization of the classical physical description" (Bohr 1955a, 85).

quantum mathematical formalism as having been created to conform to the philosophy of complementarity: "The proper mathematical tools for a comprehensive description on complementary lines have been created by the so-called quantum mechanical formalism" (1960a, 19).

As in the case of the rhetoric of inevitability, where *explanandum* and *explanans* often change places, so in the orthodox narrative past and future events are interchanged freely. An extreme example is Heisenberg's claim about the early realization of the need to introduce an observer into quantum mechanical description, which supposedly led to Bohr's theory of the atom: "There was good reason to disbelieve that the course of events was objective and independent of the observer. The consequences of these discoveries have led to Bohr's theory of atomic structure" (1979, 14). No less far-fetched are Born's presentations of history. Born claimed (incorrectly, as I will shortly argue) that Bohr fully realized the "inevitable" indeterminism of quantum transitions from the beginning—a realization that led immediately to a blurring of the separation between observer and observed.[4]

The line between such manipulation of history and deliberate deception is very fragile. Heisenberg often stated in his recollections that Schrödinger could not obtain even Planck's law without the quantum postulate—yet Schrödinger, of course, did, and published a paper at the time with a derivation of Planck's law (Schrödinger 1927b). The Göttingen-Copenhagen physicists often referred to John von Neumann's "impossibility" proof as a conclusive strike against hidden variables. Yet in 1935 Grete Hermann, Heisenberg's student, published an essay in which she argued that von Neumann's proof is faulty—no other than Heisenberg wrote a favorable preface to her essay.[5] Pauli also realized, it seems, that von Neumann's proof was deficient; he argued in his lectures on wave mechanics that "no proof of the impossibility of extending [completing quantum theory by hidden variables] has been given" (1973, 12). Nevertheless, Heisenberg and other orthodox physicists continued to use von Neumann's "proof" as an irrefutable argument for indeterminism and as a potent weapon against those who considered a deterministic version of quantum theory possible.

The magic force of the words "proof" and "von Neumann" is described vividly in Paul Feyerabend's autobiography: "He [Bohr] came for a public lecture. . . . At the end of the lecture he left, and the discussion

4. "Bohr . . . stressed, from the very beginning, the new features of the scheme, namely the indeterministic character of the transitions, the appearance of chance in the elementary processes. This means the end of the sharp separation of the object observed and the subject observing" (Born 1953a, 126).

5. Hermann's argument, as well as other criticisms of von Neumann's proof, are discussed extensively in Jammer (1974, 253–339).

proceeded without him. Some speakers attacked his qualitative arguments—there seemed to be lots of loopholes. The Bohrians did not
clarify the arguments; they mentioned the alleged proof by von Neumann, and that settled the matter. Now I very much doubt that those who
mentioned the proof, with the possible exception of one or two of them,
could have explained it. I am also sure that their opponents had no idea
of its details. Yet, like magic, the mere name 'von Neumann' and the mere
word 'proof' silenced the objectors" (1995, 78).

Discontinuities and Quantum Jumps

Discreteness is often perceived as the most basic feature of the quantum
paradigm. This was so at least in the pre-Bell era, when the orthodox
argument was that both indeterminism and antirealism follow from the
indivisibility of the quantum of action; in the post-Bell era entanglement is considered to be the most characteristic nonclassical feature of
the quantum world. Yet the disagreement concerning discontinuity between the orthodox and the opposition was not about the presence of
discontinuous phenomena in the quantum domain, but rather about
the irreducible, a priori status of discontinuities. Because the a priori
status of discontinuities cannot be conclusively proved, despite the orthodox rhetoric of its inevitability, it should come as no surprise that
the orthodox commitment to the irreducible status of discontinuity, and
belief in the real existence of quantum jumps, crystallized as the end
product of a complex historical process. As we saw in Heisenberg's
case, both theoretical considerations and psychosocial factors played a
crucial role in this process. A strong belief in, or strict commitment to,
any metaphysical presupposition acts too much like a straitjacket in a
creative, conceptually fluid phase of scientific activity. Thus it is the
existence of commitment, rather than the lack of it, that needs explanation. Often "commitment" develops to precisely those concepts most
vigorously challenged by the opposition (see chapter 13). Discontinuities and quantum jumps became the loci of the Copenhagen paradigm
because they were strongly challenged by the opposition from the advent of the new quantum theory.

The concept of discrete stationary states of an atom and instantaneous transitions between them was Bohr's most fruitful, ingenious, and
blatantly nonclassical contribution to physics (Bohr 1913). This idea enabled Bohr to lay the foundations of quantum physics, by connecting
atomic physics with Planck's hypothesis, solving the critical problem of
the stability of atoms and, unexpectedly, deciphering the whole mystery of the complex spectral "music" to boot. This incredible conception, kindling both admiration and disbelief, was problematic from the

beginning. One of the first objections, that of clairvoyance, was raised by Ernest Rutherford in a letter to Bohr: "There appears to me one grave difficulty in your hypothesis. . . . how does an electron decide with what frequency it is going to vibrate and when it passes from one stationary state to another? It seems to me that you would have to assume that the electron knows beforehand where it is going to stop" (quoted in Pais 1986, 212).

Another unsettling problem was that the observed width of spectral lines was totally incompatible with the concept of instantaneous transition. Sharp spectral lines (small $\Delta\nu$) necessitated a wave train consisting of at least 10^5 waves, and therefore a finite (noninstantaneous) duration for the emission of all the waves (Slater 1975b). As Schrödinger later put it, the radiating of such a wave train "would use up just the average interval between two transitions, leaving the atom no time to be in those stationary states, the only ones of which the theory gave a description" (1952a, 113). Though physicists could not describe the mechanism of electron transitions, contrary to what is usually assumed, they did regard the question as worthy of consideration from the beginning. Bohr himself did not exclude the possibility that eventually the mechanism of transitions would be understood (Bohr 1918).[6]

Retrospectively, two features became inextricably linked with the notion of quantum jumps: indeterminism and the impossibility of space-time visualization, or the "indivisibility" of quantum transitions. Yet these connections were, as I have mentioned, by no means clear from the beginning. Discussing the notion of quantum jumps and the probabilities of transitions in 1925, Pauli admitted that the time was not yet ripe to come to any definite conclusion on the issue of determinism (chapter 4). Claims in Bohr's later writings contrast sharply with the historical situation. Describing in 1955 the early development of atomic theory, Bohr pronounced: "The point of departure became here the so-called quantum postulate, according to which every change in the energy of an atom is the result of a complete transition between two of its stationary states. . . . *It was evident* that no explanation of the indivisibility of the transition processes, or their appearance under given conditions, could be given within the framework of deterministic description" (1955a, 86–87, my italics). Similarly, discussing the causality problem, Bohr argued: "These so-called quantum postulates . . . impl[y] an explicit renunciation of any causal description of such atomic processes. In particular, as regards its possible transitions from a given

6. In any case, questions regarding the actual nature of transitions—whether they are instantaneous or of short duration, whether they would or would not play any role in optical theory (Bohr, Kramers, and Slater 1924; Kramers and Heisenberg 1925; Bridgman 1927)—were important and meaningful.

stationary state to other stationary state, . . . the atom may be said to be confronted with a choice for which . . . there is no determining circumstance" (1939, 385).

This retrospective rewriting of history took place shortly after Bohr, Heisenberg, Pauli, and Born closed ranks at the congress at Como and the Solvay conference in Brussels in 1927. What was in 1925 only an "unsolved issue" (Pauli 1926a, 11–12), Bohr in 1929 presented as a clear-cut, uncontroversial, "conscious" theoretical choice in favor of indeterminism: "Only by conscious resignation of our usual demands for visualization and causality was it possible to make Planck's discovery fruitful in explaining the properties of the elements" (1929b, 108). Later Bohr presented his overthrow of causality as self-evident: "Since, however, no description of a state of the physical system . . . can determine the choice between different individual processes of transition to other states, we have *obviously* to do with a situation which lies beyond the scope of the classical idea of causality" (1950, 512, my italics). If initially Bohr was uncertain whether the transition process would be described in more detail[7] (recall also Slater's objections and their incorporation into the Bohr-Kramers-Slater theory), he later presented indivisibility, discontinuity, and acausality as part of the very definition of energetic transitions (instantaneous quantum jumps) in the atom.[8]

Yet these energetic transitions need not be "indivisible" or inherently indeterminate. They can be described in terms of a continuous, causal space-time process if one gives up the Copenhagen particle ontology (as illustrated by Schrödinger's description of resonance phenomena or his deduction of Planck's law, contra Heisenberg), or if one supplements the particle ontology with nonclassical additional assumptions (see, for example, the causal space-time description in Bohm and Hiley 1993, chap. 5). The Whiggish history, or the "rational reconstruction," is formulated by the orthodox with the primary aim of presenting any alternative as impossible in principle, and thus of ruling out dissidence.

I have argued that the notion of instantaneous quantum jumps was incorporated into the basic assumptions of the mathematical proba-

7. Whenever Bohr talked about the role of probability considerations in relation to transition processes before the advent of the new quantum theory (or at least before the year 1924), he was always very cautious not to draw any definitive conclusions, invariably qualifying his deliberations with the phrase "in the present state of the theory." The following quote is typical: "In the present state of the theory, it is not possible to bring the occurrence of radiative processes, not the choice between various possible transitions, into direct relation with any action which finds a place in our description of phenomena, as developed up to the present time" (quoted in Hendry 1984, 33).

8. "The basic assumption of the individuality of the atomic processes involved at the same time an essential renunciation of the detailed causal connection between physical events" (Bohr 1937b, 18).

bilistic framework of transformation theory, into Born's probabilistic interpretation, and into Bohr's original formulation of complementarity. After 1927, it became a foundational notion in the Copenhagen interpretation. No wonder the attempts to challenge the irreducibility and instantaneity of quantum transitions provoked an intense counterattack by the Göttingen-Copenhagen camp. And because the orthodox strategy was to present a free theoretical choice as an inevitable conceptual development, we should not be surprised at the inconclusive, high-pitched rhetoric of the orthodox and the opposition alike. The 1952 controversy between Born and Schrödinger about quantum jumps, which occurred in the pages of the *British Journal of the Philosophy of Science*, is a good case in point.

Throughout his life, Schrödinger sought to discredit Bohr's concept of quantum jumping. His forceful attacks culminated in the 1952 paper "Are There Quantum Jumps?" In this paper he also presented some objections against Born's collision treatment and probabilistic interpretation. But, in essence, Schrödinger repeated his old arguments, enveloped in a good deal of rhetoric. Physical science is at present in danger, he argued, of being severed both from its glorious history and from its cultural milieu. Physicists must concern themselves, not only with ingenious theoretical constructs and their agreement with experimental data, but also with the suitability of their concepts to be absorbed into a general spiritual culture of mankind: "Science is not a soliloquy. It gains value only within its cultural milieu, only by having contact with all those who are now, and who in future will be, engaged in promoting spiritual culture and knowledge" (Schrödinger 1952a, 234).

Not all concepts, according to Schrödinger, are destined to survive: "A theoretical science . . . where the initiated continue musing to each other in terms that are, at best, understood by a small group of close fellow travelers, will necessarily be cut off from the rest of cultural mankind, in the long run it is bound to atrophy and ossify" (1952a, 109–10). The "abbreviated language" of quantum jumps, Schrödinger argued, is not a suitable candidate for long-term survival. There were other "ingenious constructs of the human mind that gave an exceedingly accurate description of observed facts and have yet lost all interest except to historians," for example, the epicycles of planetary motion. The modern analogue of epicycles, Schrödinger prophesied, are the quantum jumps (1952a, 110).

It was Born, not surprisingly, who countered Schrödinger in print. He defended both his collision treatment and his probability interpretation from Schrödinger's attack. "The most important reason," we learn, for Born's entering into a public controversy with one of his "best and oldest friends" was that Schrödinger's publication "may have a confusing effect on the mind of those who, without being physicists, are interested in the general ideas of physics" (Born 1953b, 140).

Born's paper also contains a good deal of rhetoric. Its target is not Schrödinger's central case against the nature of quantum jumps but rather the "weakest, in fact quite indefensible, point in Schrödinger's arguments against the current interpretation of quantum mechanics"— his antiatomistic attitude. While accusing his opponents of "having lost the feeling of historical continuity," Schrödinger himself, argued Born, had committed a crime "that is even worse." Schrödinger's attempt to overthrow the atomistic idea, which has proved "fertile and powerful" throughout the history of physics, is "almost presumptuous, and in any case an obvious violation of historical continuity" (Born 1953b, 142).

Despite the sparring, Schrödinger and Born were each fully aware that there could be no conclusive disproof of the other's position. Thus, Schrödinger conceded, the language of quantum jumps is an extremely convenient shorthand for describing a wide range of phenomena in the atomic domain, chemistry, and thermodynamics. One can freely talk therefore "as if" these jumps occur, as long as one does not confuse this abbreviated language with reality (Schrödinger 1952a, 119). The language of quantum jumps has acquired the status of an observational language of physics, but this does not mean that such imagery should be taken literally. The situation is similar to that of the concept of valency, which is a most useful, even indispensable, shorthand in chemical research. Initially, valency had behaved "as if" it represented a "very real fact of observation." After a quantum mechanical explanation of the valency bond was given, valency continued to be retained, but only as a convenient shorthand. Similarly, in the quantum world, everything happens "as if" (but only as if) jumps are real (1952a, 119–20).

Schrödinger himself preferred to employ different theoretical notions—those that substitute a conceivable, continuous mechanism for the inexplicable and irritating "quantum jerks." Where Schrödinger envisaged the gradual evolution of changing wave configurations, Born witnessed abrupt quantum changes. Was Schrödinger's imagery absurd or obviously wrong? By no means—both Born and Heisenberg, despite their numerous objections, seemed to hold that Schrödinger's was a feasible stand, and no conclusive arguments could be brought against it. In fact, in an actual scientific situation "any one of us theoretical physicists, including Schrödinger, . . . would use the same, or at least equivalent mathematical methods, and if we should obtain concrete results our prediction and our prescription for the experimental verification would be practically the same." The whole controversy is therefore "not so much an internal matter of physics, as one of its relation to philosophy and human knowledge in general" (Born 1953b, 140).

The attitude of these physicists to philosophy is truly ambivalent. In a letter to Born, Schrödinger wrote that interpretive philosophical questions are as important as concrete scientific problems, only to reverse this opinion on the next page: actual scientific research is much more valuable than "philosophical games" (Schrödinger to Born, 16 May 1927, AHQP). Similarly, Born, after admitting that philosophical differences between Schrödinger and him did not matter in actual scientific practice, declared that philosophical questions about the real meaning of our words are just as important as the mathematical formalism (1953b, 141).

When do decisions on philosophical issues become important? From the exchange between Schrödinger and Born it follows that philosophy matters when scientists want to explain and defend their theories in front of wider audiences. Scientists then worry whether their theoretical constructs can be interpreted in such a way that these constructs will gain acceptance in wider cultural circles. Scientists aim at ensuring the longevity of their conceptions, so that their work "still will be of interest other than to historians 2000 years hence" (Schrödinger 1952a, 234). Though fully aware of the hypothetical nature of their theoretical constructs, our scientific heroes (even those with positivist sympathies), when faced with the harshest threat of all—oblivion—fight fiercely for the unobservable reality they have created.

The threat of oblivion can also come from within the scientific community, and it was as a response to such a threat, I have argued, that the reality of quantum jumps was constituted. The concept of quantum jumps not only seems to be a direct expression of experimental facts (the Franck-Hertz experiments) but is a building block of the elaborate theoretical and interpretive framework of quantum mechanics. Yet its theoretical entrenchment is no more a proof of the reality of quantum jumps than the intricate theories of electromagnetism in the nineteenth century are proof of the reality of ether.

Some kind of realism seems to be an indisputable fact of scientific life—scientists need to construct models of reality for heuristic purposes, for a means of communication, and for describing scientific experiments. Without some kind of realism "no communication about facts is conceivable, even between the most sublime minds" (Born 1953c, 153). However, we should not project from the psychosocial to the epistemic level: though scientists must presuppose reality, they need not believe in it. Often it is such conditional acceptance without commitment that creates the freedom necessary for scientific creativity.

In the last analysis, Born admitted that he and Schrödinger differed merely in their conception of "the philosophical attitude of a period, which determines the cultural background." Born consequently concluded that "the auspices of an agreement are therefore frail" (1953b, 150). Born's final argument against Schrödinger was neither scientific nor philosophical, but rather social. It is an argument, Born noted, that Schrödinger himself "is not too proud to use" (1953b, 149)— quotation of authorities. Born chose to quote from a letter from Pauli, "who is generally acknowledged to be the most critical, logically and mathematically exacting amongst the scholars who have contributed to quantum mechanics." In this letter Pauli described Schrödinger's effort as "a dream of a way back to the classical style that seems to me hopeless, off the way, bad taste . . . and not even a lovely dream" (1953b, 150).

Schrödinger must have loathed such arguments, especially those that cited the consensus of physicists against his interpretation. "Since when are scientific questions decided by majority?" he would ask Born (quoted in Born 1961, 87). In his obituary for Schrödinger, Born gave his answer—"At least since the time of Newton" was his final reply (1961, 87).

Indeterminism and Historiographical Doubts

Today physicists and philosophers of quantum physics consider entanglement, rather than acausality, the most distinctive feature of the quantum world. Indeterminism turns out to be less interesting than non-locality: one can have both deterministic and stochastic hidden variable theories, yet all of them must be nonlocal (Home and Whitaker 1992). In the previous chapter I analyzed the rhetorical strategies used by the founders of the Copenhagen interpretation to endow determinism with "finality." In this chapter I have argued that Bohr's historical recollections distort history to the same end. Born, the originator of the probabilistic interpretation, similarly constructed the course of his recollections to lead naturally to the "inevitability" of indeterminism.

That one cannot trust the recollections of participants in events is a historiographical platitude. As Einstein himself put it: "Every reminiscence is colored by today's being what it is, and therefore by a deceptive point of view" (1949a, 3). There are indeed many factual discrepancies and confusions of dates in Born's recollections of past events. Thus Born described his work with Norbert Wiener (Born and Wiener 1926) as having been done after his interpretation of the wave function (interview with Born, AHQP). There are also obvious distortions: in his Nobel lecture Born (1955b) recalled his opposition to Schrödinger's idea of a weight function, while in his original papers Born (1927b) perceived this idea of Schrödinger's as being close to his own. Born's account of his interpretation as having been conceived from the very beginning in opposition to Schrödinger, and as springing from his desire to secure a particle ontology, does not withstand an examination of the historical record (chapter 2). Interestingly, Born himself was aware at times of his lapses of memory.[9] The lapses, of course, are not cases of memory simply failing but of memory *selecting*. It is natural in retrospect to see later developments in earlier ideas, to forget the mistakes, the ambivalence, to suppress the false starts, and to bring to the fore the valid, accepted aspects of the story. Scientists turned historians are natural "rational reconstructionists." And even the authors of discoveries (and perhaps they especially) are "Whiggish" historians. In this way, tentative early attempts are described as if they had already entailed the intellectual preferences that only formed later.

Born's probabilistic interpretation in terms of particles applied initially, if at all, only to free particles. And for free particles (except for the special case of the harmonic oscillator—Schrödinger 1926e) wave packets disperse, as Heisenberg, Born, Lorentz, and even Schrödinger himself understood,[10] and the Born-Heisenberg particle interpretation is the most natural. Yet, as I discussed in chapter 2, the crux of Born's efforts lay elsewhere: in describing quantum mechanically the states and energetic changes of atoms. And for this—the more interesting and more complicated case of the bound system—the matter wave interpretation à la Schrödinger is the more natural, as Bohr emphasized from the beginning (chapter 6), and as both Heisenberg (1958) and Born (1969) later conceded. The reason Born suppressed this fact in his recollections, singling out only the more obvious case of colliding electrons, is that Born's history is an ideological one, emphasizing particlehood in contrast to Schrödinger's wave ontology—a threatening alternative to the Göttingen-Copenhagen version of the quantum world. This ideology was reinforced by the deception that the particle interpretation was there all along, and inevitably so. Together with indeterminism, it formed "a natural interpretation" in Feyerabend's (1975) sense. This is a common technique of using history for propaganda. The Copenhagen spirit was diffused by accommodating history to ideology. Because the opponents (Schrödinger, Einstein) did not write detailed recollections on this topic and because many historical studies take the map drawn

9. "Doch ist meine Erinnerung und die Ereignisse vor 35 Jahren etwas verblasst" (However, my memory of events [that took place] thirty-five years ago is somewhat faded) (Born 1961, 85).

10. Heisenberg argued this point against Schrödinger's wave ontology in his uncertainty paper (Heisenberg 1927b; see chapter 4) and Born did so in a paper published in *Nature* (Born 1927b). For Lorentz's and Schrödinger's discussion of the dispersion of wave packets, see Lorentz to Schrödinger, 27 May 1926, and Schrödinger to Lorentz, 6 June 1926, both in Przibram (1967).

by the Göttingen-Copenhagen physicists as a faithful guide to pivotal developments, one-sided and biased accounts are formed.

Ideological history has its price, of course. By concentrating on the more straightforward interpretation of $|C_n|^2$ as a density of scattered particles, Born omitted the most problematic, yet most valuable part of his contribution (the probabilistic interpretation of stationary states): the core around which pivotal developments would take place (chapter 2).

Born's probabilistic interpretation was, as I have argued, a process in flux, during which Born's intellectual preferences formed gradually, while his enthusiasm for Schrödinger's theory faded (chapter 2). There was heavy pressure from Born's younger colleagues (especially Heisenberg) not to succumb to the seductive classical beauty of Schrödinger's theory. This pressure was not the only and perhaps not even the main reason for Born's about-face on the physical significance of Schrödinger's approach. The realization that wave packets disperse, the insight that position space does not have a real ontological preference, and the brilliant extension of Born's ideas by Pauli, Jordan, and Dirac—all are substantial reasons for Born's change of mind. Still, the direction of the modifications in Born's ideas and Born's later description of the development of his ideas are not fully comprehensible without taking into account the Göttingen-Copenhagen alliance against Schrödinger. Born's militant public stand on the issue of indeterminism after 1927 (despite his private concession to Einstein that this issue is undecidable) also belongs to the ideological part of this story.

Before 1927, neither Born nor the other Göttingen-Copenhagen physicists took any definite stand on the issue of indeterminism. Although they were all open to the possibility of an ultimately probabilistic physics, this openness did not imply preference. The transition to a "committed" indeterminism was conditioned, for Born and Heisenberg, by theoretical possibilities and competitive dialogues over conflicting tastes in interpretation within the scientific community. "Commitment" is perhaps too strong a characterization. Thus at times Heisenberg (1931a, 1931b) adopted a relaxed attitude toward the causality issue, claiming that by an appropriate redefinition of causality, its validity could still be maintained.

CHAPTER 11

The Myth of Wave-Particle Complementarity

*During the summer of '25 . . . I asked him [Pauli] this question: "Was light [a]
wave or a particle?" And I can remember this kind of cynical laugh of his.*

Interview with Hendrik Casimir, 28 January 1964, AHQP

It ain't exactly real, or it is real, but ain't exactly there.

Leonard Cohen, "The Future"

*This idea [of wave-particle synthesis] seems to me so natural and simple,
to resolve the wave-particle dilemma in such a clear and ordinary way,
that it is a great mystery to me that it was so generally ignored.*

J. S. Bell 1987, 191

Introduction: The Dramatic Historical Narrative

The story of the development of quantum mechanics builds dramatically around the wave-particle dilemma. We follow the piling up of
contradictions between the corpuscular and wave aspects of matter and
radiation, one after another, until the resolution of the ultimately unbearable conceptual tension by Bohr's principle of "wave-particle complementarity." This exciting narrative appears in many textbooks (Born
1969), philosophical accounts (Sklar 1992), and personal recollections
(Heisenberg 1958). Thomas Kuhn singled out wave-particle complementarity as a prime example of a revolutionary paradigmatic transformation (Kuhn 1970, 11–12).

The photoelectric effect and the Compton effect, so we often read,
constitute direct proof of the existence of light quanta: these experiments can be accounted for only by assuming the corpuscular nature of
radiation.[1] Similarly, the Ramsauer and Davisson-Germer experiments,
and the theoretical contributions of de Broglie and Schrödinger, point

---

1. "The corpuscular nature of light itself is proved in the most obvious way by the laws
of frequency change in the scattering of X-rays [the Compton effect]" (Born 1964, 87).

unequivocally to the wave nature of matter. These developments, supposedly, caused acute distress, for they "seemed to contradict each
other without any possibility of compromise" (Heisenberg 1958, 37).
The unhappy question was: "How could the same thing be a wave and
a particle?" (Heisenberg 1958, 35). "How could this be understood?"
(Sklar 1992, 164). "How are those contradictory aspects to be reconciled?" (Born 1969, 94).

When we read papers dealing with the wave-particle issue before the
rise of the Copenhagen philosophy, we hardly find feelings of despera-
tion or distress. A patient suspense of final judgment seems to be a
more fitting characterization of the attitude of physicists.[2] To provide
a unified theory of matter and radiation seemed more important to
creative physicists—Einstein, de Broglie, Jordan, Born, Debye, Dirac,
Pauli—than to establish the ontological status of their constructs. This
wise resignation from the pursuit of ultimate answers allowed these
physicists to borrow freely from each other's powerful techniques,
even when their conceptual presuppositions seemed fundamentally
opposed to each other (Darrigol 1986, 198).

Impressive theoretical advances were possible, despite the wave-
particle dilemma, because the interpretation of "crucial experiments"
was not constrained in one particular way, as the received narrative
implies. There were many non-Einsteinian explanations of the photoelectric effect that ingeniously avoided the hypothesis of light quanta
(Stuewer 1970).[3] Similarly, there were alternative explanations of the
Compton effect—a prominent alternative to light quanta was provided
by the Bohr-Kramers-Slater theory, as I have discussed earlier (chapter 2). It was in order to avoid Einstein's concept of light quanta, that
Bohr introduced "further peculiarities" into the space-time description
of the microdomain, while dealing with the Compton effect in terms of
a virtual radiation field. Bohr did not accept even the Bothe-Geiger experiments as crucial evidence for light quanta (chapter 6), though later,
constructing a narrative of wave-particle complementarity, he would
cite the Compton effect as "proving" their existence. Nor did the wave
aspect of radiation unequivocally constrain theorists to wavelike explanations. Papers by Duane, Epstein and Ehrenfest, and Jordan interpreted the diffraction of material particles by crystals purely in terms
of particles (Darrigol 1992b). (In fact, these papers could be interpreted
as providing a physical explanation of how crystal periodicity might
produce the illusion of the wave nature of the electron.)

2. See Wheaton (1983) for a description of the early treatment of wave-particle duality.
3. Compare this with a typical statement: there "is absolutely no possible way of accounting for photoelectric effect . . . except by adopting the idea of the photon" (Tolansky
1968, 40).

The dramatic narrative assumes, and naturally leads to, the thesis that wave and particle concepts are mutually exclusive, and to a resolution of the "paradox" through Bohrian concepts of complementarity. Yet the majority of working physicists did not consider wave and particle characteristics irreconcilable. Even William Henry Bragg, who in anecdotal accounts of history canonized the incompatibility between waves and particles by saying that physicists use particles on Mondays, Wednesdays, and Fridays and waves on Tuesdays, Thursdays, and Saturdays, was not "averse to a reconcilement of a corpuscular and a wave theory." Instead, Bragg said: "I think that someday it [reconcilement] may come" (quoted in Wheaton 1983, 166 n). The work of Einstein and de Broglie was inspired by the goal of finding similar foundations for the structure of matter and of radiation; Born's work was likewise guided by the search for a unified description of matter and radiation. Born conceived his probabilistic interpretation of the ψ-function in analogy with the connection Einstein had proposed between the intensity of the electromagnetic field and the density of light quanta. Born (1926c) said explicitly that he hoped such an approach, if successful, would be a step toward a unified theory of matter and radiation.

While it remained unsolved, the wave-particle "paradox" was patiently put aside in the hopes that eventually, when a consistent atomic mechanics was found, the problem would resolve itself. For Heisenberg, the powerful new transformation theory accomplished this task and allowed a particle interpretation in the spirit of a generalization of Born's statistical interpretation. In Heisenberg's opinion, Bohr, who did not participate in the mathematical development of the new quantum theory and did not experience firsthand its wonderful power, remained trapped in the visualizable language of wave-particle duality.[4] As Heisenberg put it: "These paradoxes were so in the center of his [Bohr's] mind that he just could not imagine that anybody could find an answer to the paradoxes, even having the nicest mathematical scheme in the world" (interview with Heisenberg, 25 February 1963, AHQP).

While rejecting Schrödinger's vision of a pure wave interpretation, Heisenberg was equally reluctant to acknowledge wave-particle duality as unavoidable. The fact that one needed contradictory, wave and particle, concepts for the intuitive interpretation of quantum mechanics, Heisenberg thought, merely indicated that something essential was still missing from the present formulation of quantum theory (Heisenberg 1926b, 993; these words were written before the development of

---

4. "Bohr had not had this experience, so I would say he was not so much impressed by the new developments" (interview with Heisenberg, 25 February 1963, AHQP).

transformation theory). Heisenberg continued to be critical of wave-particle duality even after he, yielding to Bohr's pressure, appended to the uncertainty paper the postscript in Bohr's spirit. "If one begins with a statement 'there are waves and particles,' everything can be made contradictory free" was his disapproving sketch of Bohr's philosophy in a letter to Pauli written in the spring of 1927 (chapter 6).

As I have argued, contrary to the usual historical account that the heated controversy between Bohr and Heisenberg eventually ended in complete agreement due to Pauli's skillful intervention, a genuine unanimity of opinion between the two men never occurred. Heisenberg stated clearly in an interview with Kuhn that he never accepted Bohr's dualistic approach. This is the reason the papers of Jordan and Wigner (1928) on second quantization made him very happy—they demonstrated not that one needs *both* waves *and* particles but that one can do it "either way." "We have," Heisenberg emphasized, "one mathematical scheme that allows many transformations, but . . . just one mathematical scheme. . . . The fact that we can use two kinds of words . . . is just an indication of the inadequacy of words." And when Kuhn asked about the γ-ray thought experiment, in which Bohr—contrary to Heisenberg's first, mistaken account—used both particle and wave features of radiation, Heisenberg insisted: "For explaining the γ-ray experiment, it was useful to play between both pictures. . . . But it was not absolutely essential. You could actually use both languages independently" (interview with Heisenberg, 25 February 1963, AHQP).

Heisenberg's recollections are confirmed by his work at the time, especially his Chicago lectures (Heisenberg 1930). While these lectures were intended to spread the Copenhagen spirit and are often interpreted as endorsing all of the Bohrian doctrines, including wave-particle complementarity, a careful reading finds that Heisenberg's principal disagreement with Bohr persists. What Heisenberg proudly announced was not wave-particle complementarity but rather, contra Bohr, *symmetry*, a "complete equivalence of the corpuscular and wave concepts, which is clearly reflected in the newer formulations of the mathematical theory." While taking a stand at odds with Bohr's, Heisenberg paid lip service to Bohr's philosophy: there is nothing in the book that is not contained "in previous publications, particularly in the investigations of Bohr"; the purpose of the book will be fulfilled if it contributes to the diffusion of *Copenhagener Geist der Quantentheorie* (1930, preface). Such statements contributed to the mistaken conflation of Bohr's and Heisenberg's positions, and to a further proliferation of contradictions in Heisenberg's presentations.

The mathematical scheme of quantum mechanics "seems entirely adequate for the treatment of atomic processes." It is only for visualization, for the mathematically uninitiated, that one has to use "two in-complete analogies—the wave picture and the corpuscular picture" (Heisenberg 1930, 11). These analogies are "accurate only in limiting cases." We use them in a limited way, because we attempt "to describe things for which our language has no words." Wave-particle duality is of our own making: "light and matter are single entities and the appar-ent duality arises in the limitations of our language" (1930, 10). An ade-quate analysis of the limitations of wave and particle concepts "cannot be carried through entirely without using the mathematical apparatus of quantum theory" (1930, 11–12). Nor can the mutual exclusiveness of wave and particle concepts be consistently maintained in thought experiments. In an analysis of the γ-ray thought experiment, light is treated as a wave and a particle *simultaneously*. As Heisenberg put it: "It is characteristic of the foregoing discussion that simultaneous use is made of deductions from the corpuscular and wave theories of light, for on the one hand, we speak of resolving power, and, on the other hand, of photons and their recoils" (1930, 22–23).

The above sentence reveals why Heisenberg was so reluctant to adopt Bohr's "correction" to his discussion in the uncertainty paper. It also explains why Bohr moved swiftly, in his analysis of simple thought experiments, to rigid and movable diaphragms, abandoning the ca-nonical γ-ray thought experiment: this experiment violates consistency as defined by Bohr, who held that lack of contradictions in the use of wave and particle concepts is only ensured by mutual exclusivity of experimental arrangements. We can also understand why mathematical physicists, such as Bopp and Heisenberg himself, maintained that one cannot obtain a contradiction-free interpretation of quantum theory through thought experiments that rely on the concepts of classical phys-ics, such as waves and particles.

Mathematical Physicists and the Wave-Particle Dilemma

Wave-particle complementarity is a "foundational myth" of quantum mechanics (Diner et al. 1984, vii).[5] By calling it a myth, I do not deny that there is a genuine physical (and possibly philosophical) problem connected with the wave-particle duality of matter and radiation. Yet the very construction of what the problem is, and its alleged definitive resolution by complementarity, is tendentious, ignoring much of the fascinating research connected with this issue.

5. The editors of these research papers emphasized, in the spirit of de Broglie and Ein-stein, the "coexistence" (Diner et al. 1984, vii), rather than the "mutual exclusivity," of these concepts.

Philosophically, it can be argued that the orthodox solution is deeply unsatisfactory. To say that microscopic objects exhibit wave and particle properties because they obey a "complementarity between waves and particles" (Bohr) or because they are "wavicles" (Bohr's careless followers) is not much better than to say that sugar is sweet because it has an essential quality of "sweetness." The foundational status of wave-particle complementarity rests on Bohr's controversial, and as I have argued unfounded, doctrine of the necessity of classical concepts. If one gives up the foundational status of the later doctrine, the foundational status of wave-particle complementarity dissolves as well. The wave-particle dilemma consequently becomes a problem for experimental investigation—the way, in fact, it is treated in the recent physics literature (Ghose, Home, and Agarwal 1991, 1992; Ghose and Home 1992; Scully, Englert, and Walther 1991).

If one gives up Bohr's prohibition on deducing features of quantum reality from the characteristics of the mathematical formalism, a rich variety of solutions to the wave-particle dilemma become possible, depending on the local theoretical problem one is faced with. Mathematical physicists, from the very origins of quantum physics, suggested different solutions to the wave-particle dilemma, connected with the characteristics of their mathematical theorizing, though often being careful not to take definitive interpretive stands.[6] Mathematical results teased the imagination of physicists in more ways than one. No wonder Dirac, whose outstanding contributions shaped mathematical elaborations of the wave-particle issue, felt, contra Bohr, that "the final word was not yet said about waves and particles" (1963, 53).

In the context of nonrelativistic quantum mechanics, physicists often relate to Born's corpuscular statistical interpretation, and its extension by transformation theory, as a solution of the wave-particle dilemma: a substratum of particles is guided by wavelike probabilistic lawfulness.[7] Yet this interpretation is incompatible with Bohr's thesis of mutual exclusivity. As Born put it: "Both pictures are necessary for every real quantum phenomenon" (1964, 105).[8] Pauli and Born, when they did not publicly support Bohr, interpreted wave-particle duality through Born's probabilistic interpretation, with the associated mathematical notion of the interference of probabilities. Such a statistical interpretation "prevents contradictions between wave mechanics and the mechanics of point particles" (Pauli 1973, 13). The statistical calculus of

6. De Broglie: "The present theory should be taken as a form whose physical content is not entirely given" (quoted in Darrigol 1986, 209). One finds similar expressions in papers by Born, Einstein, and other physicists.

7. Maddox (1988) calls it a "delicious compromise" between Bohr and de Broglie.

8. Born's public support of Bohr's position resulted in the confusion and inconsistencies that characterize Born's writings on the wave-particle issue.

particles in quantum mechanics is characterized by a "peculiarity" similar to a wave theory, as Schrödinger, Pauli, and Jordan early realized—the so-called interference of probabilities (Jordan 1927a, 647). Thus, in the two-slit experiments, the probability of a particle's passing through the first slit when the second slit is closed is $|\psi_1|^2$, while the probability of its passing through the second slit when the first slit is closed is $|\psi_2|^2$. The probabilistic pattern when both slits are open is not $|\psi_1|^2 + |\psi_2|^2$, as it would be in the case of classical particles, but $|\psi_1 + \psi_2|^2$—an expression that, when expanded, contains a supplementary "interference" term in addition to $|\psi_1|^2$ and $|\psi_2|^2$.

This situation is completely analogous to an explanation of Thomas Young's two-slit experiment when the difference of phases adds up in a "constructive" and "destructive" manner to produce a typical interference pattern.[9] As it is in optics, it is not the probabilities that sum up, but rather the amplitudes—in a general formulation the particle and wave properties apply to every quantum mechanical process. One can, if one likes, maintain that both wave and particle attributes are in some sense essential to quantum mechanical description. One can, however, consider the particle substratum more basic, maintaining that the interference of probabilities is merely a formal analogue of a wave description, expressing the nonclassical nature of the essentially corpuscular quantum world (Landé's position—see the discussion later in this chapter).

Proponents of the Copenhagen interpretation vacillate in their writings between these two options. Such vacillation contributes to conceptual confusion about the wave-particle issue. The particle-kinematic probabilistic interpretation, which is based on the presupposition that it is meaningful to talk about the probability of a perfectly sharp position at a specific time, is not compatible with the literal applicability of de Broglie–Schrödinger matter wave packets in three dimensions, as is often discussed in the orthodox writings. As Born himself pointed out, a particle represented by a wave packet "cannot have any sharply defined position, since it is only a group of waves with vague limits" (1927b, 9). A more consistent stand would be to deny the "reality" of matter waves in the three-dimensional case as well (the Göttingen-Copenhagen physicists denied the reality of Schrödinger's waves in the 3N-dimensional case), confining all discussions to a probabilistic

9. The expression $|\phi_n(q_0)|^2 dq$ is the probability that a coordinate q lies in the interval $(q_0, q_0 + dq)$ when the atom is in the nth stationary state. In general, let us denote by $|\phi(\beta_0, q_0)|^2 dq$ the probability that q lies in $(q_0, q_0 + dq)$ when another mechanical variable has the value β. Similarly, the probability that Q takes a value in $(Q_0, Q_0 + dQ)$ while q has the value q_0 is given by $|\psi(q_0, Q_0)|^2 dq$. Then the probability that Q lies in the interval $(Q_0, Q_0 + dQ)$ while $\beta = \beta_0$ is given not, as in the classical case, by $dQ \cdot \int \phi |(\beta_0, q)|^2 \cdot |\psi(q, Q_0)|^2 dq$ but rather by $dQ \cdot |\Phi(\beta_0, Q_0)|^2$, where $\Phi(\beta_0, Q_0) = \int \phi(\beta_0, q)\psi(q, Q_0) dq$ (Jordan 1927a, 647).

treatment in terms of particles. Yet the appeal of wave imagery is too powerful and its utility for "pedagogical" purposes is too appealing, so even Heisenberg and Pauli often preferred to use the matter wave presentation of individual particles. (See the discussion later in this chapter.)

As opposed to the partial analogy between waves and particles contained in the idea of a wave packet, the Born-Pauli-Jordan idea of the interference of probabilities allowed an overarching analogy between matter and radiation. The fact that the interference of probabilities in Schrödinger's formulation corresponds exactly to the rules of multiplication for matrices with infinitely many rows and columns confirmed the great generality of this new probabilistic framework of quantum mechanics (Jordan 1927a, 647). All existing versions of quantum mechanics—Heisenberg's matrix mechanics, Schrödinger's wave mechanics, Born-Wiener's operator calculus, and Dirac's c-number calculus—were united in the single, powerful formulation of probabilistic transformation theory. That the theory was probabilistic did not seem too high a price compared with its assets of unity, generality, and problem-solving power. The unity and generality of a comprehensive probabilistic quantum theory was the most powerful argument for the indeterministic character of the quantum world and for the duality expressed in the interference of probabilities. For Heisenberg, all his previous intuitive discussions with Bohr about the duality of waves and particles were at once superseded by the unity and beauty of the newborn mathematics.

Schrödinger thought otherwise. Since 1925, when he became familiar with de Broglie's and Satyendra Nath Bose's ideas, he had realized that Bose-Einstein counting undermines the idea of an individual distinguishable particle. Particles in a gas had to be replaced by "energy excitation states" (Schrödinger 1926g), and the corpuscular picture of a gas had to be replaced by de Broglie waves. This insight spurred Schrödinger's theoretical efforts, culminating in the crowning achievement of the creation of wave mechanics (Wessels 1979). For the rest of his life Schrödinger, considering the concept of an individual particle bankrupt, advocated an underlying wave ontology, rather than the synthesis, or "coexistence," of waves and particles in the spirit of de Broglie and Einstein (Bitbol 1996; Beller 1997a). Understandably, Schrödinger did not embrace the corpuscular statistical ontology offered by Heisenberg and Born.

Heisenberg also realized very early that the Bose-Einstein statistics undermines the idea of an individual particle and used this conclusion initially to argue against the possibility of intuitiveness, or *Anschaulichkeit*, in the microdomain (Heisenberg 1926b). He later minimized the implications of the Bose-Einstein statistics in his interpretive writings when confronted with Schrödinger's competing, wave theoretical endeavors.

In 1925–27, that period of "human confusion," support for Schrödinger's position came from unexpected quarters. Though he attacked
Schrödinger publicly for deviating from, and challenging, Bohrian
guidelines (especially on the issue of quantum jumps), Jordan's theoretical efforts were much closer in spirit to Schrödinger's than to Bohr's
ideas.[10] As noted by several authors, matrix theorists were familiar
with, and even actively participated in, an elaboration of the wave theoretical point of view (Beller 1983; Darrigol 1986; Kojevnikov 1987; see
also chapter 2). Jordan, a founder of second quantization, is the most
prominent case. Following de Broglie and Einstein, Jordan made many
important advances using the assumption of a complete analogy between matter and radiation as a heuristic principle. In Jordan's hands,
the fruits of the matter wave analogy included a theoretical proof of
electron diffraction—before the experiments by Davisson and Germer
that demonstrated the diffraction of an electron beam—as well as
Jordan's early conjecture that matter could be created and destroyed
(Darrigol 1986, 219).

As early as the Born-Heisenberg-Jordan paper, Jordan had obtained
light quanta by quantizing the Maxwell fields (this part of the paper
was written by Jordan alone). As soon as Jordan learned about Schrödinger waves, he contemplated the possibility that matter as well could
be represented by quantized ψ-functions. His presumption of an analogy between matter and radiation led naturally to the idea that classical
individual particles might be replaced by wave excitations in the case
of matter as well. Jordan had very little encouragement from Heisenberg; Born, initially sympathetic to the idea, soon withdrew his support
(Darrigol 1986, 219).[11]

Jordan's quantized matter waves in three-dimensional space reinforced Schrödinger's wave theoretical position, by removing the multidimensionality of the ψ-function as an argument against Schrödinger's
wave ontology. There was a meaningful similarity between Jordan's
matter waves and the quantized waves Schrödinger used to explain
Bose's statistics before his discovery of wave mechanics. If for Heisenberg second quantization implied the equivalence, or mutual translat-
ability, of the wave and particle models, for Jordan the wave theoretical
substratum was the more basic of the two. Like Schrödinger, Jordan
was willing to give up the primacy of atomistic ideas. In this theoretical context Jordan held that the discontinuity of both matter and light
can be derived from wave quantization: "The existence of atoms is
no longer a primary basic fact of nature; it is only a special part of a

10. Jordan's theoretical work, and his crucial role in the history of quantized matter
waves, is analyzed extensively in Darrigol (1986).

11. Born may have withdrawn his support because of his change of heart regarding the
physical significance of Schrödinger's wave ontology (chapter 2).

much more general and comprehensive phenomenon—the phenomenon of quantum discontinuities (1944, 144–46, quoted in Darrigol 1986). Schrödinger, as we know, did not perceive second quantization as implying equivalence between wave and particle pictures in the spirit of Heisenberg, for by this procedure one "cannot avoid leaving indeterminate the number of particles dealt with. It is thus obvious that they are not individuals" (1950, 112).

The fascinating history of wave-particle concepts cannot be fully understood without introducing the history of quantum electrodynamics (QED), which is beyond the scope of this book. Let me simply mention two important points. First, different scientists deduced different philosophical conclusions from the same, or very similar, mathematical procedures. While not discussing these issues explicitly, their experience and knowledge of QED informed their popular writings, which were confined, as it were, to the nonrelativistic theory of quantum mechanics. This state of affairs produced contradictions, most prominently in Heisenberg's case. The second point is that even for the same author, the relative primacy given to particle versus wave concepts changed from one theoretical context to another. Thus, in their early comprehensive program for QED, Pauli and Heisenberg (1929) assumed the "symmetry," or equivalence, of wave and particle schemes. Later, disappointed by the difficulties they encountered, Pauli preferred to emphasize the dissimilarity between matter waves and light waves (matter waves are not directly observable—ψ is a symbolic quantity). Similarly, Dirac, whose efforts were initially guided by a preference for particle concepts, in his later years gave much more support to the symmetrical interpretation of second quantization (Darrigol 1986).

Important advances using second quantization were possible without a resolution of the wave-particle dilemma: Fermi and Fock, for example, used formal theoretical arguments in their work (a charge-symmetry argument, an argument of nonconservation of particle number). In conclusion, no paradigmatic consensus was achieved, and indeed none was needed, to ensure outstanding theoretical progress (Darrigol 1986; Schweber 1994). No unchanging philosophical meaning can be attached to such creative theoretical efforts, even if these efforts can be reconstructed as being directed toward the same foundational issue—the wave-particle dilemma.

Ambiguity and the Wave-Particle Issue

The proliferation of opinions, or perspectives, on the wave-particle issue was connected with the ambiguity of the designating terms. There was no agreement on the necessary and sufficient attributes of a particle. Einstein and Schrödinger argued that it is unreasonable to give up the joint applicability of position and velocity variables as belonging to the very definition of a particle. For Schrödinger, the indistinguishability of particles, implied by the new quantum statistics, signified the total bankruptcy of the concept of a particle, and the continuing use of particle concepts offended his theoretical sensibilities. What Schrödinger found so objectionable in the idea of a particle was that "it constantly drives our mind to ask for information which has obviously no significance. . . . An adequate picture must not trouble us with this disquieting urge; it must be incapable of picturing more than there is." According to Schrödinger, the problem with the concept of a particle was that it "exhibits features which are alien to the real practice" (1950, 111). In the microdomain, the only "tolerable image" of even an isolated particle is "the guiding wave group" (1950, 115). For Bohr, this was an indication of the limitation of classical concepts, not an indication that they should be given up. Feynman's verdict was crisper: "It [the electron] is like *neither*" (Feynman, Leighton, and Sands 1969, 37-1).

Difficulties in the discussion of the wave-particle issue were further aggravated by disagreement about which (wave or particle) attributes are "essential" and which are merely artifacts of interaction.[12] Here ambiguity was connected with a fruitful theoretical freedom, exploited by physicists from the early days of quantum theory until today. We have already mentioned that Duane, Jordan, and other physicists did not feel compelled to consider the diffraction of light (or matter) to be an indication of its wave nature—the diffraction was merely an artifact of the quantized structure of the grating used in the experiment. Similarly, theorists who wished to avoid ascribing discontinuities to radiation (Debye and Ehrenfest—see Darrigol 1986) located all the discontinuities in the interacting matter.

Even localized detection of light quanta on a photographic plate, though generally held to be an unequivocal sign of their particlehood, need not be considered so—localized detection events "can be regarded as originating from the quantized energy levels of the atomic constituents of the detector" (Ghose and Home 1992, 1438).[13] What some quantum theorists consider today a more adequate indication of single particle-like behavior are "single photon states" of light.[14] For an ideal

12. From a verificationist point of view, this distinction is meaningless, yet theoretically it is difficult to abstain from discussing abstract cases. Thus Bohr often talked of the propagation of radiation in free space, which, in his opinion, must be pictured by a wave theoretical model.

13. "One observes discrete events giving rise to an interference pattern even with strongly attenuated sources of classical light" (Ghose and Home 1992, 1438).

14. Such "single photon states" are Fock-space states that are eigenstates of the "photon number operator," which corresponds to the eigenvalue unity (Ghose and Home 1992, 1436).

single photon state, the probability of joint detection of more than one photon is zero—single photon states imply the notion of single particle-like behavior.[15] With this definition of particlehood, Ghose and Home argue, Bohr's thesis of mutual exclusivity is contradicted: one can have experiments in which single photon states exhibit a tunneling effect (tunneling is considered exclusively a wave phenomenon). Other physicists argue the applicability of Bohr's thesis of mutual exclusivity by considering interference, rather than tunneling, as the hallmark of wavelike behavior (Scully, Englert, and Walther 1991).[16] Clearly, the lively discussion continues. The thesis of the mutual exclusivity of waves and particles is found applicable to some situations, yet inadequate in others, depending on the definitions of the terms used.

Theoretical preferences relating to the wave-particle issue are closely connected with intellectual temperament and personal research experience. Matters of taste often dictate what seems fundamental and what derivative. De Broglie, considering quantization a mystery, introduced wave-particle duality as a way to deduce and elucidate Bohr's quantized energy levels in an atom. It was a seductive project: even Bohr, despite his claim of further "irreducibility" of the "atomic postulate," initially praised the de Broglie–Schrödinger approach for its ability to decode quantization (chapter 6). Others, such as Duane, took quantization as given and constructed their views about the wave-particle issue accordingly.

The ambiguity of wave and particle terminology resulted not only in theoretical freedom but in an abundance of ad hoc moves and inconsistencies. Born's case is typical. Discussing the Bohrian doctrine of complementarity, Born defended the thesis of the mutual exclusivity of waves and particles by an ad hoc extension of the definition of a particle (a discussion strangely at odds with Born's own probabilistic interpretation of quantum mechanics). Claiming that it is misleading to assert that at the detector in the two-slit experiment (in Born's case, a photoelectric cell), the corpuscular nature of light is revealed simultaneously with the wave aspects of the interference pattern, Born proposed the following definition of a particle: "To speak of a particle means nothing unless at least two points of its path are specified experimentally" (1969, 101).

15. Single photon states are unique in this sense: for all other states (classical or nonclassical), "the probability of a double detection is different from zero even when the average number of photons . . . is less than unity" (Ghose and Home 1992, 1436).

16. These authors argue that in interference-type experiments (Bohrian two-slit experiments, for example), the quantum formalism implies the disappearance of the interference pattern whenever one tries to obtain "which path" (particle) information, thus supporting the mutual exclusivity of waves and particles.

It is not clear why two points are necessary for the definition of a particle, unless one intends with these two points to specify the particle's *path*—something forbidden by the Copenhagen version of quantum theory. Born, in fact, inconsistently talks about particle paths: "If we propose to carry out the 'demonstration of a corpuscle,' we must settle the question *whether its path* has gone through the upper or the lower of the two slits of the receiver" (1969, 101, my italics); or similarly: "If pure interference is to be observed, we are necessarily precluded from making an observation of any *point of the path* of the light quantum before it strikes the screen" (1969, 102, my italics).[17]

The frequent use of the notion of the "path" of a particle, or a particle "trajectory" (Pauli 1973, 14), in the discussion of such thought experiments—despite the Copenhagen claim that continuous space-time motion is incompatible with the uncertainty relations—demonstrates that such visualizable interpretations cannot be carried through without inconsistencies, just as Heisenberg and Bopp asserted. And with this realization, wave-particle complementarity, as Bohr conceived it, loses much of its power and appeal.

Bohr had his own definitions of wave and particle attributes. He never accepted a complete analogy, or basic symmetry, between matter and radiation, maintaining all his life a peculiarly unshakable commitment to classical wave theory and somewhat resenting assertions of the reality of light quanta. Thus Bohr claimed that "there can be no question of replacing the wave picture of light" (1933, 5) and that "interference patterns offer so thorough a test of the wave picture . . . that this picture cannot be considered as hypothesis" (1933, 4). When one deals with the interference picture, "light quanta with such definite entities of energy is something which of course does not come into the picture at all. It is just that wherever light energy is released into material, then it is in the quantity $h\nu$" (1937d [lecture 3], 269; see also Stachel 1986, on Bohr's attitude toward light quanta, or photons).

According to Bohr, "The tangible content of the idea of light quanta is limited, rather, to . . . conservation of energy and momentum" (1929a, 113).[18] This idea that the particle nature of radiation is essential only in

17. We see here how arbitrary is the condition of having "two points," for Born talks of "any point of the path." If all one needs is *two* points, this condition is satisfied by the definite position of a light emitter, or electron gun, and of a slit, unless one assumes a strange transmutability in the *same* experimental arrangement of electrons or photons, which left the emitter as waves and passed through the slits as particles.

18. Born used this definition while discussing the Compton effect (rather than the previous case of a two-slit experiment): "The corpuscular description means at bottom that we carry out the measurements with the object of getting exact information about momentum and energy relations" (1969, 97).

the interaction of radiation and matter was at the core of Bohr's idea of the mutual exclusivity of wave and particle pictures: "It is not possible in a single picture to account for the various properties of radiation. For certain properties the idea of wave propagation is quite essential. For the study of the energy and momentum transferred in the individual processes, the photon idea, close to the particle concept, is equally essential" (1957b [lecture 3], 609).

As I have previously argued, Bohr did not assign a realistic significance either to the wave picture of matter or to the particle picture of radiation (a point also made by Murdoch 1987, 78). These pictures, as opposed to realistic models of waves and particles in the classical realm, are needed for "visualization," for adapting oneself intuitively to a nonintuitive quantum world. This being the case, it is not clear why, and in what sense, they are "equally necessary" (Murdoch 1987, 78). Nor is it clear, one might add, why it is important that they be mutually exclusive: why should imprecise and limited analogies be consistent with each other?

Bohr's idea of the mutual exclusiveness of waves and particles had its roots in the Como lecture. What he considered contradictory, and therefore mutually exclusive initially, were the ideas of a single, infinitely extended harmonic wave and a precisely localized, free material particle (chapter 6). Yet their contradictory natures did not bother Bohr—they were "abstractions," unrealizable in physical experimental situations. The contradiction between these two ideas, Bohr argued, is resolved by the de Broglie–Schrödinger wave packet.

As Bohr's emphasis on the usefulness of wave imagery was replaced gradually by an operational emphasis on measurement, his method of avoiding contradictions changed: moving away from the notion of a wave packet, he turned to discussions of the mutual exclusivity of experimental arrangements for disclosing wave and particle properties, respectively. Bohr adopted wave-particle complementarity by analogy with kinematic-dynamic (space-time and causality) complementarity. In the case of kinematic-dynamic complementarity, which followed in a straightforward way from Heisenberg's uncertainty relations, the mutual exclusivity of the respective experimental arrangements was demanded for the consistency of the quantum mechanical scheme.

This misleading analogy between wave-particle and kinematic-dynamic complementarity resulted in the widespread misconception that there must be a precise connection between the two. Not surprisingly, different commentators interpreted this connection in different terms—some correlated energy and momentum with particles, and space-time with waves (Born 1969, 97), while others identified particles with space-time. Yet the two kinds of complementarity are not reducible to each other, as an analysis of the Como lecture shows. In the Como lecture, as I have argued, "complementarity between space-time and causality" is an umbrella concept covering several basically different physical situations. In elucidating the applicability of conservation laws in the Bothe-Geiger experiment, Bohr applied causality to particles, and space-time notions to waves. In discussing the limits of space-time models for an atom in a definite stationary state, he correlated space-time with particle models, and causality with the notion of a wave (a definite stationary state is represented by a constant value of energy and by a single harmonic wave). The frequent assumption that one kind of complementarity reduces to the other is unfounded and leads to confusion.

Ideological and Pedagogical Uses of Wave-Particle Complementarity

If Bohr's wave-particle complementarity is neither unambiguous nor necessary in theoretical research, what is its function? To what uses is it put and what aims does it serve?

Orthodox visualizable discussions of wave-particle complementarity have clear ideological and pedagogical objectives. This is the main reason quantum theorists, most notably Heisenberg and Born, sometimes supported Bohr's discussions of wave-particle complementarity in public. Their support for an accessible, Bohrian interpretation became especially strong when they targeted their expositions toward wider audiences and not only the mathematically initiated: "The existence of this mathematical theory [quantum mechanics] shows that the whole structure is logically coherent. But this proof is rather indirect and convincing only for those who understand the mathematical formalism. It is therefore an urgent task to show directly for a number of important cases why, in spite of the use of two such different pictures as particles and waves, a contradiction can never arise" (Born 1936, 47).

The status of wave-particle complementarity is closely tied to the Bohrian doctrine of the necessity of classical concepts. If one gives up this controversial, if not unfounded, doctrine, the basic role of wave-particle complementarity fades away. The necessity of classical concepts, the overthrow of the concept of reality (the impossibility of a quantum ontology), and the finality of indeterminism—all are tied to wave-particle complementarity in the overall rhetoric of the inevitability of the Copenhagen orthodoxy. Not surprisingly, those physicists who opted for some version of realism—Bohm, Schrödinger, Einstein, Landé—rejected Bohrian complementarity. Wave-particle duality is also extensively used in the rhetoric of the inevitability of indeterminism. Born's assertion is typical: "It is clear that the dualism, wave-corpuscle, and the indeterminateness essentially involved therein, *compels us* to abandon any attempt to set up a deterministic theory" (1969, 102, my italics). The philosopher Norwood Hanson (1958, 1959) followed the orthodox guidelines in arguing that wave-particle complementarity necessarily excludes a deterministic alternative to the Copenhagen interpretation (for a full discussion, see chapter 14). Wheaton opened his valuable historical description of the wave-particle issue in the early days of quantum theory by emphasizing its importance to the "overthrow" of determinism: "The first years of this century witnessed the final rejection of determinism in physical theory; there is no more compelling example of this than the synthesis forged in the early 1920s between theories of matter and theories of light" (1983, 3).

The most extensive use of wave-particle duality was the simple derivation and legitimation of the uncertainty relations.[19] The easiest, most accessible way to demonstrate the uncertainty relations is by identifying a particle with a limited wave field, the argument Bohr presented in the Como lecture, and consequently the way many textbooks introduce the uncertainty formula. Heisenberg, who considered the uncertainty relations fundamental to arguments for the self-consistency of quantum theory, was willing to use Bohr's accessible dualistic explanation of the uncertainty relations for wider audiences and for pedagogical purposes. As he later revealed, he used Bohr's complementarity because "it did not do any harm" to his own explanation, yet he did not believe "it was necessary" (interview with Heisenberg, AHQP).

The reason wave-particle complementarity is not necessary for an elucidation of the uncertainty relations is that Heisenberg's formula follows from the mathematical formalism of Dirac-Jordan transformation theory, while wave-particle complementarity does not. As Jammer put it: "Complementarity is an extraneous interpretive addition to it [to the formalism]" (1974, 60–61). Moreover, wave-particle complementarity is always discussed for a very limited, physically uninteresting case—that of a free particle. Bohr elaborated the de Broglie–Schrödinger idea of a wave packet and the Planck–Einstein–de Broglie relations ($E = h\nu$, $p = hk$), which Schrödinger had used to argue the inapplicability of particle concepts in the microdomain, into an argument for wave-particle complementarity, uncertainty, and indeterminism.[20] In fact, the

19. Conversely, the use of wave and particle visualization means the uncertainty formula must be taken into consideration: One can apply the wave or particle picture by taking into consideration limitations imposed by the uncertainty relation (Heisenberg 1958, 43).

20. Initially, Bohr used the Planck-Einstein formula $E = h\nu$ to argue the inapplicability of particle concepts—see his Bohr-Kramers-Slater theory, in which, due to the "irrationality" of this formula, Bohr was eager to avoid the corpuscular concept of light quanta at any price.

wave-particle duality expressed in the idea of a wave packet is not compatible with Bohr's later elaborations of complementarity, which emphasize the mutual exclusiveness of wave and particle attributes. The continuing confusion between wave-particle duality and wave-particle complementarity gives the impression that most physicists accept wave-particle complementarity, while they only endorse wave-particle duality. Born and other physicists sometimes followed Bohr in his rhetoric of the inevitability of wave-particle duality: "The lasting result of Bohr's endeavors is the simple consideration given above [analysis of the wave packet] which shows with irrefutable logic that the Planck–de Broglie laws of necessity imply the duality [of] particles-waves" (1953a, 128).

As already noted, the Göttingen-Copenhagen physicists later ascribed a realistic wave meaning to a single particle (a three-dimensional wave packet) but continued to treat the ψ-function for higher dimensions as an abstract, purely mathematical concept. This artificial divide between three and more dimensions has important implications. If wave-particle complementarity applies only to free particles, the concept is too limited to be of general significance and wave-particle complementarity becomes marginal, if not superfluous. If, for reasons of homogeneity and consistency, one applies the abstract probabilistic particle interpretation to a single particle as well, then the wave packet in the three-dimensional case is as "abstract" as in the multidimensional case and simply signifies the probability of a particle's having a certain position value.

If one extends the basic particle substratum, guided by probabilistic lawfulness, to the case of a single particle, the whole argument for wave-particle complementarity dissolves. We do not have in this case reality of matter waves even for a single particle, viewing it instead as the three-dimensional case of the same probability calculus that is used in higher dimensions. The last stand is implied in the following remark by Feynman: "Although one may be tempted to think in terms of 'particle-waves' when dealing with one particle, it is not a good idea. . . . For if there are two particles, the amplitude [probability] to find one at $\vec{r}_1$ and the other at $\vec{r}_2$ is not a simple wave in 3-dimensional space, but depends on *six* space variables $\vec{r}_1$ and $\vec{r}_2$" (Feynman, Leighton, and Sands 1969, 3:3–4). This point of view is indeed consistent with Born, Heisenberg, Pauli, and Jordan's particle-kinematic probabilistic interpretation. Yet, following Bohr, the orthodox physicists often invoked the three-dimensional wave packet as more than a formal analogy. Pauli, in popular presentations, used the idea of a wave packet not merely as a visualizable illustration but as a physical explanation for the uncertainty relations: the uncertainty relations are valid because in wave imagery no packets exist that contradict the relation $\Delta x \Delta (1/\lambda) \sim 1$

(the relation between the spatial limitation of a group of superposed waves and wavelength λ; Pauli 1950).

If wave-particle complementarity is an arbitrary addition to quantum theory, we can easily understand why it generated so much sterile verbiage.[21] Born again appears on the scene with a rhetorical sword in hand (see my previous discussion of Born's confrontation with Schrödinger on quantum jumps), this time instigated by Landé's (1965) criticism of wave-particle complementarity. Landé argued that "duality" might sometimes be "helpful for heuristic reasons" but that it is totally "unphysical." He stressed eloquently and repeatedly that the "crucial" evidence for matter waves, supposedly impossible to explain otherwise—diffraction phenomena—is adequately explained by Duane's "third rule" of the quantization of linear momentum for a diffracting lattice, which is a body periodic in certain space directions and which therefore can change its momentum component only by discrete amounts. This condition yields a set of discrete angles of electron deflection that determine and explain the diffraction pattern without the assumption of de Broglie waves. Similar considerations can, according to Duane, explain all of the other "mysteries" that allegedly can only be accounted for by the "fiction" of the dual nature of matter (Landé 1965, 123).

Landé (1965) also did not accept Heisenberg's conciliatory formulation of the "symmetry," rather than the "duality," of waves and particles, based on second quantization: a photon whose role is merely that of a quantum number attached to the periodic components of the continuous Maxwell field hardly deserves to be regarded as a particle. Similarly, the fact that atomic probabilities for the distribution of particles obey a rule of wavelike interference rather than of simple (classical) addition is not a reason to conclude that particles have a "wave nature," but rather a stimulus to search for a simple and logical foundation from which such interference (or equivalent rules for matrix multiplication) for probability amplitudes can be deduced.[22] Such a simple and logical foundation, from which Landé succeeded in deducing the interference of probabilities, contains the non-quantum postulates of symmetry, order, and coherence (Landé 1955).

Landé's derivation of the interference of probabilities was largely ignored by representatives of the Göttingen-Copenhagen alliance, but his attack on the Copenhagen credo was countered at once. Rosenfeld dismissed Landé's work as "making a muddle of a perfectly clear situ-

21. In contrast to the idea of a wave-particle synthesis, which, I argued earlier, served as a fruitful heuristic principle.

22. The basic innovation in the traditional view in physics and philosophy, according to Landé, was not "wave-particle duality" or "reality" but the irreducible statistical lawfulness in the quantum domain.

ation" (1956, 133). Born and Walter Biem (1968) published a highly critical rebuttal of Landé's views, using the characteristic rhetorical tactics of discrediting the opponent and appealing to authority. Landé "does not realize the historical origin of the dualistic interpretation" and does not "correctly describe its physical meaning." He is driven by "prejudice" and "dogma," and he "ignores important physical discoveries" (1968, 51). Dualism is a "discovery, and not an invention" (1968, 54) of Einstein himself, made when Einstein was young and drew "irrefutable inferences" (1968, 51), as opposed to Einstein's deterioration into speculation in later years. While every physicist must accept Duane's rule, which "describes correctly all experiments of momentum exchange" on periodic structures (1968, 51), the rule is "obscure" without de Broglie's idea of the connection between particles and matter waves. Besides, the dualism Landé himself introduces ("particulate nature of matter and wave nature of light") is even more unsatisfactory. And while Landé's derivation of the quantum mechanical probabilistic rules "can be interesting in itself," there is no need, Born and Biem reprimanded, to accompany it "by attacks on supposed enemies" (1968, 55).

Landé's (1968) answer was a fitting rhetorical counterattack: wave-particle duality "received its death-knell" from none other than Born himself, through his "admirable" statistical interpretation of the ψ-function in term of particles; to present Einstein "as a champion of dualism is utterly unhistorical"; Landé's own views are no longer those "of a lonely Don," but are shared "by many prominent physicists and philosophers of science" (1968, 56). In the end, Born and Biem had no choice but to retreat: "Clearly, it is possible to formulate a quantum mechanics of particles avoiding all wave-like terms." Still, they asked, "Why the effort?" (1968, 56).

〜

CHAPTER 12

Complementarity as Metaphor

Perhaps our ability to convince others depends on the intensity with which we can persuade ourselves of the force of our own imagination.

Words attributed to Niels Bohr by Heisenberg 1971, 131

We have at the end only to take recourse to painting with words just like one paints, as [an] artist paints with colors just trying to use them in such a way [as] to be able to give to one another just an impression of certain connections of certain harmonies.

Niels Bohr 1937d, 354

Bohr felt that whenever one came with a definite statement about anything, one was betraying complementarity.

Interview with Leon Rosenfeld, 22 July 1963, AHQP

Introduction

Bohr was an avid storyteller. A whole generation of physicists was raised on Bohr's stories. As Pais reminiscences: "Sooner or later, for the purpose of illustrating some point . . . Bohr would tell one or more stories" (1991, 6). The most inspiring, never-ending story, which Bohr never tired of repeating, was about the "great interconnections" in science and life as revealed by complementarity. But what is complementarity? It is not a principle, as both Bohr and Heisenberg often stressed. Even less is complementarity a model. A model of quantum theory should contain something essentially quantum. Complementarity does not. Complementarity is a metaphor, powerful enough to cut across many domains, inspiring enough to construct a new sensibility.

Complementarity is not a rigorous guide to the heart of the quantum mystery. Nor do Bohr's numerous analogies between quantum physics and other domains, such as psychology or biology, withstand close scrutiny. Complementarity does not reveal preexisting similarities; it generates them.[1] Complementarity builds new worlds by making new sets of associations. These worlds are spiritual and poetic, not physical. Complementarity did not result in any new physical discovery— "it is merely a way to talk about the discoveries that have already been made" (interview with Dirac, AHQP).

Only great poets and great prophets succeed in imposing their private associations on the whole culture. Einstein considered Bohr "a prophet," and Freud the author of a "huge mythology." Einstein had little sympathy for either. Neither Freud nor Bohr was intimidated by Einstein's criticism. In a shrewd reply to Einstein, Freud was anything but apologetic. Freud had no intention of demarcating psychoanalysis from science: "But does not every natural science lead intimately to this—a sort of mythology? Is it otherwise today with your physical science?" (quoted in Erikson 1982, 168).

As is suitable for a prophet, Bohr talked in fables and parables. Offending some sensibilities, Heilbron compared Bohr to a "guru" (Heilbron 1987, 221). As gurus do, in his later years Bohr inspired by personal contact.[2] As Otto Frish recalled, "after dinner, we sat close to him—some of us literally at his feet, on the floor—so as not to miss a word" (French and Kennedy 1985, 353). Bohr had the exclusive authority to reveal the "hidden harmonies" in nature. When one did not understand Bohr, the reaction was not inquiring criticism—it was a feeling of awe for the "deep and subtle" philosophy of Bohr.[3] Bohr expected the new framework of complementarity he had built to be taught to children in schools, just as the heliocentric theory is taught (interview with Bohr, AHQP). Even Bohr's close collaborator and co-author of complementarity, Pauli, thought the "imperialism of complementarity" was going too far. Objectivity, or at least the illusion thereof, demands some intersubjective agreement. No metaphor can form a basis for "unambiguous communication," to use Bohr's expression. Ambiguity is a necessary part of a metaphor's suggestiveness. Strip complementarity of its imaginative, imprecise associations and not much

---

1. Bohr's disciple Kalckar wrote: "The connection between Bohr's work and the whole of his personality is so close that one can almost speak of an identity. The turn he gave to the trend of modern physics and through which it received its far reaching epistemological consequences, arose so directly from and harmonized in such a rare degree in his own mind, that one dares to use of him the phrase which one would otherwise reserve for the great artists: that he created a world from within" (1967, 229).

2. "Only those who knew him personally could experience the immense inspiration exuding from his intuitive grasp of physics and his humane personality" (Pais 1991, 29).

3. Using the expression "deep and subtle" has become almost obligatory when writing about Bohr.

will remain except a fancy formulation of the uncertainty principle. Bohr certainly intended complementarity to mean more than that.

The ambiguity, opaqueness, and obscurity of Bohr's writings is legendary. Bohr himself, despite the heavy rhetoric of "unambiguity of communication" that underlay the inevitability of complementarity, considered obscurity a virtue.[4] Bohr's ambiguity was enveloped, by Bohr and others, in a fog of profundity. "Deep truth" cannot be adequately expressed. It is almost as hard to write about complementarity as "about religion . . . being almost as long as life itself" (quoted in Blaedel 1988, 27). Presenting complementarity to a French audience, Bohr announced that "the situation is very difficult to express in words of any language" (notes for talk given at the Institut d'Henri Poincaré, 18 January 1937, AHQP).

Bohr extended complementarity to psychology, biology, and the theory of culture. In this he followed his teacher Harald Høffding, who perceived many analogies between physics and psychology (Faye 1991). Bohr elaborated Høffding's ideas, construing analogies between "disturbance" in introspection and in physical interaction. Initially, this was no more than a hunch—Bohr spoke hesitantly of "suggestive analogy" (1929b, 100), or of "more or less fitting analogies" (101). In a letter to Høffding, Bohr revealed that he had "the *vague idea* that there might be a possibility of proving a similar complementary relation between those aspects of the description of the individual psychological processes which relate to the emotions and those which relate to the will as that which quantum theory has shown to obtain, with respect to elementary processes in physics" (Bohr to Høffding, 1 August 1928, quoted in Faye 1991, 58–59, my italics). Many of Bohr's analogies did not undergo any substantive change over the years. Rather, what was suggestive and vague became, merely by virtue of repetition, rigorous and compelling: "It is not a question of weak parallels. It is a question of investigating as accurately as we can the conditions for the use of our words" (Bohr 1958e, 704).

When Bohr first explored the similarities between physics and psychology, he felt that we "can hardly escape the conviction that we have acquired a means of elucidating general philosophical problems" (1929b, 101). It is this conviction that sustained repetition without substantive elaboration. The initially novel and fresh metaphors became familiar, obvious. "Vague" analogies become "beautiful examples," "striking analogies" (Bohr 1958e, 704).

According to Heisenberg, Bohr was a "natural philosopher," not "a

---

4. Bohr especially liked a fable about the talk by a rabbi that was understood neither by his listeners nor by the rabbi himself (Pais 1991).

mathematical physicist." For a mathematical physicist, to understand or explain means to construct a suitable set of models and to specify the rules for connecting the abstract theoretical entities with empirical data. Mathematical physicists explicate the meaning of operators, eigenfunctions, observables, projections in Hilbert spaces. Bohr had no use for any of these notions. For Bohr, as for a pre-Newtonian natural philosopher, "any explanation or analysis only means to use analogies from simple experience" (Bohr 1937d [lecture 6], 353). Vico's description of what is distinctive about proper philosophizing fits Bohr especially well: "Specifically philosophic quality . . . [is the] capacity to perceive analogies existing between matters lying far apart. . . . It is this capacity which constitutes the source and principle of all ingenious, acute and brilliant forms of expression" (quoted in McMullin 1991, 57).

Bohr was a philosopher of "harmonies," of symbolic meanings. As Pauli did not fail to notice, Bohr's correspondence between macro and micro, based on an analogy between a planetary system and a microscopic atom, retained the original medieval notion of harmonies between macrocosm and microcosm. Yet, if the correspondence principle, by an ingenious mathematical elaboration, especially in the hands of Kramers and Heisenberg, led to the new quantum theory, complementarity never left "common language." By not leaving common language, Bohr could not even attempt to construct a quantum ontology. Bohr and his followers presented this idiosyncratic choice as a prohibition in principle. Not surprisingly, many of Bohr's thought experiments intended to demonstrate complementarity contain nothing quantum in their analysis (Beller and Fine 1994).

If there is a single idea that inspires Bohr's analogies, it is the breakdown of the classical concept of motion, and the loss of continuity (infinite divisibility) and of the visualizability *(Anschaulichkeit)* associated with it. It is this idea that inspired, often mistakenly, Bohr's intuitive discussion of the interconnection between reality, acausality, and loss of visualizability in the quantum domain. This idea also inspired Bohr to see a general epistemological similarity between atomic physics and psychology: "We may say that the trend of modern psychology can be characterized as a reaction against the attempt at analyzing psychical experience into elements which can be associated in the same way as are the results of measurements in classical physics" (1938b, 27). The loss of visualizability in physics inspired many of Bohr's far-fetched analogies: "emotions and volitions" are similar to the quantum of action because they are incapable of being represented by visualizable pictures. Improvising on the complementarity theme, Bohr gave free rein to his imagination. Many of his analogies—for example, between the quantum of action and the concept of life, between the flow of thinking and the wave nature of matter, or between the unity of personality and the individuality of material particles—contain no more objective content than the correspondence between dreams and winning lottery numbers outlined in an essay by his grandfather.[5]

Many of Bohr's analogies that initially seem appealing fall apart when probed more closely. In his early writings, Bohr often draws a parallel between atomic measurement and introspection. In both cases, apparently, it is inherently impossible to separate the observing from the observed. Bohr's conclusion was that atomic interaction is in principle "unanalyzable" and "unsurveyable"—a prohibition not many working theoretical physicists would choose to comply with. Many of Bohr's examples of mutual exclusivity are far from convincing. The often repeated example of complementarity between "thoughts" and "feelings" is contradicted by contemporary psychological research (see Lazarus, Kanner, and Folkman 1980 on the connection between emotion and cognition). The mutual exclusivity of reason (associated with the masculine) and emotions (associated with the feminine) has also been subjected to penetrating critique, by feminist scholarship. Though some of Bohr's associations remain private and inexplicable, others, such as the preceding example, resonated within the culture of his time. Heisenberg, who joined Bohr in the dissemination of his metaphors, understood well their persuasive power. Though to his colleagues Heisenberg preferred to talk in unambiguous mathematical language, he appealed to the general "mental climate" when lecturing to popular audiences.[6] Bohr's numerous "harmonies" were those "useful (sinnvoll) interrelations 'belonging together' within the human mind" (Heisenberg 1979, 68).[7]

Bohr's writings on complementarity are metaphorical in a strong sense. Complementarity is an "artificial word," which "serves only briefly to remind us of the epistemological situation" in quantum physics (1937c, 293). Because the situation is unprecedented, one can only turn to "quite other branches of science, such as psychology or . . . to thinkers like Buddha and Lao Tse" (1937b, 20). But these disciplines are themselves ambiguous and in need of more rigorous explication, so one has to turn back to quantum physics, where matters can be explicated, so Bohr intimated, more precisely. Clearly, this state of affairs cannot be understood by any "substitutive view" of metaphor, in which the literal

5. Pais described Bohr's grandfather's "witty essay" in which "the dream table establishes a correspondence between a specific type of dream and the lottery number to be picked" (1991, 35).

6. "Every scientific theory arises in a certain mental climate . . . the author of the theory may be only vaguely conscious of it" (Heisenberg 1979, 65–66).

7. I would translate sinnvoll as "meaningful," rather than "useful."

meaning of the primary subject can replace the metaphorical meaning of the secondary one. Even the "interactive" view of Max Black (1962) is only of limited help. According to this view the primary and secondary domains (for example, physics and psychology) interact so as to produce novel understandings of both, but Black assumes that the primary domain is understood to a considerable degree. No such well-understood primary domain exists in Bohr's analogies. Rather, we encounter here a hermeneutical circle. Bohr's complementarity should be approached as a foreign culture, as a newly encountered, unfamiliar symbolic world. Clifford Geertz's characterization of the anthropological approach is a fitting one to describe the encounter with the strange framework of complementarity: "Hopping back and forth between the whole conceived through the parts which actualize it and the parts conceived through the whole which motivates them, we seek to turn them by a sort of intellectual perpetual motion into explications of one another" (1976, 236).

In this hermeneutical web of metaphors, "hopping back and forth" often interchanges what is foundational and what is derivative, *explanandum* and *explanans*. As I will shortly argue, Bohr's language of "correspondences," conjoining certain concepts and thus creating "harmonies," was inspired by powerful, though often misleading, images. His understanding was that of a poet, a mystic, an artist—Bohr's own words (this chapter's epigraph) characterize his thought especially well: "We have at the end only to take recourse to painting with words just like one paints, as [an] artist paints with colors just trying to use them in such as way [as] to be able to give one another just an impression of certain connections of certain harmonies" (1937d, 354).

The Web of Correspondences and Harmonies

Bohr's thought teems with "correspondences" and "harmonies." If, in the old quantum theory, such correspondences served as powerful heuristics (Jammer 1966; Darrigol 1992a), after 1927 Bohr advanced the idea of correspondences, or harmonies, to legitimate an unfamiliar and abstract quantum theory. The idea of harmony between the possibilities of observation and theoretical definition pervades, as I have argued, the Como lecture. Later, when Bohr elaborated the lessons of complementarity, he argued that features of the quantum formalism "correspond" to the freedom to choose the appropriate measuring devices.[8] Similarly,

8. "The possibility of disposing of the parameters defining the quantum mechanical problem just corresponds to our freedom of constructing and handling the measuring apparatus, which in turn means the freedom to choose between the different complementary types of phenomena" (Bohr 1948, 452).

"emphasis on permanent recording under well-defined experimental conditions . . . *corresponds* to the presupposition, implicit in the classical physical account, that every step of the causal sequence of events in principle allows of verification" (my italics). It is this "correspondence" that allows one to view quantum mechanics as fulfilling "all demands on rational explanation with respect to consistency and completeness" (Bohr 1958c, 6). By presenting numerous "correspondences" between the formalism and the "conditions of experience," Bohr created the illusion that the quantum formalism is a direct confirming instance of complementarity, thus enforcing the acceptance of an unfamiliar formalism and a new philosophy of physics concurrently.

Heisenberg, when presenting the lessons of quantum theory to wider audiences, sometimes adopted Bohr's language of correspondences. Arguing that the quantum formalism represents our knowledge rather than the objective course of events, Heisenberg pointed out the correlation, or reciprocal "image," between the act of registration in observation and the collapse of the wave function: "The discontinuous change in the probability function, however, takes place with the act of registration, because it is the discontinuous change in our knowledge in the instant of registration that *has its image* in the discontinuous change in the probability function" (1958, 54, my italics). Earlier, Born had advanced his statistical interpretation of the wave function and his statistical solution of the collision problem also by asserting a preestablished harmony between the possibilities of theorizing and the possibilities of experimentation (Beller 1990).

Bohr's writings are permeated not only with harmonies between theoretical advances and experimental possibilities but with what he considered necessary connections between key concepts, such as causality, visualizability, objectivity, and the distinction between subject and object. These connections, firmly entrenched in Kantian philosophy and in the classical idea of motion, Bohr fully embraced. His philosophy of complementarity, as I have argued, sprang from his basic intuition of the breakdown of the classical idea of motion (overthrow of visualization, causality, and reality) and was aided initially by the incorrect idea of disturbance (the impossibility of demarcating object from subject, the inseparability of phenomena and their means of observation, and the subsequent "indivisibility," or "wholeness," of the quantum interaction; chapter 7). In the classical idea of motion, space-time concepts, conservation laws, and visualizability (*Anschaulichkeit*) are intimately connected: from given initial space-time conditions and the conservation laws (causality), one can calculate the subsequent dynamical evolution of a system, as well as form a model (a picture, a visualization) of its behavior in space-time. Bohr's initial conflation of the ideas of causality and reality, and of causality and visualization, had its source in the classical idea of motion—Bohr thought intuitively that the breakdown of the classical idea of particle motion implied the breakdown of both causality and our concept of reality.[9]

Bohr characterized classical description as "causal pictorial" (1960c, 11), and he retained this intuitive, yet incorrect, idea of a necessary connection between causality and visualizability [10] throughout his life, even as his ideas about reality and acausality changed (chapter 7). If visualizable classical physics is "causal," then it is natural that the nonvisualizable quantum physics be noncausal: "In conformity with the non-pictorial character of the formalism, its physical interpretation *finds expression* in laws of essentially statistical type" (Bohr 1958c, 3, my italics).

Bohr's metaphorical framework of harmonies and correspondences is sustained by such expressions as "finds logical expression," "is suited," "it is not surprising that." Consider two characteristic examples: "Since, . . . we cannot neglect the interaction between the object and the instrument of observation . . . , *it is not surprising that* in all rational applications of the quantum theory, we have been concerned with essentially statistical problems" (Bohr 1929b, 93, my italics). And many years later: "We are dealing [in quantum mechanics] with a mathematical generalization of classical physical theories which by its non-pictorial character *is suited* to embrace the indivisibility of quantum phenomena" (Bohr 1957a, 669, my italics).

The two examples sound innocently similar. Yet they are worlds apart. The first is taken from Bohr's early writings at the time when all of his considerations were informed by the idea of disturbance. It is with the help of this idea that Bohr justified the connection between the indivisibility of measurement interaction and the necessary overthrow of causality and reality (chapters 7 and 9). As Bohr himself later repudiated the misleading idea of disturbance, no intuitive ground remained for his analysis of the inevitability of indeterminism and the overthrow of objectivity in the quantum domain. Before 1935, the idea of disturbance provided the illusion that Bohr's argument was grounded in a solid, experimental state of affairs. No illusion of this sort could be retained when the idea of disturbance was discarded and Bohr declared that quantum theory was merely a tool for the description and prediction of measurement results. In the post-1935 framework Bohr had to

9. Supplemented by his doctrine of the indispensability of classical concepts, this intuition led Bohr to deny the possibility of an objective causal theory, such as Bohm's.

10. One can have a nonvisualizable deterministic description, as well as a visualizable nondeterministic description.

postulate, rather than explain, the inseparability between measuring and measured, as well as the indivisibility, or wholeness, of quantum phenomena.[11] Bohr's wholeness often meant nothing more than an operational definition, where the meaning of a concept is intrinsically tied to the procedure of measurement. How is then the "nonpictorial" character of quantum physics suited to embrace the indivisibility of quantum phenomena? If the phenomena are defined relationally, or contextually, there is little reason for discarding causality a priori. Not surprisingly, in Bohm's version of quantum mechanics wholeness does not lead to acausality. In fact, Bohr retained the intuition of a correspondence between wholeness and acausality from his older, discredited framework of disturbance.

Thus Bohr's complementarity is built on a clustering of associations, which are nowhere grounded. Bohr's liberal interchanging of *explanandum* and *explanans* leads to his peculiar assertion that Heisenberg's uncertainty relation follows from complementarity: "The ultimate reason that in no conceivable measurement conjugate quantities can be fixed with a greater accuracy than that given by (5) [the uncertainty formulas] is indeed the complementary character of the pictures employed" (1939, 391). Complementarity, begot by repetition, assumes a life of its own and begins to serve as a basic explanatory concept, from which other aspects of the Copenhagen philosophy seem "inevitably" to follow. When he was in a Copenhagen mood, Heisenberg too freely interchanged *explanandum* and *explanans*. He often argued, using disturbance imagery, that the atomic constitution of matter leads to the uncontrollability of the measurement interaction. Yet sometimes his reasoning was reversed—it is the uncontrollability of the interaction that necessitates the finite divisibility of matter: "The existence of elementary particles is only justified by this fact [uncontrollability of interaction]" (1952, 73).

The Copenhagen framework is built on "correspondences" and "harmonies" among a cluster of intuitively interrelated ideas, despite the fact that the meaning of these ideas and the nature of their interrelation vary over time. It is this clustering that misled Bohr's readers into regarding his conceptual framework as basically stable. Because his reasoning is essentially metaphorical, Bohr often uses such expressions as "harmonizes," "corresponds," "is suited to," or "finds proper expression," rather than "follows from" or "is deducible from." Here lies a

---

11. One could, of course, turn to the quantum formalism and fully acknowledge its new features of nonseparability and nonlocality. Yet Bohr, who refused or was unable (see the discussion of Bohr's attitude toward mathematics below) to leave the "common language," avoided this option altogether.

striking difference between the analogical use of the correspondence principle and the metaphorical use of complementarity. The correspondence principle guided physicists, by suitable analogies, to explore the new quantum realm in ways suggested by regularities existing in the well-explored (primary) classical domain. No such primary domain exists in the complementarity framework. The complementarity principle was a metaphorical tool of legitimation—it led to no new physical knowledge.

"Wholeness" as Metaphor

"Hunger for wholeness" permeated the early twentieth century's Weimar culture (Gay 1968, 70–101). This hunger was not confined to the Weimar republic. In Denmark many intellectuals inherited the craving for "wholeness," "unity," "irrationality," and "unanalyzability" from Kierkegaard's existential philosophy. The young Bohr admired Kierkegaard (see Holton 1970 and Jammer 1966 on Kierkegaard's impact on Bohr). The wholeness, interconnectedness, and unity of science was, albeit from a different angle, a central concept in Kantian philosophy, which, as a result of the "back to Kant" movement, dominated academic philosophy in German-speaking countries.[12]

From his neo-Kantian teacher and friend Harald Høffding, Bohr learned that "true knowledge does not consist of accumulated experiences but is insight into the interrelationships between experiences" (Faye 1991, 9). In a Kierkegaardian vein, Høffding attempted to create "unity and harmony of the opposing views" (Faye 1991, 12) and a holistic notion of science that integrates physics and psychology. Bohr's early analogies between quantum physics and psychology, and his general notion of wholeness, are due to Høffding. It is from Høffding, according to Bohr's own words, that he learned about the "relativity of all our concepts" (1932, 200, 203). Høffding's impact was one of those meaningful mental processes that cannot, according to Bohr, be fully explained or analyzed: Høffding's "influence and guidance could be absorbed almost without being felt by the one who received it" (1931a, 178). It was because of Høffding's discussions about the "balance between analysis and synthesis" that Bohr maintained throughout his life the centrality of the idea of "wholeness": "While the whole may be built of individual parts, the appearance of each individual part is influenced in turn by the whole." Høffding's discussions had for Bohr and "for many of the listeners at his lectures as well as the even more numerous

12. See Beller (forthcoming) for Kant's impact on Einstein's philosophy and Chevalley (1994) for an analysis of the Kantian context of the Bohrian use of the terms *Anschauung* and "symbol." Bohr's usage of "wholeness" also belongs to this context.

readers of his book, a significance much deeper than what any one of us could easily explain" (Bohr 1931a, 177).

Høffding's general idea of wholeness permeated Bohr's writings on complementarity from his Como lecture until the end of his life. Like the terms "complementarity" and "causality," the word "wholeness" in Bohr's writings has many different meanings and is exemplified by different analogies in different contexts. Some of these analogies conflict with each other, while others are misleading (see the discussion below). None of these analogies adequately reflects the post-Bell notions of "inseparability" and "nonlocality."

In what follows I will discuss several of Bohr's metaphorical uses of the idea of wholeness. In these discussions Bohr constructed an illusion of explanation, by using outmoded traditional (classical) principles and ideas. Bohr's early intuition of wholeness drew strength from his misleading notion of disturbance. Wholeness, based on finitude and atomicity, is expressed, according to Bohr, by the prototype of the quantum jump—"an individual process, incapable of more detailed description, by which the atom goes over from one so-called stationary state to another" (1929a, 108).

As Bohr later argued, the essential characteristic of the quantum interaction is its uncontrollability, not its finitude. I have argued that the uncontrollability follows from the mathematical formalism of quantum mechanics (chapter 9)—it cannot be deduced from a general philosophical principle. In Bohr's pre-1935 writing, "uncontrollability" figures in his discussions of the uncertainty principle—uncontrollability prevents a circumvention of uncertainty. While Bohr avoided discussing the mathematics of quantum mechanics, he appealed to the age-old metaphor of "hiding" and elusive Nature, who prevents the inquirer from getting too close to her and "penetrating" her secrets. Thus, Bohr argued, we cannot know the position of particles in an atom in a given stationary state, because the use of measuring instruments will imply "an exchange of energy between the atom and the instruments which completely *hide* the energy balance" (1932, 201, my italics). This idea drew its metaphorical strength from an analogy with biology: "In every experiment on living organisms there must remain some uncertainty as regards the physical conditions to which they are subjected, and the idea suggests itself that the minimal freedom we must allow the organism will be just large enough to permit it, so to say, *to hide its ultimate secrets from us*" (Bohr 1933, 9, my italics).

Initially, disturbance (we erase phenomena while trying to observe them) was the metaphorical tool used to express wholeness—nature hides its detailed working by our interference, though no knowledge without such interference is possible. Later, Bohr distanced himself from the disturbance notion and introduced a "distinction in principle between the objects we want to examine, and the measuring instruments" (1935b, 218).[13]

Yet the idea of disturbance, though he later criticized it, remained his most vivid, most potent metaphorical tool for discussing quantum uncertainty and wholeness without recourse to mathematical formulas. This is the reason Bohr and Heisenberg, while denying its viability at times, regressed into using the idea of disturbance.[14] Even when he rejected the notion explicitly, Bohr implicitly retained the intuition of disturbance and of its "hiding" aspect in his later (especially post-1935) discussions of individuality or wholeness. This intuition is apparent in Bohr's notes for a talk he gave at the Institut d'Henri Poincaré in Paris in 1937: "Individual phenomena in quite a new sense in physics. When trying to analyze, phenomena disappear. They appear only under conditions where it is impossible to follow their course" (18 January 1937, AHQP). The example that Bohr provided was the two-slit experiment—if we could determine, by momentum transfer, through which slit the particles had passed, we would necessarily have "latitude" in the position of the measuring diaphragm and would thus exclude the appearance of an interference effect.

This argument is repeated in an obscure manner in the Hitchcock lectures (Bohr 1937d, 323) and is much more lucidly elaborated in "Discussion with Einstein" (Bohr 1949, 217). Bohr claimed that we cannot say what is going on with a particle between its passage through a slit and its arrival at a detector; measurement does not allow it because under measurement the "phenomenon . . . disappears entirely" (1937d, 323).[15] This reasoning makes sense in an antirealist approach, where it is meaningless to discuss behavior and properties independently of measurement—here "analyze" means "to measure": "we actually cannot at all analyze in such an arrangement . . . what is going on from the time the particle comes in until it is caught. We cannot control that without wiping the phenomenon out entirely. The phenomenon is in that sense an individual phenomenon, just like the individual process of transitions between the states of the atom" (Bohr 1937d, 324).

13. Note how different the idea of "irrational," "hiding" Nature is from Einstein's argument for the comprehensibility of nature: his belief that the "subtle, but not malicious" God would not construct a world that fundamentally hides part of its workings.

14. "The electron has been pushed by the light quantum, *it has changed its momentum and its velocity*, and one can show that the uncertainty of this change is just big enough to guarantee the validity of the uncertainty relations (Heisenberg 1958, 47–48, my italics).

15. The problem is not that the "phenomenon disappears" but that the laws of quantum mechanics do not allow one to obtain the specifics of the previous situation by calculation. Again arguments for consistency are presented as arguments for "inevitability."

Bohr's conception of wholeness as eradicating phenomena if we try to probe into them in more detail goes back to William James's analysis of mental phenomena. Bohr himself connected this idea with the idea of wholeness expressed by James: "If you have some things . . . they are so connected that if you try to separate them from each other, it just has nothing to do with the actual situation" (quoted in Faye 1991, xvii). According to Weizsäcker, "in the winter of 1931–1932 Bohr was constantly reading James" (1985, 186). Yet this Jamesian idea of wholeness is poorly suited to a characterization of quantum entanglement or inseparability. Bohr's and James's understanding of wholeness better fits the idea of a chemical compound: if we separate its constituent parts (we *can* do it), the phenomenon—the compound—disappears. A chemical compound is radically different from its constituting elements—water is different from hydrogen and oxygen. Such was the understanding of wholeness by Vygotsky, who argued for a holistic approach to thought and language.[16] Jamesian and Bohrian wholeness, while it may capture the idea of a classical interaction, is unable to provide insight into a quantum wholeness that evades classical analogies.

Bohr's wholeness plays an essential role in his metaphorical web of seemingly interconnected ideas. Intuitively, but wrongly, Bohr argued for a necessary connection between wholeness and complementarity (the impossibility of a unified ontology) and between wholeness and indeterminism. Thus, in "Mathematics and Natural Philosophy," Bohr stated: "The essential indivisibility of proper quantum phenomena *finds logical expression in the circumstance* that any attempt at a well defined subdivision would require a change in the experimental arrangement that precludes the appearance of the phenomenon itself. Under these conditions, *it is not surprising* [again we meet the language of correspondences] that phenomena observed with different experimental arrangements appear to be contradictory when it is attempted to combine them in a single picture. Such phenomena may appropriately be termed complementary" (1956, 559, my italics). And on the existence of a "natural" connection between wholeness and determinism Bohr wrote: "The indivisibility of quantum phenomena *finds its consequent expression* [!] in the circumstance that every definable subdivision would require a change in the experimental arrangement. . . . Thus, the very

16. "Two essentially different modes of analysis are possible in the study of psychological structures. . . . The first method analyzes complex psychological wholes into *elements*. It may be compared to the chemical analysis of water into hydrogen and oxygen, neither of which possesses the properties of the whole. . . . Psychology . . . analyzes verbal thought into its components, thought and word, and studies them in isolation. In the course of analysis, the original properties of verbal thought have disappeared (Vygotsky 1962, 3, Vygotsky's italics).

foundation of a deterministic description has disappeared" (1955a, 90, my italics).

Yet quantum wholeness does not necessitate either complementarity or indeterminism. Bohm's causal ontology of the quantum world is an eloquent counterexample to Bohr's categorical assertions. Even though both Bohr and Bohm emphasized the contextual nature of measurement, their notions and uses of wholeness are radically different. Bohr, turning away from mathematics, talked about the "essential unsurveyability" of the measurement interaction. Bohm considered measurement to be a special case of the quantum process and subjected it to detailed mathematical analysis. Bohr, using the notion of wholeness, denied the possibility of a space-time description of the two-slit experiment or a space-time description of energetic transitions in the atom. Bohm provided causal, realistic space-time descriptions of both (Bohm and Hiley 1993; Cushing 1996). Bohr based his conclusions on the breakdown of the classical idea of motion and on the alleged impossibility of coming up with entirely new ideas. Bohm construed a new conceptual tool for describing nonclassical space-time behavior—the "quantum potential." Bohm's wholeness is a mathematical tool for the exploration of the quantum world. Bohr's wholeness is a weapon of prohibition.

The connection between wholeness and indeterminism is also prominent in Pauli's writings. As with Bohr, so too with Pauli it is often not clear what follows from what, what is being assumed and what deduced. In his later years, Pauli proposed an argument, according to which inseparability between the observer and the observed phenomena is a direct outcome of the statistical nature of quantum mechanics: "If two observers do the same thing even physically, it is, indeed, really no longer the same: only the statistical averages remain in general the same. The physically unique individual is no longer separable from the observer" (quoted in Faye 1991, 196). Like Bohr's metaphorical allusions to wholeness, Pauli's argument is intuitively appealing yet conceptually fragile. The notion of the "same experimental arrangement" in which different experimental results are obtained, or of an "observer" that "does the same thing" as another observer, is never adequately defined. This notion presupposes, rather than argues, the impossibility of "hidden variables" and the inevitability of indeterminism. Similarly, Bohr's arguments, such as the following one, simply beg the question: "The very fact that repetition of the same experiment . . . in general yields different recordings pertaining to the object, immediately [!] implies that a comprehensive account of experience in this field must be expressed by statistical laws" (1958c, 4).

Pauli's argument from acausality to wholeness is also unconvincing.

There is nothing essentially quantum in this argument; the "whole-ness" seems to apply to any statistical theory, classical or quantum. Pauli's wholeness, as well as Bohr's, is not a crisply defined concept. For both Bohr and Pauli it is a tool for the legitimation of the Copenhagen credo. For Pauli, the notion of wholeness possessed considerable appeal as a mystical idea. Pauli attempted to connect the physical realm and the unconscious, breaking the limits of rational knowledge. He specu-lated that the humanistic and irrational concept of *anima mundi* (world soul) should replace the spiritually empty and conceptually bankrupt mechanical conception of the world as a great machine (Laurikainen 1988, 42–43). Pauli welcomed the statistical nature of quantum me-chanics because it opened the way for "irrational causes"—those con-nections between things that cannot be subjected to rational analysis. It also opened a possibility that "supernatural will," in the sense of Schopenhauer, permeates the world and determines its actuality (Lauri-kainen 1988, 55). While Pauli had some initial reservations about the Jungian idea of synchronicity (coincidences that are experienced as mystical or supernatural and that are not described by the ordinary concept of causality), he nevertheless was driven to broaden "the old, narrower idea of causality (determinism)" to the more general form of "connections" in nature. In this effort, he was especially keen about "correspondence," or "meaningful coincidence," between the comple-mentarity or wholeness of quantum physics and the idea of the uncon-scious (Laurikainen 1988, 204). He advanced imaginative, metaphorical ideas of wholeness and acausality to a mystical abyss that Bohr felt no inclination to approach.[17]

The post-Bell understanding of quantum inseparability, or whole-ness, relies on a mathematical analysis of previously interacting, but now separated atomic systems. It was the "conservatives," Einstein and Schrödinger, who provided the first adequate analysis of quantum inseparability by analyzing mathematically the features of the corre-sponding wave functions in such situations.[18] Bohr's translation of the EPR mathematical analysis into his own terms obscured rather than re-vealed quantum wholeness (chapter 7). Quantum inseparability was buried in a "tranquilizing" metaphorical web spun through positiv-ist maneuvers and the repeated enunciation of "inevitable" comple-mentarity and acausality. Only after the mathematical analysis of Bell

17. For Born "wholeness" had a well-defined meaning: in quantum physics there is "wholeness" because to predict a situation one needs to know the wave function every-where, including the boundary, where conditions are determined by the specifics of the experimental situation (1950, 101).

18. Einstein, Podolsky, and Rosen (1935); Schrödinger (1935). See Fine (1986) for a de-scription of Schrödinger's crucial role in the elaboration of the idea of "entanglement."

(1966) did progress again become possible. Some scholars attempt to construe Bohr as "anticipating" Bell. But "anticipation" is a biased historiographical notion, with Whiggish assumptions built into its very meaning.

Some philosophers argue that because Bohr did not accept nonlocality, he must have embraced inseparability (entanglement): Bohr must have denied that "separate systems possess separate real states" (Folse 1989, 264). Yet nothing in Bohr's writings warrants such a conclusion.[19] Bohr's frequent assertion that in atomic physics "wholeness is going beyond the ancient idea of the limited separability of matter" is too vague to be interpreted in terms of post-Bell inseparability. Bohr used this expression in different conceptual circumstances, from discussing the initial idea of quantization, through propounding the idea of disturbance, to declaring a priori the unanalyzability of the quantum interaction. Bohr's intuitions are congenial neither to nonlocality nor to nonseparability. Thus, in the case of the double-slit experiment, Bohr claimed that if we were able to give a more detailed deterministic description, we would "really be lost," for "how could the phenomena depend on whether this hole was open or closed?" (1930, 144).

That Bohr did not anticipate Bell is witnessed by the unease and eagerness with which the implications of the Bell theorem were (and still are) explored. Some scholars see the explanation of quantum interconnectedness as "the grand task" of our "golden age of metaphysics" (Shimony 1989; Teller 1989). Many understand that it cannot be done in Bohr's way, by keeping the classical "words and pictures without keeping the meaning of the words and of the pictures" (interview with Heisenberg, February 1963, AHQP). Those who, unlike Fine (1989) and van Fraassen (1989), are not content to accept the foundational status of quantum correlations[20] turn to the mathematical models of the quantum formalism looking for significant interrelations. For Bohr, "any explanation is an analogy." Mathematicians and mathematically inclined philosophers think otherwise. They assume that quantum theory transcends all classical analogies and look for answers in the characteristics of the mathematical structure itself. Such structural explanations are aimed at analyzing invariant connections in different models, displaying how the elements of a model fit together. Discarding all classical analogies, one gets an understanding from within, by demonstrating

19. Bohr's assertions of the wholeness of experimental arrangements are, in a fact, statements about the contextual nature of measurement. For Bohr (though not for Bohm), contextuality is different from entanglement because of the Bohrian "unsurveyability" of the measurement interaction.

20. One need not ask why there is quantum holism any more than in Newton's time one needed to ask why there exists a gravitational force.

how the *explanandum* gets built into a mathematical model (Hughes 1989). This is exactly what Bohr refused, or was unable, to do.

Bohr: Mathematics and Common Language

When John Slater came to work with the great Bohr in Copenhagen, he discovered, to his amazement, that Bohr had no use for mathematics: "I had supposed, when I went to Copenhagen, that although Bohr's papers looked like hand-waving, they were just covering up all the mathematics and careful thought that had gone on underneath. The thing I convinced myself of after a month was that there was nothing underneath" (interview with Slater, AHQP). What was "underneath" was Bohr's intuitive analogy-based approach to science. It enabled Bohr to guide his mathematically competent students to elaborate rigorously Bohr's analogies by use of the correspondence principle, leading to the formulation of a new quantum theory. After the erection and elaboration of the new quantum mechanics, the metaphorical approach enabled Bohr "to familiarize" himself with the "epistemological lesson" of the theory without mastering its mathematical formalism. There is a mythology about Bohr's "superhuman intuition"—his direct communion with the "deep truths" of nature achieved without the mediation of mathematics.[21]

Niels Bohr's brother Harald, a great mathematician, once said that Niels could get along without mathematics because of his great intuition. Heisenberg declared that "mathematical clarity in itself had no virtue for Bohr." "Physical understanding," according to Bohr, "should precede mathematical formulation" (quoted in Blaedel 1988, 111). According to Pauli, Bohr himself admitted that his "interest in physics was not so much that of a mathematician as that of a craftsman and of a philosopher" (1964, 1052). Many scholars who analyze Bohr's work notice his peculiar attitude toward mathematics. And, strangely, they present Bohr's choice to dispense with the power of mathematics as a legitimate stand, if not an advantageous one (Kaiser 1994).

I argue otherwise. There is ample evidence that Bohr's mastery of mathematics was very limited. Lacking advanced mathematical skills, Bohr could not build a new quantum ontology but instead had to use "common language" and simple analogies. This personal trait, if not weakness, was canonized into the universal doctrine of the indispensability of classical concepts and the impossibility of a quantum ontology. Bohr claimed that the quantum world, no matter how much it transcends the classical world, could be apprehended only by simple

21. Some philosophers of science accept this mythology (Honner 1987).

analogies with familiar domains, most of them nonmathematical. Theories of classical physics and their mathematical structures seem to have been, in a rather limited way, the only tools at Bohr's disposal. Bohr made no contributions to relativity theory (relativity played a minor role in Bohr's development; Pais 1991, 14), nor is there any evidence, I will shortly argue, that Bohr really mastered the formalism of quantum mechanics.

During the 1930s, the physicists at Bohr's institute in Copenhagen joked that Bohr used only three mathematical symbols: $\gg$ (much greater than), $\ll$ (much less than), and $\approx$ (about the same as; Weizsäcker 1985, 186). Bohr's brother Harald, despite being younger than Niels, had helped Niels in school (Blaedel 1988, 17). Apparently, Harald also helped Niels with his epoch-making, yet mathematically quite simple 1913 paper. According to Oskar Klein, Bohr's later assistant, all the mathematics of Bohr's great 1918 paper on atomic theory was done by Bohr's assistant at the time, Kramers: "Bohr had no idea how to do the mathematical part, even though the physical ideas, so to say, he had." In fact, Bohr was somewhat embarrassed when Kramers arrived to work with him. According to Klein, Niels asked Harald: "What should I do with this learned mathematician who is proposing to work with me?" (interview with Klein, 20 February 1963, AHPQ; see Dresden 1987, on Kramers's critical role in the elaboration of Bohr's early atomic theory).[22]

Pauli, in a 1927 letter, asked Klein to explain the contents of the letter to Bohr—for he, Klein, was the "only one in Copenhagen capable of following mathematical calculations" (Pauli to Klein, 18 February 1929, PC). The only interpretive paper of Bohr's that has calculations is the joint paper with Rosenfeld on the measurability of electromagnetic field quantities (Bohr and Rosenfeld 1933). For this paper, Bohr

---

22. This state of affairs inevitably led to questions about the rationale for collaboration and tension about the distribution of credit. Thus Bohr wrote to Kramers: "One may expect to obtain a large number of significant results in the nearest future; hence it is absolutely necessary that both of us are perfectly clear about the form of our collaboration, and that this be arranged in a way that is reasonable and justifiable for both of us. . . . in talking with you last night about calculations, I got the feeling that perhaps you do not think that the manner of continuing our collaboration . . . is the wisest for you, but that you might prefer to try, more independently, to work out some problem." Kramers felt uneasy, yet reassured Bohr about the importance of Bohr's contribution: "If I only knew well and clearly what the matter is. The whole thing appears to me as some vague half ethical and half practical question . . . just as little as I can refrain from working a little independently, just as little can I refrain from later asking your advice, for you can further the matter so very much by the philosophy that you can put into it" (Bohr to Kramers, 15 November 1917, Hellerup, Niels Bohr Archive; Kramers to Bohr, 11 November 1917, Niels Bohr Archive, quoted in Kojevnikov 1997).

provided the intuition, Rosenfeld the mathematical calculations: "Bohr wanted somebody who knew mathematics and was willing to go through it; Rosenfeld sort of sacrificed himself" (interview with Felix Bloch, 15 May 1964, AHQP). The division of labor was also clear during the Bohr-Rosenfeld lecturing tour to Russia—Bohr lectured on "matters of principle," Rosenfeld on the more "technical" aspects (draft of lectures, AHQP).

Similarly revealing is Bohr's reaction to von Neumann's remarks on the issues of acausality in quantum mechanics and hidden variables at a conference in 1938. While Bohr (1939) expressed his admiration for von Neumann's mathematical skills, he considered all von Neumann's mathematical virtuosity superfluous. Bohr insisted that his own elementary analysis of simple experimental cases demonstrated the same essential point as von Neumann's elaborate mathematics.

According to James Franck, Bohr was really "something different" from other quantum physicists: "Namely, Bohr was an amateur, and if [a] man like Wigner or so . . . came with too much involved mathematics, he even left, and said 'I can't understand it'" (interview with Franck, AHQP).

Harald provided mathematical help to Niels throughout their lives. As early as 1912 Niels wrote to Harald: "I have thought of you often these days; for, I had to use some mathematics, and thought of asking for your advice" (Niels Bohr to Harald Bohr, 12 June 1912, BCW, 1:555). And, later, one could hear Bohr saying: "I can't work out the math, but here comes Harald" (Blaedel 1988, 18). It was probably Niels Bohr's remoteness from mathematical labor that allowed him to make such bizarre statements as "after all, general relativity is just common sense" (1937d, 333). In numerous notes prepared by Bohr for lectures on quantum philosophy, nowhere do we find any quantum mechanical calculations—at most only formulas expressing the uncertainty relations. What we frequently encounter instead are rough mechanical drawings of springs, shutters, and bolts for discussions of thought experiments (notes for talks at Caltech, at Institut d'Henri Poincaré, and in Japan, 1937, AHQP). Discussing simple thought experiments, Bohr informed his audience that he "tried not to enter into any details of the mathematical methods" (1937d, 267). Yet all of the available evidence, published as well as unpublished, shows that Bohr never developed any mathematical treatment of these thought experiments. Nor did Bohr systematically elaborate his brief (too brief) analogies between quantum physics and other domains.

According to Heisenberg, a consistent, rigorous interpretation of the quantum formalism is not possible without recourse to its mathematical aspects: "One may say that the concept of complementarity, introduced by Bohr . . . has encouraged physicists to use an ambiguous rather than unambiguous language, to use classical concepts in a somewhat vague manner. . . . When this vague and unsystematic use of language leads into difficulties, the physicist has to withdraw into the mathematical scheme" (1958, 154).

This sort of withdrawal, I have argued, was not open to Bohr, who never seemed at ease with the esoteric game of symbols that his younger colleagues had mastered effortlessly. Analyzing the use of analogy by alchemists before the "mathematization of the world picture," Gentner and Jeziorski, following Vickers (1984) wrote that alchemists were "owned by their analogies, rather than owning them" (1993, 448). Can we not make a similar assessment of Bohr?

Metaphorical Appeal and Conclusion

In later years Bohr based his notion of scientific objectivity on nonambiguity of communication. Yet the metaphorical nature of complementarity implies its essential ambiguity. It is not clear how Bohr's emphasis on nonambiguity of communication is compatible with his insistence that the "definition of a concept stands in complementary relationship to its practical implication." Even less clear is how such nonambiguity is possible when "improvisation . . . points to a feature essential to all communication" (Bohr 1955b, 79).

Duality between the rhetoric of nonambiguity and the essential ambiguity of improvised, associative thinking produces another tension in Bohr's writings. It is the frequent vacillation between the two extremes of presenting his complementarity philosophy as exceedingly complicated or as transparently obvious. Thus Bohr announces that "the difficulty is, of course, very, very deep by nature" (1930, 136). Explaining complementarity and claiming that "phenomena are essentially the result of the interaction with the measuring instruments," Bohr warned his listeners that "this is very important . . . very complicated" (1937d, 262). Yet a few lines later he called it "very simple": "It is something very simple—the situation can be explained and can be detailed by looking a little closer into some of the very simplest experiments" (1937d, 263). No wonder Bohr shuttles between the opinion that physicists have readily accepted complementarity—"it is something very simple which at the moment by physicists [sic] appear quite clear"—to the opinion (just two sentences later!) that "all physicists will have to work very hard to get accustomed to [it]" (1932, 201).

Bohr was notorious not only for the incomprehensibility of his talks but for his extreme difficulties with writing. As Franck recalled: "Bohr had so great trouble to write little essays one has to write in school about everything, that finally he became always ill and stayed home when the day came to write essays" (interview with Franck, AHQP). Heisenberg singled out Bohr's difficulty with writing as one of the reasons for Bohr's early withdrawal from active scientific research. According to Heisenberg, when he came to Copenhagen in 1925, Bohr had already stopped doing physics. Heisenberg found Bohr overwhelmed with administrative duties: Bohr "had many visitors and then writing letters. . . . Since it took him so much time to write even a single quite trivial letter, it's obvious he could not find time for many other things" (interview with Heisenberg, AHQP).

Part of the Bohrian myth is that he thought very clearly and only expressed himself obscurely. Bohr's struggle with language is often presented as an attempt to encapsulate his worked out, but not yet spelled out, ideas. The ideas themselves, so the myth goes, are sharp, crisp, unambiguous; only their "expression" is vague. Bohr himself strengthened this impression: "I must have expressed myself most confusingly" (1938b, 31). This separation between Bohr's thought and his language seems especially out of place in view of his emphasis on the holism of mental phenomena. The distinction between Bohr's linguistic formulations and unexpressed thoughts is peculiar when advocated by Bohr's followers, who subscribe to his dictate that no phenomenon is a phenomenon unless it is an observed phenomenon. From this perspective, what access to Bohr's thought could one have except through Bohr's talking and writing?

From the early days of his scientific career, Bohr dictated his papers and scientific correspondence to numerous assistants. Their role was to help Bohr with the formulation of his ideas, to provide mathematical help, and to keep the discussion within the confines of relevancy.[23] Bohr's assistants did not, of course, simply record Bohr's fully backed ideas—rather the ideas were created in a dialogical, yet unequal interchange between Bohr and his helpers. Bohr's assistants often had no choice but to help Bohr to formulate ideas they themselves disagreed with. Reminiscences by Kalckar give the flavor of this process, as well as of Bohr's personality: "Sometimes, during work on a particularly difficult point in a treatise, where one had not succeeded in reaching any unanimity in the final formulation, and when, therefore, in order to get on, one had accepted his suggestion without really approving it in ones heart, he might stop walking around the table. He would look pene-

23. Rosenfeld recalled helping to draft Bohr's replies to Pauli's letters: "I would stand by as usual, in my role of helper, putting a hand to the more mathematical considerations and generally signifying by approbation or dissent whether the statements diffidently proposed by Bohr seemed or not to meet the point under debate" (1967, 119).

trating at one and say: 'Now, you mustn't be unhappy about it. Don't think I've gone quite crazy, but . . . ' And then he would go over the disputed argument from every side . . . until he felt that his young pupil was again with him" (1967, 230). In fact, in each paper signed by Bohr only, an informed reader can perceive the intellectual inclinations of the assistant who helped in its composition. The two versions of the Como lecture, written only weeks apart, form an eloquent example. The manuscript of the lecture (Bohr 1927c), with its emphasis on the wave nature of matter, bears the imprint of Darwin's predilection and preference for Schrödinger's wave mechanics. The *Nature* version (Bohr 1928), written with Pauli's help, has a much stronger operational emphasis.

The obscurity and vagueness of Bohr's writing reflects the associative and metaphorical nature of his thought. Numerous anecdotes tell about Bohr's talking before thinking, or starting a sentence without knowing how to complete it. According to witnesses, Bohr would sometimes say: "I do not know how to finish this sentence." [24] Bohr's brother Harald, his loyal supporter, explained the opaqueness of expression: "Niels talks about things he means to explain later" (Pais 1991, 45). But most things are never explained later—Bohr offers confident promises that are not, and cannot be, fulfilled. Among such promises belongs Bohr's claim that his complementarity philosophy provides a resolution of the mind-body problem. Reading Bohr, one is naturally led to assume that somewhere in his writings or unpublished notes, one can find the fully worked out solutions. Yet the solutions do not exist. Bohr scholars therefore charitably and ingeniously reconstruct from Bohr's short remarks what he might have had in mind. Each scholar inevitably brings in his or her own, not Bohr's, set of associations. Perhaps because of Bohr's open-ended analogies, Pais coined the term "complementarism" to distinguish them from what he supposed to be Bohr's logically impeccable and unambiguous complementarity in physics (1991, 438–47).

The analogies of complementarity are partial, incomplete, vague, and contradictory. This is not surprising, for, according to Bohr, the main conclusions of his complementarity philosophy follow from the indivisibility of the quantum of action, yet there is no rigorous analogue of this indivisibility in nonquantum domains. Consequently, no meaningful structural network of analogies can be established between, say, quantum physics and psychology. Pauli, the coauthor of complementarity, must have understood the limitations of Bohr's analogies: "It

---

24. To which Dirac apparently replied: "I was told never to start a sentence which one does not know how to finish" (quoted in French and Kennedy 1985, 249).

seems to me that Bohr's analogies, insofar as they pertain to *something psychological*, are an arbitrary construction without any deeper significance" (Pauli to Marlus Fierz, translation in Laurikainen 1988, 200, my italics).

Metaphors often underpin both common and specialized scientific language (Lakoff 1993). Terms that seem to be purely abstract have an underlying metaphorical meaning (Turner 1987). Such supposedly neutral terms as "causality," "disturbance," and "interference" may trigger a leap to desired, unfounded conclusions. When Donnan wrote to Bohr that his complementarity reasoning, following from the disturbance idea, was absolutely compelling, he was seduced by the vivid attractions of metaphorical imagery (Donnan to Bohr, 1935, AHQP).

Yet the wide appeal of complementarity is based on its existential message—an affirmation of the "irreducibility" of life, an invitation to sustain and to accept contradictions,[25] a quest for unity that transcends the fragmented, painful human existence. The appeal of complementarity is in its proclaimed goal of "harmonizing all situations" (Bohr 1958e, 699)—from contradictions in physics (waves versus particles) to contradictions in spiritual life (the pragmatic versus the mystical; Bohr 1960c, 14–15). When Weizsäcker found himself enchanted by complementarity, it was because it promised to bridge his scientific and personal lives (1985, 184–85). Bohr's discussion of existential matters resonated strongly in the tender souls of his young disciples: "Best of all was when the conversation turned to the so-called eternal questions. Nowhere did Bohr's influence as an educator have a more profound significance. . . . Nowhere was the overpowering intellectual and emotional impact of his personality more irresistible" (Kalckar 1967, 235).

When Born was distressed by the Cold War and devastated by the use of the atomic bomb on Japan, he turned to the philosophy of complementarity because of its conciliatory message (chapter 8). For Bohr himself, his numerous analogies had a profound emotional significance. He was constantly struck ("striking analogies"—Bohr 1955a, 93) by contrasts around him, and he commented on them sometimes in a sage and sometimes in a humorous way.[26] Bohr had an intense lifelong attachment to complementarity, as Kepler had to his cosmic geometri-

25. According to Kierkegaard, the ability to sustain conflicts rather than trying to resolve them belongs to a higher "stage" of life.

26. "The mutually exclusive relationship between different psychological phenomena, denoted by such words as 'thoughts' and 'feelings,' or 'reason' and 'instinct,' remind us in fact *most strikingly* of the complementarity relationship between atomic phenomena described by kinematical and dynamical concepts respectively" (Bohr 1939, 403, my italics).

cal construction of the plan of the Creator. As Nietzsche lived his philosophy, so did Bohr. After 1927 Bohr would talk for hours about the significance of the "great connections" he had discovered.

Part of the appeal of complementarity has to do with ideas so deeply entrenched in western European culture that one wrongly considers them unquestionable, obvious, universal. Complementarity between "thought" and "feeling" is, I have argued, such an example. Another, especially "liberating" (Pais 1991, 447) example is that between "love" and "justice." As the "mutual exclusiveness" of emotion and reason, so the complementarity of love and justice draws its appeal from contrasting the just and the rational (male) with the emotional and the intellectually inferior (female). The eloquent "argument" behind the complementarity is that of Herbert Spencer (1873): "The love of the helpless inevitably affects her [woman's] thoughts and sentiments, and this being joined in her with the less developed sentiment of abstract justice" (quoted in Easlea 1981). The greatest scientists, Bohr and Einstein included, accepted uncritically many of the stereotypes of their culture. Bohr felt that a "specially drastic example" is the one in which "the role of men and women are reversed" (1938b, 30).

The power of complementarity as a weapon of legitimation flows from its emotive, existential, and cultural appeal. Complementarity was indeed more warmly embraced in Europe than on the more pragmatic American soil (Cartwright 1990, 417–24). Bohr used his brief analogies with other fields less to explore and elaborate the unity of knowledge than to prepare audiences to accept complementarity in physics: "I just want to explain why I think it has importance to emphasize the analogy. First of all it is interesting when one as [sic] regards the situation in atomic physics. . . . it is, of course at any rate, very alarming that we should be in this situation . . . but it is interesting to remember that there are other fields of knowledge where similar situations must be looked into and have been looked into" (1937d [lecture 6], 352–53).

Bohr, despite his modest (and rhetorically pointed) declarations that he "has been forced to enter on such problems" only for his "own clearness of mind" (1937d [lecture 1], 263–64), made deliberate efforts to shape his presentations in the most persuasive way. Thus he informed his former assistant Kramers about his talk at the Copenhagen Society for Philosophy and Psychology: "I learned a great deal from the ensuing discussion. In particular, I know better which points nonphysicists resent, and I also believe that for this very reason I found on this occasion better words than previously to answer the objections" (quoted in Faye 1991, 64).

When Bohr adjusted his talks to his audiences, he did not present a systematic, rigorous elaboration of the idea of complementarity.[27] With the passage of time, Bohr's presentations of complementarity became more, rather than less, ambiguous. This ambiguity allowed very different audiences to experience kinship with the philosophy of complementarity—from Jordan in his Nazi days (Beyler 1996; Heilbron 1987; Wise 1993) to today's New Age advocates, left intellectuals (Aronowitz 1988), and feminists (Wilshire 1989). A rigorous clarification and analysis would undermine Bohr's metaphorical framework. Davidson's words (though relating to another matter) aptly characterize Bohr's metaphors: "It is hard to improve intelligibility while retaining the excitement" (1984, 183).

In one of his later talks Bohr referred to complementarity by telling an amazing "little joke": "Churchill said, about the Battle of Britain, that never in history have so few been able to do so much for so many. Now we can certainly turn this around in physics and say that hardly, in the history of science, have so many physicists . . . worked so many years and so hard on so small and simple [a] point" (1958e, 705). Was it the epitome of Bohr's numerous vacillations between complementarity as difficult and profound and complementarity as simple and self-evident? Or might Bohr have been teasing the collective of future scholars who have invested so much effort and ingenuity in presenting his ambiguous set of associations as a rigorous philosophical framework?

27. Bohr was aware of the frustration his audiences often experienced: "You might most of you think that I am using a lot of words and that just plain words and how will I define these words? But now of course that is right, that is just the whole difficulty of speaking about any field without that it were very daily experience" (1937d, 354).

CHAPTER 13

乀丿

Hero Worship, Construction of Paradigms, and Opposition

Even to the big shots, Bohr was the great God.

Richard Feynman 1980, 129

Scientists are inclined to take their own outlook for the natural way of looking at things, while the outlook of others, inasmuch as they differ from theirs, are adulterated by preconceived and unwarranted philosophical tenets, which unprejudiced science must avoid.

Erwin Schrödinger 1955, 14

Introduction

In our times, the dethroning of heroes is commonplace. We are reassured to learn that our cultural and scientific heroes—the overpowering Newton and incomparable Mozart—are in fact flesh and blood, endowed with the frailties, ambitions, and vices of simple mortals. In the history of science we also have come to recognize hero worship as a major distortion of the past.

Yet hero worship is not merely a historiographical sin. It is a powerful tool for legitimating orthodoxy. Hero worship does not merely deify and mystify the hero. Hero worship erects psychological barriers against criticism and delegitimates opposition. Hero worship effectively conceals the dialogical network in the emergence of novelty by centralizing insights, dispersed among many participants, into the hands of a few great scientists. In this way, the "lesser" figures are painlessly eliminated from history. If, in the evolutionary narrative, the scientific community becomes superfluous, in the revolutionary account, the scientific collective is no more than a dutiful army of puzzle solvers. The communicative, interactional nature of scientific creativity is as alien to the revolutionary as to the evolutionary narrative (see chapter 15).

In this chapter I describe what can be called the hero worship of Bohr, and I indicate how Bohr's authority promoted uncritical acceptance of the Copenhagen philosophy. I also describe the strategies by which the orthodox portrayed their insights as revolutionary by discrediting the opposition and constructing a simplistic notion of past science.

Bohr and Hero Worship

One cannot overestimate the impact of the authority figure in the evaluation and acceptance of ideas. Bohr's unprecedented authority not only promoted the widespread, uncritical acceptance of the Copenhagen philosophy but obtained a favorable reception for his dubious and poorly developed ideas outside of his area of competence (Heilbron 1987). Bohr's deliberations on complementarity in biology, psychology, and anthropology are good examples.[1]

Bohr's authority was based on his outstanding scientific achievements in the past, his formidable institutional power, and his unique personal charisma. Bohr could provide simultaneously intellectual stimulation and help in advancing careers, spiritual fulfillment and down-to-earth fun, material benefits and psychological counsel.[2] Bohr became a father figure whose unique status many young scientists were eager to honor and whose authority not many dared to challenge.

Many physicists, among them Pauli and Jordan, used the words "father figure" when referring to Bohr. Pais (1991) calls Bohr a "father figure extraordinary." Jordan referred to Bohr as "the father of the great world-wide family of quantum physicists" (1972, 214). Bohr was a father figure for "physicists belonging to several generations" (Pais 1991, 3). Bohr's assistant Kalckar called Bohr a "fatherly friend" (1967, 227).

Scientific heroes are often described as being endowed not only with supernatural reasoning faculties but also with superhuman personal qualities. Any event or trait, no matter how irrelevant, or even discrediting, is construed as reinforcement of the hero's unique stature. Often such a construction is accompanied by genuine affection, as in Pais's (1991) depiction of Bohr. Pais presents Bohr's ignorance as formidable intuition—Bohr did not know what an isomer was but knew that the

1. As Landé commented: "It is a matter of social psychology how people perceive the ideas uttered by great men outside of their field of expertise" (interview with Landé, AHQP).

2. All this was offered at Bohr's institute in Copenhagen. As early as the 1920s, Bohr could support Heisenberg's and Pauli's visits to Copenhagen much better than the young German physicists could expect to be financed in Germany. About Bohr's connection with funding agencies, including the Rockefeller Foundation, as well as about the institute as a place of fun (sports, the opposite sex), see Pais (1991). Bohr's expertise in securing funding is explored in Kojevnikov (1997). Many recollections by Bohr's colleagues and disciples reveal a unique, powerful charismatic personality.

arguments of a lecturer pertaining to isomers were "all wrong" (1991, 7). Reminiscences, irrelevant to the scientific problems at hand, Pais presents as "warming up exercises," while Bohr's inability to deliver a coherent public lecture Pais interprets as an "unrelenting struggle for truth" (1991, 11).

To a certain extent, Bohr was a tragic figure. He laid the foundations of the quantum theory of an atom and inspired and supervised the erection of the new quantum theory. Yet he was unable, with his heavy administrative duties and limited mathematical knowledge, to participate actively in further developments when the field became too mathematical (chapter 12). While Bohr presented himself as a dilettante who had to approach "every new question from a starting point of total ignorance," Pais graciously remarks that "Bohr's strength lay in his formidable intuition and insight rather than in erudition" (1991, 7).

Bohr's "formidable intuition" and "subtle reasoning" were often used by the orthodox to certify the Copenhagen interpretation as final and to disarm those who sought an alternative. The legend that Bohr had some sort of access to nature's secrets, qualitatively different from that of other mortals, directly discouraged critical dialogue. This legend is supported by another, peculiar claim—unlike other theoretical physicists, Bohr did not need to calculate in order to obtain "the truth." Blaedel's is a typical statement: "Perhaps his intuition allowed him to grasp things when others needed calculations" (1988, 11). Bohr's personal limitation is thus uncritically transformed into a strength. At the same time, strong legitimation is given to Bohr's metaphorical presentation of complementarity by singling out his associations as privileged (chapter 12). Needless to say, none of the cases in which Bohr turned out to be wrong—the Bohr-Kramers-Slater theory, the existence of spin and of the neutrino, complementarity between "vitalism" and "mechanism"[3]—are mentioned when Bohr's intuition is being worshipped or used as a weapon against dissent.

It became almost obligatory, when writing about Bohr, to refer to the "subtlety" of his thinking. Followers and opponents alike characterized Bohr's thought as subtle. Yet, peculiarly, it is rarely a specific argument that is singled out for this description. Rather, the word "subtle" is used when the author encounters a difficulty in understanding Bohr, failing to penetrate the structure of the argument, to achieve a clear and coherent reading of Bohr's writings. Thus Sklar, characterizing Bohr's philosophy as "exciting and subtle," simultaneously complains that "his so-called Copenhagen Interpretation is not easy to summarize neatly" (1992, 172) and that "complementarity is a difficult notion to fully pin

---

3. Bohr avoided this terminology after the discovery in 1953 of the DNA structure by Watson and Crick.

down" (173). Hailing Bohr's philosophy as "extraordinarily ingenious," Sklar calls it "not a view of the world easy to understand" (1992, 175).

Even Bohr's opponents felt the pressure. Thus Bohm and Hiley, describing "Bohr's very subtle thought," mentioned that his ideas "do not appear to have been well understood by the majority of physicists" (1993, 15). It seems that the more one feels at a loss to extract a clear and coherent message from Bohr's philosophy, the more often one calls it "subtle" and "ingenious." This terminology creates a qualitative gap between oneself and the hero, excusing inability to understand and simultaneously preventing criticism of the hero's authority. No wonder Bohr's notion of "deep truth" was used to disarm those young visitors to Bohr's institute who thought they had found contradictions in his reasoning.[4] Any doubts physicists had about Copenhagen wisdom were silenced by an appeal to Bohr's unreachable depth of discourse, beyond the usual "simplistic" cannons of argumentation. When the "simple" truth seemed to contradict the orthodoxy, the inaccessible "deep" truth prevented further questioning.

This is not to deny ingenuity to Bohr's reasoning on complementarity, in which arguments of consistency are subtly disguised as those of inevitability. Some of the consistency arguments are very clever; other arguments owe their subtlety to skillful ad hoc maneuvering.[5] One can easily criticize the lack of rigor in such arguments, for Bohr and his followers often transcended the framework of nonrelativistic quantum mechanics while attempting to demonstrate its consistency. In the case of the Bohr-Einstein argument about the time-energy uncertainty relation (see Bohr 1949, 224–28), Bohr was forced to invoke ad hoc a formula from general relativity theory. In the case of Heisenberg's γ-ray thought experiment, Bohr used the formulas of classical wave theory while simultaneously denying that theory's validity by appeal to a photon concept. No wonder Einstein was not swayed by such arguments— it is an illegitimate mixture of classical and quantum concepts similar to the setup he found so unsatisfactory in Planck's initial derivation of the distribution law. Some physicists, such as Heisenberg and Bopp, understood that the frivolous mixing of classical and quantum concepts in the analysis of thought experiments was inconsistent, Bohr's ideology of complementarity notwithstanding. Yet Bohr's arguments were accepted uncritically by many physicists, historians, and philosophers; these arguments were construed as the "triumphant victory" of Bohr, heroically countering the threatening moves of the opposition.

4. According to a saying of Bohr's, there is a "simple truth," the opposite of which is falsity; and the "deep truth," the opposite of which is also a truth (1949, 240).

5. Bohr's uses of correspondence arguments in his early creative years display an outstanding ingenuity and skillfulness. My criticism here refers to the philosophical legitimation of the Copenhagen ideology.

The fact that many physicists were willing to accept Bohr's authority on fundamental issues of quantum theory is not entirely surprising. Bohr's arguments for the consistency of quantum mechanics (his response to EPR) and of quantum electrodynamics (Bohr and Rosenfeld 1933)[6] were perceived by physicists as a vindication of the powers and consistency of their tools, as a green light to go on confidently (Beller and Fine 1994; Pais 1991). Few bothered to carefully study either the EPR paper or Bohr's response to it.[7]

More striking are testimonies of blind acceptance of Bohr's authority even in such matters as the physics of the atomic bomb. According to Feynman's reminiscences about Bohr's visit to Los Alamos, nobody there dared to subject Bohr's proposal for improving the bomb to critical scrutiny. Feynman, who was unaware at the time that he was talking to the great Bohr himself, criticized Bohr's ideas freely. After the meeting Bohr told his son Aage that "Feynman had been the only person at the meeting who had been willing to say that an idea of his was 'crazy'" (Schweber 1994, 403). According to Feynman, Bohr said: "Next time when we want to discuss ideas, we are not going to do it with these [big shots] . . . who say everything is yes, yes Dr. Bohr" (Feynman 1976, 29).

Hero worship, and the associated suppression of criticism, need not always be sober and serious.[8] Thus Hendrik Casimir, a physicist from Bohr's circle, wrote a comical poem for Bohr's fiftieth birthday. He described Bohr's famous theory that a defensive, as opposed to aggressive, shooter has an advantage because supposedly making a voluntary decision takes more time than reacting in a purely mechanical way. The poem concludes with an attempt to give Bohr a gun so he can prove his theory "experimentally" by defending himself:

> So the three of us went to the center of town
> And there at a gunshop spent many a crown
> On pistols and lead, and now Bohr had to prove
> That in fact the defendant is quickest to move.
> Bohr accepted the challenge without ever a frown;
> He drew when we drew . . . and shot each of us down.
> This tale has a moral, but we knew it before:
> It's foolish to question the wisdom of Bohr.
> Casimir's translation (Casimir 1983, 98–99); original German
> (Casimir 1967, 113)

6. See Schweber (1994) for an excellent discussion of Freeman Dyson's use of the Bohr and Rosenfeld paper.

7. "When I asked George Uhlenbeck, who was an active physicist in his mid thirties when EPR appeared, what he recalled about physicists' reactions, he replied that no one he knew paid any attention" (Pais 1991, 430).

8. I discuss the humorous critique of complementarity in my "Jocular Commemorations: The Copenhagen Spirit" (Beller 1999).

The moral of the story is clear. Nothing positive awaits those who challenge Bohr's authority. But is not such an interpretation inconclusive and tendentious, misrepresenting the joyful and humorous atmosphere in Copenhagen? Clearly not, for let us quote Casimir's own sober reading of his poem: "The moral of the story that one should not doubt the wisdom of Bohr applies to more important things than shooting gunmen in Westerns" (1967, 113; 1983, 98–99).

The exuberant celebrations at Bohr's institute, such as a feast held by the graduate students' club on the occasion of the twenty-fifth anniversary of the institute, are also instructive about the nature of Bohr's authority. Students, standing on chairs, beer in hand (with Bohr similarly on a chair), sang the hymn "Fathers of Selena," which hailed the "noble Bohr" who "knows the way amidst all false tracks" (Pais 1991, 6). Those false tracks no doubt referred to directions of thought at odds with the Copenhagen orthodoxy.

Bohr's unpublished correspondence discloses the overwhelming guilt experienced by those physicists who dared to challenge him. Thus Bopp argued in a letter to Bohr that one cannot obtain a contradiction-free interpretation of quantum theory by relying on thought experiments that explicitly use the concepts of classical particle mechanics—exactly the concepts that the new physics gave up. Bopp's contention strikes, of course, at the very heart of Bohr's lifelong enterprise. Bopp prefaces his argument with the following words: "A young Japanese colleague said once: we must do something, that our parents do not understand, and this is very painful" (Bopp to Bohr, 4 February 1962, AHQP).[9] These words, again, reveal the enormous parental authority of Bohr.

Many of Bohr's correspondents could not transgress the psychological barrier of even beginning to criticize Bohr. Thus Jesse Du Mond wrote to Bohr of his numerous unsuccessful attempts to grasp the meaning of Bohr's writings: "This is of course not the first time I have read and tried to grasp your point of view. . . . regarding your viewpoint I have never been able to get a clear answer from others. I hasten to say that I do not mean to imply any shortcomings in your own written exposition and am very ready to admit that the difficulty is entirely the fault of my own slowness and dullness" (Du Mond to Bohr, 7 March 1961, AHQP).

Yet Weizsäcker's reminiscences about his first encounter with Bohr constitute perhaps the best evidence for the overpowering, almost disabling impact of Bohr's authority. After the meeting with Bohr, Weizsäcker asked himself: "What had Bohr meant? What must I understand

---

9. "Ein junger japanischer Kollege hat einmal gesagt: wir müssen etwas tun, was unsere Eltern nicht verstehen, und das schmerzt uns sehr."

to be able to tell what he meant and *why was he right*. I tortured myself on endless solitary walks" (Weizsäcker 1985, 185, my italics). The question was not: Was Bohr right? or To what an extent was Bohr right? or On what issues was Bohr right? but, quite incredibly, What must one assume and in what way must one argue in order to render Bohr right?

The Issue of Consistency

Scholars of Bohr's thought are bewildered by the number of contradictions in his writings. As I have argued, the inconsistencies are genuine, stemming from many conflicting voices, from confusion between consistency and inevitability arguments, from a mixture of metaphorical and model theoretical argumentation, and from a mixture of classical and quantum concepts. Moreover, unbridgeable gaps exist between Bohr's pre-1935 and post-1935 philosophies.

Encountering these inconsistencies, scholars respond in ingenious ways and develop different interpretations of Bohr's thought—anything short of admitting the contradictions. The unshakable belief in his consistency more often than not follows directly from explicit hero worship: Bohr's mind is too "sublime," "subtle," "too relentless," or "too scrupulous" to produce inconsistencies (Honner 1987, 12). And even a historically sensitive philosopher, Don Howard, who insists on applying "the critical tools of a historian" in order to understand Bohr, defines understanding Bohr as reconstructing "from Bohr's words a coherent philosophy of physics" (1994, 201). Howard's imaginative reconstruction, as he does not fail to realize, transcends Bohr's words significantly.[10]

One might criticize such reconstructions of Bohr's thought as historiographical sins. And historiographical sins of this sort enhance hero worship not merely by refusing to admit inconsistencies but by ascribing to the hero figure the later insights and achievements of others. Such are the arguments for Bohr's alleged foreknowledge of Bell's nonlocality (Folse 1989).

Even the most competent and friendly readers find Bohr's philosophy obscure and inconsistent. Physicist and philosopher Abner Shimony admits: "I must confess that after 25 years of attentive—and even reverent—reading of Bohr, I have not found a consistent and comprehensive framework for the interpretation of quantum mechanics" (1985, 109). We might wonder what reason he has to remain "reverent" after

---

10. In fact, Howard uses von Neumann's analysis based on pure cases and mixtures, which Bohr himself considered superfluous, and which cannot be construed as a natural extension of Bohr's deliberations.

repeatedly (for twenty-five years!) finding Bohr's writings unsatisfactory and whether anybody but a "hero" would receive such charity. Yet perhaps the most extreme expression of the passion to save Bohr's consistency by any means comes from Honner: "If Bohr's thought is not found to provide a consistent framework for the interpretation of quantum mechanics, then perhaps one's expectations of 'interpretation' should be revised" (1987, 23). Clearly, the soundness of Bohr's reasoning is not judged by objective standards, rather Bohr's authority defines what the legitimate standards of reasoning should be.

As befits a hero, Bohr's weaknesses in argumentation are transformed into assets. Bohr's obscurity, Folse explains, followed from his greatness—"a work of genius resists categorization" (1994, 119). Bohr's difficulty in writing followed, according to Chevalley (1994), from his philosophy of the ambiguity of language. According to Honner, the opaqueness of Bohr's philosophy followed from "his concern for precision of expression"; repetitiousness was a proof of depth: "The articles are often repetitious, but it only should deepen our convictions that they [Bohr's examples] are not merely analogies" (1987, 169).

Philosophers often apply double standards when judging the orthodox and the opposition. A weakness in the opposition's stand might be a reason for total dismissal. A similar deficiency in the case of the winners is downplayed and rationalized. Hooker, one of the most sympathetic and penetrating commentators on Bohr's thought, after (reluctantly) criticizing Bohr's doctrine of the indispensability of classical concepts, defends him: "Bohr was only driven to adopt indispensability [of classical concepts] by his efforts to understand the conceptual significance of quantum theory." Moreover, while there is no evidence that Bohr ever considered abandoning this peculiar, untenable doctrine, Hooker feels compelled to ascribe such open-mindedness to him: "Though Bohr held strongly to BC15 [the doctrine of the indispensability of classical concepts], he would in principle give it up [!] had circumstances demanded it" (1991, 491).

Opposition, Paradigms, and Past Science

There are numerous ways to delegitimate the opposition and to discredit its stand. When skillful rhetorical techniques, disguised as a disinterested search for truth, are used by powerful authority figures, their effect is potent. It is difficult enough to produce a well-developed alternative to the deeply entrenched and elaborated quantum orthodoxy; it is intimidating, if not paralyzing, when all such alternatives are confidently ruled out by the "unbearable weight" of Bohr's authority and by such scientific heroes as Heisenberg and Pauli.

The orthodox exaggerated the difficulties of the opposition stand while ignoring their own. As the opponents realized, not without some bitterness, their criticism of the Copenhagen interpretation was simply "brushed off" with accusations of not "understanding Bohr" (Landé 1965, 123). As in politics, so in science, the orthodox misrepresented, trivialized, and caricatured the opposition's stand.

Perceiving threats from the technically sophisticated and proficient theoretical explorations of Einstein, Schrödinger, Landé, and Bohm, the orthodox translated them into "simple" language. This tactic ensured that few would bother to look into the original arguments, learning about the opposition stand through a presentation by the orthodox. Such translation not only distorted the original arguments, hiding their strengths, but also tailored them for the upcoming attack. Bohr's simple exposition of the EPR argument misrepresented its original structure and hid its radical message about either the incompleteness or the nonseparability of quantum mechanics (chapter 7 and Beller and Fine 1994). Diverting the issue from the focus of the opposition's criticism to a position on which the opponent may be weaker is a strategy often used. Thus, in a controversy between Born and Schrödinger, Born (1953b) advanced arguments against Schrödinger's (1952a) attempts to eliminate particlehood, when the bulk of Schrödinger's paper was directed against quantum jumps.

The orthodox did not merely misrepresent the opposition's stand— they trivialized and often deliberately caricatured it. Schrödinger's attempts to develop a wave ontology, which originated in his work on vibrations in a gas before the new quantum mechanics and were reinforced by his researches in cosmology and statistical mechanics in the 1930s and 1940s, were the result of penetrating mathematical analysis and philosophical reflection on his own work and on the work of other scientists (Wessels 1980; Bitbol 1996; Darrigol 1992b; Beller 1997a). Schrödinger was accused by the orthodox of disregarding quantum discontinuities (chapters 2 and 4); Schrödinger, of course, did not deny discontinuities, but he believed they could all be deduced mathematically from a wave theoretical scheme. He was accused of lacking the elementary understanding that wave packets disperse (Heisenberg 1927b, 73); yet he recognized the inevitable dispersion of wave packets even before Heisenberg's paper.

Orthodox quantum physicists often used the argument that ψ-waves are not real—Schrödinger was amazed by it: "Something that influences the physical behavior of something else must not in any respect be called less real than the something it influences—whatever meaning we may give to the dangerous epithet 'real'" (1950, 110).

The Göttingen-Copenhagen physicists discredited Schrödinger's interpretive aspirations by pointing out that classical waves propagate in the usual three-dimensional space, rather than in a 3N-dimensional space (Born 1953b, 142–43). Yet this criticism was misleading and outdated. Since 1927, Schrödinger had replaced his original "naive" wave interpretation with that of the "second quantization approach" (according to this approach, one can translate any statement about N-particles in 3N-dimensional space into a statement about N-level excitations of the three-dimensional vacuum state). In fact, Schrödinger's own work (even before the advent of wave mechanics) foreshadowed the modern concept of quantized fields (Darrigol 1986). And though he had some reservations about the very "abstract" character of these "waves," Schrödinger's preference for this scheme over the orthodox model was unequivocal: "I believe the discrete scheme of proper frequencies of second quantization to be powerful enough to embrace all the actually observed discontinuities in nature" (1952b, 27).

Schrödinger's attempt to substitute a wave ontology for a particle ontology was not an expression of a naive regressive nostalgia for the old way of doing physics, as the Copenhagen propaganda implied. Nor did his attempts at interpretation stem from purely philosophical presuppositions, or "prejudices." Since the 1930s, Schrödinger's preference for a wave ontology had been embedded in his research program: to unify general relativity and quantum mechanics (Rüger 1987). Following Eddington, Schrödinger hoped to deduce the discrete structure of matter from a unification of atomic physics and cosmology. The discreteness of a material substratum, conceived as a continuous wave field, follows from its being enclosed in a finite volume.[11] This research program also illuminated the lack of individuality and the prevalence of holism in the quantum mechanical domain: a single electron is inseparable from all others in the universe. The statistical nature of quantum theory arises precisely because, illegitimately, the orthodox approach implicitly presupposes such an individuality (Ruger 1987). From the perspective of the unification of quantum physics and cosmology, Schrödinger's search for a (nonclassical) model of reality was not merely natural—it was unavoidable. The search for such a realistic alternative by other physicists was aborted, in Schrödinger's view, mainly because the orthodox Copenhagen interpretation was "administered fairly early and authoritatively" (1950, 111).

The orthodox delegitimated the opposition by presenting it as "conservative" and "dogmatic." Heisenberg, intimating that Einstein "had difficulty" understanding the Copenhagen interpretation, labeled him

11. The discreteness is deduced because the vibrating field, enclosed in a finite volume, can have only discrete proper modes.

a "dogmatic" realist, as opposed to Heisenberg's own, "nondogmatic" brand of realism (1958, 81). According to Heisenberg, those who questioned the Copenhagen orthodoxy were dealing with "cracks in old bottles instead of rejoicing over the new wine" (1958, 139). The orthodox often discredited the opposition by accusing it of incompetence, indulgence in wishful thinking, or even personality disorders. Those who did not agree with the orthodoxy were "unable to face the facts." Those who allowed the possibility of hidden variables were presented as dreamers who had lost touch with reality: "To hope for hidden variables is as ridiculous as hoping that $2 \times 2 = 5$" (Heisenberg 1958, 132). Those who seek an alternative to the Copenhagen interpretation are simply outdated "grumblers": "This group of distinguished men . . . may be called philosophical objectors, or, to use a less respectful expression, general grumblers" (Born 1953a, 129).[12]

The most powerful technique for discrediting the opposition was by identifying its views as part of an outmoded, conservative stand. For this reason, the orthodox painted themselves as intellectual revolutionaries. They constructed the stand of the opposition and the image of past science simultaneously, thus caricaturing the opposition and trivializing past knowledge.

It is tempting, of course, to present one's own contributions as revolutionary, where "revolutionary" is synonymous with novel, bold, original, radically new.[13] Intensely ambitious individuals are especially prone to using revolutionary rhetoric. Heisenberg hoped that his own contributions to quantum physics would "revolutionize" the twentieth century as Copernican ideas had transformed the Renaissance (1979, 21). Bohr, who wished "to realize his wishes as to the future of physics," expected that his complementarity would be taught in elementary school alongside Copernicus's ideas (interview with Bohr, AHQP; Rosenfeld, quoted in French and Kennedy 1985, 323). Yet originators of new theories and the judgment of posterity do not always agree about what exactly was revolutionary in their work. Planck, for example, considered his discovery of the numerical value of h, rather

12. The opposition did not spare the sharp words either, though these found expression more often in private correspondence than in published writings. Einstein called the Copenhagen philosophy a "tranquilizer" for an uncritical follower, and Schrödinger defined it as "philosophical extravaganza dictated by despair" (Landé 1965, 124). More eloquently, Schrödinger referred to believers in the Copenhagen philosophy as "asses": "With very few exceptions (such as Einstein and Laue) all the rest of the theoretical physicists were unadulterated asses and I was the only sane person left" (Schrödinger to Synge, quoted in Moore 1989, 472).

13. In classical times one used the word *mutatio* for novelty. In the sixteenth and seventeenth centuries one used the word "new" rather than "revolutionary."

than discreteness of the quantum of action, as constituting a "revolution."[14]

Exactly where one chooses to demarcate between the old and the new depends on the local theoretical and sociopolitical context. While in the 1930s Heisenberg presented relativity theory and quantum theory as together opening a new era in physics (the quantum overthrow of Kantian causality is a direct continuation of the Einsteinian overthrow of Kantian space-time), both Einstein and relativity disappeared from Heisenberg's speeches and writings during the Third Reich (see the collection of his essays, Heisenberg 1979). In the postwar years, when Einstein was perceived as the most prominent opponent of the Copenhagen interpretation, the line again was drawn differently—relativity, which still preserves the notions of causality and of an invariant objective reality, belongs to the prerevolutionary past—it is quantum theory that opened a genuinely new age.

Whether one chooses to present a contribution as revolutionary, and what feature one singles out as the "most revolutionary," also depends on local theoretical and sociopolitical circumstances. Heisenberg's reinterpretation paper (1925) is usually accepted today as the basis of the new quantum theory and as the inauguration of the conceptual revolution in physics. Yet initially neither Heisenberg nor Born and Jordan, who extended Heisenberg's ideas (Born and Jordan 1925a; Born, Heisenberg, and Jordan 1926), presented the new quantum theory as revolutionary. Quite the contrary—because of the highly abstract character of the new quantum formalism, they preferred to emphasize its connection with the past and with cherished classical ideas (the identical form of the canonical equations, the identity of the perturbation methods, energy conservation). No indeterministic conclusions were initially deduced from the formalism (chapter 2). The rhetoric of a return to classical purity and beauty was abruptly silenced when something still more classically pure and beautiful appeared—Schrödinger's competitive wave theory. I have argued that Heisenberg and other orthodox physicists developed the arguments for a revolutionary overthrow of causality as a response to Schrödinger's competing theory of quantum mechanics—a continuous, causal alternative. This example indicates how the choice is made of which ideas to label "conservative."

This analysis also suggests why some elements, rather than others, are chosen to serve as foundational pillars for a new paradigm. It also points out the intimate connection between a challenge from the op-

14. His idea was that h, together with other fundamental constants—e (electron charge), k (Boltzmann's constant, and c (the speed of light)—would provide absolute universal units of measurement, independent of the "accident" of human life on Earth. For details, see Klein (1977).

position and the construction of the scientific past. It was because of Schrödinger's challenge and Einstein's critique in the late 1920s and early 1930s that the Göttingen-Copenhagen physicists chose acausality and indeterminism as the focal points of their emerging quantum paradigm (rather than the more recently proposed nonlocality).

As this new paradigm emerged, its founders constructed a profile of the opposition and a description of past science simultaneously. The ideas of the opposition were projected as most characteristic of the overthrown past—in this way opponents were automatically presented as conservatives; disposing of the old and discrediting the opposition went hand in hand. The opposition became simpleminded and reactionary; the past became monolithic. The diversity, ingenuity, fluidity, and epistemological resilience of past science was thus forced into a few rigid, simplistic categories. In this way classical physics became uniformly deterministic. Yet probabilistic ideas were introduced into classical physics as early as the beginning of the nineteenth century (see the papers in Krüger, Gingerenzer, and Morgan 1987). Statistical methods were the mainstream in quantum physics since Planck's and Einstein's work at the turn of the century, before the new quantum theory. This was the reason Born did not initially regard his probabilistic interpretation of the wave function as signifying a revolutionary departure, the way it was later construed by the Copenhagen physicists (interview with Born, AHQP).

Nor were classical physicists naive about the possibility of exact predictability. Since Poincaré's work at the turn of the century, physicists were aware of the existence of chaotic systems—nonlinear dynamical systems with sensitive dependence on initial conditions.[15] Born himself was cognizant of this phenomenon—he wrote a paper with an unequivocal title: "Is Classical Mechanics in Fact Deterministic?" (Born 1955a). The prominent orthodox argument that quantum physics gives up causality in principle, while classical physics gave it up only in practice, was itself crystallized in this tendentious construction of the past. It is the very meaningfulness of this distinction between indeterminism in principle and indeterminism in practice that chaos theorists challenge today.

In the simultaneous construction of the opposition and the past, the opponents—such as Einstein, Schrödinger, Bohm, and Landé—were presented as conservatives, unable to digest the revolutionary novelties. I have not been able to find any common criteria by which their stand was characterized as conservative—their conservatism consisting

---

15. An error in the determination of the initial conditions quickly becomes large enough to preclude the possibility of predicting the future state of the system.

merely in their disagreement with the orthodoxy. We are all familiar with the caricatures of the opponents—the image of aging Einstein, stubbornly mumbling "God does not play dice." We are also familiar with Galileo's construal of the fictional Peripatetic Simplicio. As Galileo's brilliant caricature distorts and disguises the complexity and resilience of Aristotelian thought (Schmitt 1983), so too the Göttingen-Copenhagen physicists trivialized and distorted the ideas of Einstein, Schrödinger, and Bohm. Recent scholarship, notably Arthur Fine's and Don Howard's work, reveals that Einstein was neither a simpleminded determinist (he did not hope for a completion of quantum physics by hidden variables, but for a radically new theory that would subsume the current quantum physics—Fine 1986), nor was he an unsophisticated realist—in fact, his philosophy had prominent conventionalist strands (Howard 1990, 1993). Far from holding a correspondence theory of truth, Einstein considered the concept of independent physical reality meaningless (Fine 1986; Beller, forthcoming). Nor was Einstein a dogmatic adherent to causal, continuous field theories—there was "another Einstein," and the debate between Einstein's two conflicting voices lasted to the end of his life (Stachel 1993).

Recent scholarship has also refuted the picture of Schrödinger as an adherent to realism. As did Einstein, so too did Schrödinger deny the meaningfulness of the concept of independent reality, regarding the idea of reality rather as a regulative construct (Ben-Menahem 1992; Bitbol 1996; Beller, forthcoming).

Schrödinger's defense of a comprehensive wave ontology, as well as the persistent Copenhagen caricature of Schrödinger's position, has resulted in an image of Schrödinger as a conservative, simpleminded, classical realist, unwilling and unable to sacrifice traditional concepts and accept new ones. Yet an analysis of Schrödinger's writings reveals instead a very sophisticated position, along neo-Kantian lines: the concept of reality "as such," as it objectively exists independent of all human observers, is indefensible, if not downright meaningless. Similarly, Schrödinger fully understood that the correspondence theory of truth can hardly be sustained. Still, the concept of reality, held Schrödinger, is as indispensable in science as it is in everyday life. There is no distinction in principle between a layman's and a scientist's conception of reality—both are regulative constructs, indispensable for mental (and physical) activity.

Schrödinger, no fan of duplications and divisions, dismissed those dichotomies on which representative realism rests—the dichotomy between primary and secondary qualities, between mind and matter, between theory and experiment. The division into experimental and theoretical physics is artificial—it is "mostly caused by the fact that the two kinds of activity require, each of them, such elaborated special training and skill, that they are seldom commended by the same person" (Schrödinger 1954, 124).

The objectification of measurements, the intersubjective agreement about the results of experiments, is primarily based on geometric and kinematic statements. For it is in the assessment of a geometric coincidence between the marks on the scale and the pointer that the discrepancy between the judgments of two different observers is the least possible. In this sense, empirical statements about geometric and kinematic relations clearly have a privileged epistemological status. Yet this state of affairs contributes to the fallacious naive realist view that "something is distributed in space in a definite arrangement and well defined order," this "something" is changing according to objective laws, and "this changing something" constitutes so-called objective reality (Schrödinger 1954, 144–45).

Positing this objective reality leads, according to Schrödinger, to a very awkward philosophical situation. Since "secondary" sensory experiences (sensations of color, taste, smell, sound) are removed from "objective reality," one invents a new realm for them—the mind (!), forgetting that "all that we have been talking about till now is also in the mind and nowhere else." This division creates philosophical pseudoproblems, such as how mind and matter act on each other, and the like (Schrödinger 1954, 145).

Schrödinger's draft for the James lectures (1954) allows us to form a clearer view of his aversion to the strong positivist elements of the Copenhagen interpretation—an aversion that the Göttingen-Copenhagen physicists mistakenly characterized as a form of simpleminded realism. An analysis of these lectures reveals why Schrödinger fundamentally rejected another dichotomy—Bohr's notorious division between the micro- and macrodomains, between classical "reality" that is close to common sense and the quantum mechanical "abstract formalism" that is a tool for description and prediction of measurement results. Schrödinger rejected this division not because he extended the naive realism of the macrodomain into the microdomain. Quite the opposite: he denied to the macrodomain a realistic status that the Copenhagen philosophy regarded as a self-evident fact.

Schrödinger was fully aware of the pitfalls, deficiencies, and ideological power of positivism. He called the legitimating positivist stratagems of the Copenhagen orthodoxy "unfair subterfuge" (1952d, 83), a "supreme protector" (1935, 157). The positivist approach entrenches the basic assumptions of the Copenhagen interpretation and endows controversial assertions with an aura of inevitability. The Copenhagen claim that quantum mechanics is complete is among such assertions, prohibiting further questioning. Completeness seems to arise as a straightforward deduction from positivism: "It must be impossible to add onto it additional correct statements, without otherwise changing it; else one would not have the right to call meaningless all questions extending beyond it" (Schrödinger 1935, 159).

Schrödinger was no less a philosophical "opportunist" than his Göttingen-Copenhagen opponents (compare chapter 3). He did not shy away from positivist arguments when they suited his purposes. In fact, it seems that Schrödinger was no less skillful in positivist analysis when criticizing the orthodox camp than the Göttingen-Copenhagen physicists were when fending off the challenges of the opposition. In such criticisms, Schrödinger often presupposed the verificationist meaning of quantum formulas: the uncertainty relations, for example, are not merely limits on the possible measurement values of physical variables—uncertainty restricts the very definability of the concepts used (see his criticism in chapter 6). Similarly, Schrödinger used positivist arguments when "deconstructing" the concept of a particle: "We are not experimenting with single particles, any more than we can raise Ichthyosauria in the zoo. We are scrutinizing records of events long after they have happened" (Schrödinger 1952a, 240). As Born noted shrewdly, refusing to go from an experimental event to an underlying substratum (particle matter) is a positivist argument (1953b, 144).

A letter from Schrödinger to Eddington (written in 1940, reprinted in Bitbol 1995, 121–22) reveals that Schrödinger indeed considered himself simultaneously an heir to Machean positivism and a follower of Boltzmann's descriptive tradition. The approaches of Boltzmann and Mach, at first sight irreconcilable, were in fact directed toward the same goal: avoiding doubtful presuppositions, excluding contradictory assumptions, clearing obstacles on the path to truth. While "Boltzmann's idea consisted in forming absolutely clear, almost naively clear and detailed 'pictures'—mainly in order to be *quite* sure of avoiding contradictory assumptions"—Mach "was most anxious not to contaminate this absolute reliable timber [an economical summary of the observed facts] with any other one of a more doubtful origin" (Bitbol 1995, 121). Thus Schrödinger wrote to Eddington that he could hardly be impressed by the "brave new world" of the Copenhagen positivism—a rather crude and naive version of an approach he was always familiar with. While one may and even must use pictures, one has to do so with one's eyes open to their limitations, revealed by an analysis of the experimental possibilities (Bitbol 1995, 122). It is the relative weight of positivist and model-descriptive elements in Schrödinger's arguments that changed, one can conclude, according to the theoretical challenges that he encountered.

Philosophically sophisticated and technically ingenious are also Bohmian alternatives to the Copenhagen interpretation (Cushing 1994a; Bohm and Hiley 1993). And Anthony Valentini has recently argued that de Broglie's scientific contributions have also been misconstrued in a grotesque way—de Broglie was not some cranky outsider, or second-rate theorist, but a deep thinker and an outstanding physicist, who in fact as early as 1927 (twenty-five years before Bohm) had derived all the correct "pilot wave" dynamics for a multiparticle system.[16]

As to the claim that the opponents did not have a deep working knowledge of quantum physics, anyone who is even superficially familiar with Einstein's and Schrödinger's publications immediately sees how ridiculous it is. It was the opposition—Einstein and Schrödinger—who in the mid-1930s discovered and mathematically elaborated the basic nonseparability of quantum systems (Einstein, Podolsky, and Rosen 1935; Schrödinger 1935). It was the orthodoxy, as I have argued, who diffused these arguments by operational stratagems, preventing serious exploration of nonseparability until Bell's seminal work. Einstein's characterization of Bohr as a "Talmudic philosopher" referred precisely to Bohr's circumventing, rather than directly confronting, the most fundamental problem of quantum theory. This deep physical challenge was met with the rhetoric of "sacrifice."[17] The "sacrifice" primarily meant elimination of the opposition's ideas.

16. This assessment of de Broglie's work is opposed to the received view that de Broglie developed only a primitive case, treating ψ as some classical field in three-dimensional space. Valentini also challenges the accepted story that at the fifth Solvay conference de Broglie was unable to reply to Pauli's penetrating criticism and he therefore abandoned his efforts for an alternative to the Copenhagen orthodoxy. I am grateful to Valentini who shared with me his reinterpretation of de Broglie's work in private communication before publication. This subject is treated in Valentini (1998). It must be noted that Valentini's interpretation is controversial. Scholars disagree about where de Broglie stood conceptually in 1927, and about the success with which he countered Pauli's objections at the Solvay conference (James Cushing, private communication).

17. Bohr, Heisenberg, and Born talked repeatedly about a "sacrifice" of the old. The following statement is typical: "All progress has been achieved by sacrifice" (Heisenberg 1979, 27). The rhetoric of sacrifice is noted in Heilbron (1987, 219).

CHAPTER 14

Dialogues or Paradigms?

Physics gets authority from ideas it propagates but never obeys in actual research, methodologists play the role of publicity agents whom physicists hire to praise their results.

Paul Feyerabend 1993, 261

There is a crack in everything. That's how the light goes in.

Leonard Cohen, "The Future"

Introduction

Thomas Kuhn's *Structure of Scientific Revolutions* (1970, hereafter referred to as *Structure*) is the best known and most influential critical model of science in our time. Yet the central concept of *Structure*—that of a paradigm—is incompatible with the notion of criticism. Criticism presupposes the existence of a contemporary alternative—the concept of a paradigm denies it. By definition, a paradigm excludes any diversity of opinion or open-minded discussion. It legitimates dogmatism and silences the opposition. Not surprisingly, Kuhn and sociologists of science following in Kuhn's footsteps talked in terms of "deviation," "impermissible aberration" (Kuhn 1970, 209), or "deviance" rather than legitimate disagreement.

In this chapter I argue that the notion of a paradigm has not only clear totalitarian implications but also dogmatic ideological roots. I argue that close historical links exist between the notion of incommensurable paradigms and the ideology of the Copenhagen dogma. In particular, I disclose the importance of Bohm's alternative to the Copenhagen interpretation in the emergence of a post-positivist philosophy of science.

The notion of a paradigm is also historically inaccurate. Historical research on different episodes in the history of science reveals not incommensurability but rather openness, selective borrowing, and communication. This openness and communication suggest that a dialogical approach is also appropriate to describe the growth of knowledge in fields other than twentieth-century physics.

Heisenberg's "Closed Theories" and Kuhnian "Paradigms"

The Kuhnian notion of a paradigm is peculiarly reminiscent of Heisenberg's notion of a "closed theory" (Heisenberg 1948, 1958, 1979). According to Heisenberg, progress in science occurs through a sequence of closed theories. A closed theory is true in the limited domain of its validity and is not open to modification.[1] A closed theory cannot be improved by small changes; it can only be replaced by something essentially different. The notion of a closed theory, like the notion of a Kuhnian paradigm, has a strong holistic aspect: "The connection between the different concepts in the system is so close that one could generally not change any one of the concepts without destroying the whole system" (Heisenberg 1958, 94).[2] From time to time, physicists discover that what they have held to be an unshakably true theory (classical physics, say) has only a limited range of validity. How can one talk in such circumstances of exact science at all? (Heisenberg 1979, 44). We can save the notion of an exact science by sacrificing the notion of unity, by relinquishing the Einsteinian goal of an ultimate unified theory. Science is a collection of independent "islands," of different domains, which cannot be united into a common structure. Within the domain of a closed theory we can use its basic concepts with complete confidence. How do we know that in the domain of a closed theory its basic concepts are exact? Heisenberg's answer is a careless one: "When we use concepts unhesitantly, they are exact" (1979, 44). Heisenberg, of course, does not provide us with any independent guidelines about when we can use concepts "unhesitantly" unless we already believe that they are correct and exact—a characteristic circularity of Heisenberg's arguments.

There are other obvious weaknesses in Heisenberg's notion of a closed theory—he provides no argument for identifying the correctness of a theory with its completeness, nor does he indicate any criteria that determine the limits of applicability of a closed theory (again, a characteristic circularity: "classical physics extends just as far as its concepts can be applied"—1979, 23).[3] Yet it is not philosophical inade-

---

1. "Scientific systems must be complete in order to be correct" (Heisenberg 1979, 23).

2. This holism has recently been emphasized in Kuhn (1987).

3. Heisenberg's notion of a "closed theory" has received very little scholarly attention. It was treated briefly by Chevalley (1988), who pointed out the similarities between Heisenberg's notion and the conception of a physical theory as a partially interpreted formalism, along the lines of logical positivism. It was used imaginatively by Hacking (1992a, 1992b) in his analysis of stable domains in theoretical and laboratory science.

quacy, but the *use* of the concept of a closed theory that concerns us here. A paradigmatic case of a correct closed theory is, according to Heisenberg, quantum mechanics. Under the Copenhagen interpretation, quantum mechanics can neither be modified by small changes nor supplemented by hidden variables. Therefore, argued Heisenberg (as well as Bohr and Pauli), the attempts by the opposition to challenge quantum theory and its orthodox interpretation are futile (chapter 9). It was usually in the context of combating criticisms of quantum theory that Heisenberg invoked the notion of a closed theory. The notion of a closed theory implies that the advance from an old theory to a new one demands an intellectual jump (very much like the Kuhnian Gestalt switch): "From the point of view of quantum theory . . . Newtonian mechanics cannot be improved; it can only be replaced by something essentially different. . . . This realization can preserve us from [the] mistake, not always avoided in the past, of attempting to force new fields of experience into an outmoded unsuitable structure of concepts. . . . We shall serve the future best by . . . easing the way for the newly won methods of thought rather than by combating them" (Heisenberg 1979, 26).

Neither the philosophy nor the history with which Heisenberg supported his notion of a closed theory is convincing. Quantum mechanics *can* be supplemented by hidden variables, albeit nonlocal ones, despite von Neumann's so-called impossibility proof (Bohm 1952; Bub 1997). Nor is it true that quantum mechanics cannot be modified by small changes—a notable example is the Ghirardi-Rimini-Weber theory (1986), which circumvents the measurement problem by modifying Schrödinger's equation. Nor can we reasonably hold that classical physics is a closed theory, given the host of new, nonclassical phenomena disclosed in its domain by research in chaos theory.

Heisenberg's idea of a closed theory, conceived in a context of confrontation, is an ideological notion, aimed at a defense of quantum theory and its orthodox interpretation. It is a powerful tool of legitimation in the rhetoric of the irrefutability and finality of the quantum theory. I argued in chapter 9 that the central Copenhagen pillars, such as the doctrine of the inevitability of classical concepts and operationalism, are powerful social strategies of legitimation, disguised as objective philosophical arguments. The notion of a closed theory is also ingeniously constructed to argue for the inevitability and finality of quantum theory and to encourage uncritical commitment by its practitioners, very much like the Kuhnian paradigm.

There are, of course, differences between Kuhnian paradigms and Heisenberg's closed theories. Kuhn's paradigms, when replaced, are forever buried in history. Heisenberg's closed theories are correct for all time. They serve as indestructible memorials to the greatness of their discoverers.[4] The intensely ambitious Heisenberg, not surprisingly, turns out to be the author of such an eternally true theory. As Weizsäcker put it, not without envy: "The concept originated in a reflection on what Heisenberg himself had achieved twenty years earlier when he was lucky enough to lay the foundation of the latest closed theory" (1987, 278).

Despite this difference, it is the similarities between closed theories and paradigms that are significant. Both resist improvement by small changes. Both are holistic. Both demand that advances come by way of jumps to a qualitatively new intellectual experience. Both imply "incommensurability." And both require dogmatic commitment and discourage criticism. No wonder, when Heisenberg read Kuhn's book, he found it all too familiar: "Alright, but the new paradigm is a new closed theory" (quoted in Weizsäcker 1987, 281).

Are these similarities coincidental? Or can we find close historical links between the philosophy of *Structure* and the ideology of the Copenhagen orthodoxy? And, in particular, can we uncover the roots of Kuhn's novel and controversial notion of incommensurability?

Where Did Kuhnian Incommensurability Come From?

Thomas Kuhn himself considered the notion of incommensurability the "central innovation introduced by the book [*Structure*]" (1993, 315). It is this notion that is tied to the most radical tenets of *Structure*: relativism, lack of rationality in theory choice, theory-ladenness of observation, changes in fundamental perceptions and intuitions following the acceptance of a new paradigm, lack of communication between representatives of different paradigms, Gestalt conversion from one paradigm to another. Over the years, trying to escape the relativism of *Structure*, Kuhn developed a much milder, taxonomic notion of incommensurability, identifying it with linguistic untranslatability. Kuhn emerged from his search feeling more strongly than ever that incommensurability must be a component of any historical, developmental, or evolutionary view of scientific knowledge" (1991, 3).[5] It is the initial introduction of incommensurability into *Structure* and its historical roots, philosophical underpinnings, and ideological consequences that are the focus of my analysis.

The notion of a scientific revolution, of a radical and comprehensive change, need not contain the controversial and pregnant notion of in-

4. As Galileo's Medicean planets were conceived by Galileo as an indestructible memorial to Prince Cosimo and to himself; see Westfall (1985) and Biagioli (1990).

5. For an excellent comparison between Kuhn's initial and recent notions of incommensurability, see Maudlin (1997).

commensurability. Revolution need not be a violent, Gestalt switch of meanings, practices, and allegiances, as Kuhn's own *Copernican Revolution* (1957), written before *Structure*, so impressively demonstrates.

What we see there is a gradual, sometimes irresistibly rational, transition from the "closed world to the infinite universe," an almost "inevitable" incorporation of atomism into this process, and a constant "essential tension" between tradition and innovation. As Robert Westman (1994) indicates in his valuable essay, there is neither incommensurability nor theory-ladenness of observation in *The Copernican Revolution*, though one is often tempted to read them back into the book under the influence of *Structure*. The author of *The Copernican Revolution* appears to be a faithful disciple of Alexandre Koyré: Kuhn's book is an outstanding conceptual history emphasizing theory at the expense of observation, analyzing the influence of wide philosophical and cultural currents at the expense of local everyday practice, and carefully laying out the fabric and inner logic of the evolution of concepts.[6] And while Kuhn, like Koyré, perceives revolution as the destruction of an old world and its replacement by a largely incompatible new world, nothing in Kuhn's analysis foreshadows the radicalization of the concept of revolution introduced by *Structure*.[7]

One can argue, perhaps, that Kuhn's insightful and sympathetic description of rejected knowledge, his presentation of the Copernican and Ptolemaic astronomies as having the same evidential basis, his insistence that in the end Copernicus's system suffered from the same degree of inaccuracy, inconsistency, and complexity as the Ptolemaic model, imply a rejection of the notion of truth and the introduction of relativism. Yet Kuhn is far from acknowledging such a possibility: "Each new scientific theory preserves a hard core of the knowledge provided by its predecessor and adds to it" (1957, 3).[8] At most, Kuhn's

---

6. A good example is Kuhn's analysis of the gradual development of the concept of "gravitational force" (1957, chap. 7).

7. Compare Kuhn and Koyré: "As science progresses, its concepts are repeatedly destroyed and replaced. . . . Like Aristotelianism before it, Newtonianism at last evolved . . . problems and research techniques which could not be reconciled with the world view that produced them. For half a century we have been in the midst of the resulting conceptual revolution" (Kuhn 1957, 265). Similarly, Koyré: "The founders of modern science had to destroy one world and to replace it by another. They had to reshape the framework of our intellect itself, to restate and reform its concepts" (1943, 405). Yet this process is the opposite of a Gestalt switch: "The spiritual change did not occur in a sudden mutation. . . . revolutions too need time for their accomplishments. . . . the heavenly spheres did not disappear at once. . . . the world-bubble grew and swelled before bursting" (Koyré 1957, viii). It is this evolutionary conception of revolutionary change that underlies Kuhn's *Copernican Revolution*.

8. At the very least, this "hard core" consists of observational results—we have not even a hint at the radical analysis of "seeing" provided in *Structure*, which maintains the theory-ladenness of observation.

analysis is compatible with, and perhaps influenced by, another thesis, that of the "underdetermination of theory by data." Yet this thesis, though undermining the notion of a single truth, does not exclude it in the same radical way as theory-ladenness of observation does: because underdetermination does not imply the impossibility of communication. As long as communication takes place, one can at least attempt to single out consensual truth by appeal to shared epistemic virtues. The miscommunication inherent in the notion of the incommensurability of paradigms leaves no such option. Kuhn, indeed, presents an astronomer's choice of the Copernican system over the Ptolemaic as due to neo-Platonic considerations, harmony and mathematical beauty.

One of the most fascinating aspects of *The Copernican Revolution* is Kuhn's penetrating description of the "interface" between tradition and innovation.[9] The delicate balance between tradition and innovation, conservatism and radicalism, "convergent" and "divergent" thinking, commitment and open-mindedness constitutes the focus of Kuhn's analysis in his "Essential Tension" (1959), an essay written when he began his work on *Structure*. Kuhn's struggles with the issue of conservatism versus innovation reveal an acute historical sense of the continuity of change rather than an uncritical commitment to the idea of static frameworks. Yet the "essential tension" swiftly developed, in Kuhn's hands (1963), into a celebration of "dogma" in scientific research. "Preconception and resistance to innovation" became, for Kuhn, hallmarks of the scientific identity—necessary conditions for scientific creativity, indispensable aspects of scientific education.[10] The forthcoming *Structure* disclosed a deep conceptual connection between incommensurability and the notion of a paradigm: incommensurability excludes the possibility of being suspended between two different, incompatible worlds, of creatively participating in both, of sustaining for long a creative tension between the old and the new. Such work is possible only during a short period of crisis, disarray, inconsistency. Incommensurability logically dictates total, unquestioning, dogmatic commitment.

9. See especially Kuhn's analysis of Copernicus's attempts to "justify" the earth's motion (1957, 144–55).

10. "Preconception and resistance seem the rule rather than the exception in mature scientific development" (Kuhn 1963, 348); "a quasi-dogmatic commitment . . . is . . . instrumental in making the sciences the most consistently revolutionary of all human activities"; "the commitment is actually constitutive of research"; "scientific education remains a relatively dogmatic initiation into a pre-established problem-solving tradition that the student is neither invited nor equipped to evaluate" (349). This essay was written before 1963, and probably before 1962, for Kuhn informs his readers: "The ideas developed in this paper have been abstracted . . . from . . . my forthcoming monograph *The Structure of Scientific Revolutions*, which will be published in 1962 by University of Chicago Press" (1963, 347 n. 1).

Kuhn's "encounter with incommensurability" and his discovery and elaboration of the concept of a paradigm must have been connected. Kuhn's reminiscences locate the beginning of his work on *Structure* in 1958–59. It was during this year that, after having "great trouble" developing his ideas (Kuhn 1977, xviii), he "finally had a breakthrough" (Horgan 1991, 14) achieved by finding the "missing element" (Kuhn 1977, xix)—the concept of a paradigm. It must have been at the same time that Kuhn came upon the notion of incommensurability: "My own encounter with incommensurability was the first step on the road to *Structure*" (Kuhn 1993, 314–15). Where could Kuhn have met with this notion?

Another book appeared before *Structure*, a book that undertook a radical analysis of the notion of seeing, of the theory-ladenness of observation, of Gestalt experiences in perception, of the revolutionary restructuring of an old pattern into a new one, with the resulting loss of the ability to communicate. In this book the word "paradigm" appears on the first page. It is, of course, Norwood Hanson's *Patterns of Discovery*, which appeared in 1958—an important book that would be totally eclipsed by *Structure*.

Hanson's book opens with a fictive encounter between a pro-Copernican Kepler and an anti-Copernican Brahe and with the surprising and explosive question: "Do Kepler and Tycho see the same thing in the east at dawn?" (1958, 5; compare with Kuhn's fictive speech by a Copernican convert—1970, 115). By discussing different organizations of visual data exemplified in Gestalt phenomena (a bird versus an antelope, an old Parisienne versus a young woman à la Toulouse-Lautrec, a perspex cube viewed from below versus above—such cubes decorate the cover of *Structure*), Hanson prepared the ground for his penetrating conclusion that "seeing is a 'theory-laden' undertaking" ("observation of X is shaped by prior knowledge of X"; Hanson 1958, 19). Tycho and Kepler do not see the same thing and merely *interpret* it differently; they do have in fact incompatible visual experiences: Tycho sees a mobile sun, Kepler sees a static sun. "Seeing as" and "seeing that" become indistinguishable. It is because there is no pure observation language, it is because the data themselves are different for Tycho and Kepler, that they live in incompatible worlds. We are familiar with these arguments, of course, from Kuhn's "Revolutions as Changes of World View," chapter 10 of *Structure*—arguments that are the basis of incommensurability.

Hanson's work, including his extensive discussion of the connection between perception and language, "pictures" and "sentences," "facts" and "expressions," in the spirit of Wittgenstein, deserves closer attention. So does his possible impact on the historiographical revolution of the 1970s in general, and on Kuhn's work in particular. Yet on the more narrow, though absolutely central, issue of incommensurability: the tight connection between Hanson's and Kuhn's ideas is undeniable. The evidence is not simply the similarities between the two texts, but Kuhn's own references to Hanson's work. While discussing the Gestalt changes of paradigm during revolutions, Kuhn mentions Hanson, who "has used *Gestalt* demonstrations to elaborate some of the same consequences of scientific belief that concern me here" (1970, 113). Another essay (written soon after *Structure* appeared), in which Kuhn discussed the "revolutionary reconceptualization" that permits data "to be seen in a new way," again refers the reader to Hanson's works.[11] Kuhn's use of incommensurability goes, of course, beyond Hanson's—it is the powerful synthesis of philosophical analysis, sociological incorporation of the notion of a scientific community, and rich historical knowledge and insight that makes *Structure* a masterpiece. Yet the basic similarity between Hanson's and Kuhn's notions of incommensurability leads directly to the following question: Where did Hanson get this controversial, if not bizarre idea that Tycho and Kepler lived in incompatible worlds and were unable to communicate? As we know, they lived (at least for a while) on the grounds of Tycho's castle and did, in fact, argue constantly, loudly, and passionately. What, then, could have suggested to Hanson the idea of incommensurability? What were Hanson's direct sources and aims?

Hanson's Incommensurability and the Copenhagen Dogma

As Hanson informed his readers in the introduction to *Patterns of Discovery*, "the approach and method of this essay is unusual." Disappointed with the inadequacy of discussions concerning the nature of general philosophical issues such as the status of observational facts and causality, Hanson decided to use current atomic theory as "the lens through which these perennial philosophical problems will be viewed." All discussions in his book, Hanson informed readers, "are written with the final chapter in mind"—the chapter in which he discussed the lessons of the quantum revolution. Hanson's credo was not simply Whiggish. In an outspoken way, he refused to cope with any historical

---

11. "N. R. Hanson has already argued that what scientists see depends upon their prior beliefs and training, and much evidence on this point will be found in the last reference cited in note 31" (Kuhn 1977, 263 n. 33). Similarly, "N. R. Hanson . . . has used gestalt demonstrations to elaborate some of the same consequences of scientific belief that concern me here" (Kuhn 1970, 113). There are other references to Hanson's work in *Structure*, as well as in almost all the other essays Kuhn wrote at the time he was working on *Structure*; see the papers collected in Kuhn (1977).

material that did not fit his aims: "Any argument not applicable to microphysics has been held generally suspect; conversely, arguments have been regarded as established if they help one to understand the conceptual basis of elementary particle theory" (1958, 1–3).

While discussing historical cases of incommensurability—between Tycho and Kepler, Descartes and Beckman, Mach and Hertz—Hanson referred repeatedly to the situation in quantum physics as the prime example of a new conceptual pattern's incompatibility (logical discontinuity) with the old.[12]

What is, then, the content of Hanson's final chapter? The bulk of the chapter is devoted to description and approbation of the Copenhagen dogma and argues the impossibility of an alternative to the orthodox interpretation. Hanson's writing spirals into opacity[13] as he approaches the issues of the holism of paradigms and incommensurability.

Hanson's discussion was very close to Heisenberg's writings, sometimes embarrassingly so. He dutifully repeated Heisenberg's arguments about the impossibility of visualization in the quantum domain, about wave-particle duality, and about the inevitability of acausality, following from Heisenberg's uncertainty relations. As for Heisenberg, so for Hanson the uncertainty relations are the cornerstone of the whole quantum theory—their violation is a "conceptual impossibility." This does not mean that the uncertainty principle is a tautology, or definition: "Had nature been other than it is . . . the principle might never have been formulated at all" (1958, 136).

Hanson used wave-particle duality to deduce the "conceptual impossibility" of the simultaneous determination of the positions and momenta of particles. Note that in Bohm's theory, position and momentum can be defined simultaneously, though they cannot be simultaneously measured—Bohm's version of quantum mechanics constitutes the best disproof of such "impossibility" arguments. Hanson relied uncritically on the usual Copenhagen assumptions (those that the orthodox presented as indisputable), such as the indispensability of classical concepts in a quantum theoretical interpretation. This resulted in Hanson's repetition of the usual Copenhagen rhetorical strategy, presenting

12. "To say that Tycho and Kepler, Simplicius and Galileo, Hooke and Newton, Priestley and Lavoisier, . . . Heisenberg and Bohm all make the same observations but use them differently is too easy. It does not explain controversy in research science" (Hanson 1958, 19): "It is the sense in which Tycho and Kepler do not observe the same thing which must be grasped if one is to understand disagreements within microphysics" (18). The difference between Descartes's and Beckman's idea of motion is "like Einstein, de Broglie, Bohm and Jeffreys on the one hand, and Heisenberg, Dirac, Pauli and Bethe on the other, all considering the uncertainty relations" (48–49). A similar assertion is made later (118).

13. Compare with the force, lucidity, and succinctness of Kuhn's *Structure*.

arguments for the consistency of quantum theoretical description as arguments for its inevitability (chapter 9).[14]

Yet it is not the details of Hanson's argument but his conclusion that is the focus of my considerations: because a violation of the uncertainty relations is a "logical impossibility," quantum mechanics simply cannot be modified—it can only be totally overthrown: "The uncertainty principle may of course be given up, but this would not be a reshuffling of one or two elements at the top of the pile of micro-physical knowledge: the whole structure of that pile would collapse. . . . One cannot maintain a quantum-theoretic position and still aspire for the day when the difficulties of the uncertainty relations will have been overcome. . . . To hold a quantum-theoretic position just is to accept the relations as unavoidable" (1958, 149). As Kuhn's paradigm forbids any dissent from within, so Hanson's "conceptual pattern" forbids all but wholesale change—both notions are holistic in a strong sense and both discourage local criticism.

What would be an example of such an overarching alternative? Hanson, with reservations, presented Bohm's theory as a possible candidate.[15] Observations and experiments cannot decide as yet between Bohm, de Broglie, and Einstein on one side, and Heisenberg, Born, and Dirac on the other (1958, 174). The situation in quantum physics is similar to the incommensurable ways in which Kepler and Tycho interpreted the astronomical data of their time.

In another, more technical discussion, Hanson developed the notion of incommensurability as linguistic untranslatability. This development, too, is aimed at defending the Copenhagen interpretation against criticism. Some physicists and mathematicians, notably Hermann Weyl, argued that the two central principles of the quantum theory—Heisenberg's uncertainty principle and Bohr's correspondence principle—contradict each other. According to the correspondence

14. Hanson used for his discussion Schrödinger's idea of a wave packet, with the usual Copenhagen excuse: "Though Schrödinger was wrong . . . there is an advantage in adopting his exposition provisionally" (1958, 147). Hanson then asserted, without any proof, that matrix mechanics will "issue in exactly equivalent results" (148). The excuse for not providing any proof is the usual one of the "formidable" difficulty of the formalism of matrix mechanics. Yet the equivalence of matrix and wave mechanics was used by Hanson, as by Göttingen-Copenhagen physicists, in a misleading way. The formal mathematical deduction of the uncertainty relations from the matrix formulation refers to the statistical spread of measurement results. Therefore, it leaves open the question of the exact determination of position and momentum for an individual particle.

15. Hanson mentioned doubts raised by Heisenberg, Oppenheimer, Dirac, and Bethe about the correctness of Bohm's theory (1958, 174). He reminded the reader that similar arguments by de Broglie were "pulverized" by Pauli; yet despite the hopelessness of the "hidden variables" project, "Bohm remains undaunted" (173).

principle, classical physics is a limiting case of quantum physics—as we approach higher quantum numbers, quantum results merge into classical ones. The uncertainty principle, however, implies that statements exist that are meaningful in the classical realm yet meaningless in the quantum domain (for example, the exact determination of a particle's position and momentum). How can then meaningless empirical assertions transform into meaningful ones as we approach higher quantum numbers?[16]

This technical objection seems to threaten the whole edifice of quantum physics. For it seems to imply that either the correspondence principle or the uncertainty principle must be given up. In Hanson's words, either "(1) quantum physics cannot embrace classical physics as a limiting case; or (2) quantum physics ought not to be regarded as permanently restricted such that no analogue for the classical S [meaningful sentences] is constructable."[17] Hanson clearly rejected the second option, which implies the incompleteness of quantum physics—the position of the critics of quantum theory: "Alternative (2) has been chosen by many eminent physicists and mathematicians: Einstein, Podolsky, Rosen, de Broglie, Bohm" (1958, 153). Does it follow, then, that we must choose the first option—abandon Bohr's correspondence principle? No, argued Hanson, if we properly qualify the meaning of the correspondence principle using the notion of linguistic incommensurability.

The languages of quantum and classical physics, he explained, are "discontinuous"—there is no "ultimate logical connection between them."[18] Languages of "so different conceptual structures" cannot simply mesh. When we increase quantum numbers, quantum mechanical sentences do not merge continuously into classical statements; instead, there are "formal analogies" between the expressions in the quantum and classical realms.[19] The fact that we use the same mathematical symbols, or formulas, in the classical and quantum languages obscures the fact that they are logically distinct (for example, we use the

16. A similar objection applies when we express the same problem in terms of probability distributions. "Classical theory allows joint probabilities of accuracy (in determining pairs like time-energy and position-momentum), and allows them to increase simultaneously. In quantum theory this is illegitimate. But as quantum numbers get larger, the legitimacy of these joint probabilities seems to increase; the same perplexity arises" (Hanson 1958, 152).

17. Hanson mentioned a third possibility, which he immediately dismissed: "Classical physics itself should be regarded as restricted against the construction of S, just as is quantum physics" (1958, 153).

18. Hanson might have been inspired by Bohr's assertion that an absolute "cut" exists between the classical and quantum domains.

19. On the correspondence principle as "formal analogy," see the excellent treatment by Darrigol (1992a).

signs + and − in both number theory and valence theory, but there is no logical identity between them). So understood, the correspondence principle is not threatened by the consequences of the uncertainty formulas (1958, 49–157).

This linguistic untranslatability Hanson projected back into history, to Kepler and Tycho and their supposed inability to communicate: "The conceptual differences here reflect earlier examples: Kepler and Tycho at dawn . . . , Beckman's problem and Descartes' problem, Kepler's law[s] as they were for him and then for Newton, . . . for Mach . . . and Hertz" (1958, 156). This statement reveals where the strange idea of Kepler's and Tycho's inability to communicate comes from. Yet Hanson was not merely projecting the then current situation in physics ahistorically into the past. He was projecting a certain version—the orthodoxy—that was to serve as a "lens" through which he would interpret the whole history of physical thought. We see the strategy by which, intentionally or not, the stand of the orthodox, through supposedly disinterested philosophical discussion, is canonized into an overarching theory of the growth of knowledge. Hanson first used quantum theory to interpret the past, to "describe" the essence of the "miscommunication" between Tycho and Kepler; then he projected the reconstructed past onto the situation in quantum physics: "It is the sense in which Tycho and Kepler do not observe the same thing which must be grasped if one is to understand the disagreements within microphysics" (1958, 18). History itself, by this trick, comes down with all its force on the side of the current orthodoxy.

But is not my assumption that Hanson was defending the orthodoxy too extreme? If my reading of *Patterns of Discovery* admittedly involved some interpretive effort, Hanson's zeal to protect the orthodoxy is crudely obvious in his paper "The Copenhagen Interpretation" (Hanson 1959), written in the year *Patterns* was published. Hanson spelled out his intentions clearly: his paper "aims . . . to argue for orthodox quantum theory as it now stands" (1959, 1). In this paper Hanson's defense of the irrefutability of the Copenhagen dogma reached a high pitch: "There is . . . no working alternative to the Copenhagen Interpretation. Ask your nearest synchrotron operator" (1959, 5). Following Heisenberg faithfully, he attacked all those physicists who dared question the Copenhagen dogma. He joined Heisenberg in discrediting "yesterday's great men who have not offered one scrap of algebra to back up their grandfatherly advice (Einstein, Schrödinger, von Laue, de Broglie, Jeffreys)," as well as those physicists who offered simply "pure mathematics" (Janossy) and "hidden variable" physicists who offered a formalism that does not preserve the mathematical properties of symmetry that have been "the power and glory" of quantum theory (1959, 10).

Hanson's defense of Copenhagen was triggered by a draft of a favorable review of Bohm's theory by Paul Feyerabend (1960). The existence of the Bohmian alternative convinced Feyerabend, as it would Bell after him, that the Copenhagen philosophy of indeterminism and subjectivity was not a logical necessity (Heisenberg's, Bohr's, and Hanson's rhetoric notwithstanding) but a deliberate theoretical choice. The Bohmian alternative did indeed inspire Feyerabend's philosophy of pluralism. It was this pluralism that Hanson rose to strangle at its very birth. Feyerabend regarded Bohr's interpretation not as a necessity but only as a metaphysical addition to the "bare" theory. Thus, according to Hanson, Feyerabend concluded that we are free to invent alternatives to Copenhagen. Hanson objected strongly to this reasoning: "Let us suppose that Feyerabend is correct: would it follow that admitting this implies that we are . . . free to invent and consider other metaphysical interpretations? Not at all!" (1959, 5).

Hanson's conclusion was sharp: "Certainly no reinterpretation yet suggested by philosopher or physicist presents a case for abandoning Bohr's views." Bohm was Hanson's direct target: "There is little practical warrant for the alternative interpretations which have, since Bohm, been receiving prominence" (1959, 1). Then Hanson judged even the question of an alternative interpretation unwarranted, defending the "puzzle solvers" who unquestioningly follow the Copenhagen line: "No one should think that because most quantum physicists are unperturbed by the type of question brought to prominence by Bohm, that therefore they are unreflective, resigned, . . . predicting machines" (1959, 15).

My discussion of Hanson's philosophy and its connection with Kuhn's ideas reveals why Kuhn was such a "reluctant revolutionary." It sheds light on what seemed a puzzling reaction by Kuhn to the relativistic way the tenets of *Structure* were later interpreted. As for Hanson, so for Kuhn incommensurability and the existence of different theoretical options did not imply relativism—there is only one preferred paradigm, to the exclusion of all others. We can now better appreciate Kuhn's objections to arguments that translated his notion of incommensurability into irrationality of theory choice. We can also see, having analyzed the connection between Hanson and Heisenberg, the source of the striking similarity between Kuhn's "paradigms" and Heisenberg's "closed theories."

Unlike Hanson, Kuhn, as far as I know, never mounted a direct defense of the Copenhagen dogma. Yet revolutions are revolutions and they demand their victims. Kuhn had little compassion for the defeated. As for the alternatives to Copenhagen, Kuhn, like Hanson, had little doubt who deserved to disappear into the dustbin of history: "Those who rejected Newtonianism proclaimed that its reliance upon innate forces would return science to the Dark Ages. Those who opposed Lavoisier's chemistry held that the rejection of chemical 'principles' in favor of laboratory elements was the rejection of achieved chemical explanation. . . . A similar . . . feeling seems to underlie the opposition of Einstein, Bohm, and others, to the dominant probabilistic interpretation of quantum mechanics" (1970, 163).

For Kuhn, Bohm's theory is not a genuine alternative—it is an irritable nuisance, if not a betrayal. Feyerabend, who discussed *Structure* with Kuhn (and whose "far-reaching and decisive" contributions to the final draft of *Structure* Kuhn acknowledged—1970, xii), recalled: "I remember very well how Kuhn criticized Bohm for disturbing the uniformity of contemporary quantum theory" (1970, 206). Not surprisingly, in his critique of *Structure*, Feyerabend mentioned a conceptual similarity between the notion of a paradigm and the situation in quantum theory: "Does he [Kuhn] want every subject to imitate the monolithic character of, say, the quantum theory in the '30s?" he asked (1970, 199).[20]

Ironically, Bohm himself, wrote a very favorable review of *Structure* when it first appeared, and he harbored some hopes (naive, we must call them) that through *Structure* scientists would "become aware of the role that paradigms actually play in the life of scientific research in order that they shall be able more easily to realize the need for a change" (1964, 379). Years later, a disillusioned Bohm had become aware of the tight dogmatic implications of Kuhnian paradigms, and he changed his initial enthusiasm for *Structure* into sober criticism (Bohm and Peat 1987).

Paradigms and the History of Science

The notion of incommensurable paradigms was, I have argued, conceived by Hanson in the image of the ideology of the quantum orthodoxy. Through pretense, disguise, or unreflective belief in disinterested philosophical analysis, philosophers can easily objectify the interests of certain power groups, scientific or political. Kuhn subsequently took the notion of incommensurability as the focal point of a comprehensive

20. Feyerabend himself argued eloquently for incommensurability between Aristotelian and Galilean physics in *Against Method* (1975); he claimed, however, that such cases of "incommensurable" change are "rare" in the history of science (Feyerabend 1981a, vol. 2, ix).

and far-reaching theory of scientific development. Incommensurability, I have argued, however, did not grow organically from Kuhn's historical scholarship—it was swiftly, perhaps hastily, superimposed on Kuhn's emerging *Structure*, most likely as a result of an encounter with Hanson's work. It is perhaps because of this abrupt introduction that Kuhn struggled with the meaning of incommensurability throughout his life: "Even before *Structure* appeared, I knew that my attempts to describe its central conception [incommensurability] were extremely crude. Efforts to *understand* and refine it have been my primary and increasingly obsessive concern for thirty years" (Kuhn 1993, 315, my italics).[21]

Having uncovered the ideological underpinnings of incommensurability, we should hardly be surprised that historians of science find it increasingly difficult to build their narratives in Kuhnian terms. Past scientific developments do not imitate the paradigmatic example of quantum theory—not even, as I have argued throughout this book, the development of quantum theory itself. Rather than dogmatic commitment to a rigid set of ideas, scientific creation—be it during the Copernican, chemical, or quantum revolution—is often characterized by the ingenious mingling and selective appropriation of ideas from different "paradigms."

As we learn from Robert Westman's essay (1994), Copernicus himself combined heliocentric and geocentric ideas, considering at first a model very similar to Tycho Brahe's. Tycho's geometric system also was the product—historical and conceptual—of an interface between two "incommensurable paradigms"—the Ptolemaic and the Copernican. The pre-*Structure* Kuhn himself defined the Tychonic system as "an immediate by-product" of Copernicus's *De Revolutionibus*, despite the fact that "Brahe himself would have denied this" (1957, 205).

In fact, during the sixteenth century some Wittenberg astronomers accommodated certain Copernican theses to the traditional geostatic framework. Far from finding "communication" between Kepler and Tycho impossible, astronomers incorporated Keplerian orbits into the Tychonic system[22]—a development that the post-*Structure* Kuhn

21. During these years, Kuhn gave up the notion of Gestalt conversion, yet retained his feeling that following a revolutionary change, a scientist experiences a totally different world, "incommensurable" with the old one.

22. "The use made of Kepler's elliptical hypothesis shows the same kind of selective appropriation that can be found already in the assimilation of Copernicus's planetary models in the sixteenth century. Not everyone who accepted Keplerian ellipses in the seventeenth century also adopted the terrestrial motions, nor was anyone 'converted' by new planetary theory in the SSR [*Structure*] sense of conversion. The astrologer and French Royal Professor of Mathematics Jean-Baptiste Morin (1583–1653), for example, managed to accommodate Keplerian ellipses in a Tychonic frame. In this he followed a

perceived as a conceptual impossibility: "What is regularly ignored . . . is the *elementary fact* that elliptical [Keplerian] orbits would have been useless if applied to any geocentric astronomical scheme" (1971, 138, my italics).

Turning to another revolution—that in chemistry—we again find much greater flexibility and selective appropriation than the notion of paradigm change entails. Lavoisier himself initially combined the phlogiston theory with his emerging ideas on the role of air in combustion and calcination (work by Perrin, as discussed in Mauskopf in press).[23]

Another field of research—relativistic gravitational physics—has been explored by John Earman (1986, 1993). What might look from textbooks like a monolithic paradigm is in fact a pluralistic field, characterized by a diversity of opinions and approaches.

Finally, I have argued throughout this book that in the quantum revolution, mutual borrowing of techniques from competing camps and even of "incommensurable" imagery characterized the creative efforts of Heisenberg, Schrödinger, Dirac, and Born. The emergence of Heisenberg's uncertainty principle was inspired by Schrödinger's wave imagery, which Heisenberg publicly denounced; Schrödinger's solution of the radiation problem was inspired by Heisenberg's matrix approach, which Schrödinger characterized as "frightening and repulsive."

As we have noted, Heisenberg and Schrödinger were closer on certain issues than Heisenberg and Bohr (chapter 6); just as Kepler was closer to the "anti-Copernican" Tycho than to the Copernican Giordano Bruno (Westman 1994, 95).

Paradigms and Holism

The notion of holism underpins Kuhn's incommensurability. Paradigms are holistic because "they require a number of interrelated changes of theory to be made at once." Only at the price of incoherence could these changes be made piecemeal, one step at a time. Kuhn illustrated his point by describing his first encounter with Aristotelian science, which at first seemed "dreadfully bad." Only when Kuhn understood Aristotle on his own terms and realized how the pieces of Aristotelian physics "fit together," giving each other "mutual authority and support," did he understand the nonarbitrariness and holism of Aristotelian physics (1987, 8).

---

strategy similar to those mid-sixteenth-century Wittenberg astronomers who accommodated Copernicus's equantless devices into a geostatic reference frame" (Westman 1994, 105–6).

23. I am grateful to S. M. Mauskopf, who shared with me his manuscript on the historiography of Lavoisier's studies, "The Chemical Revolution."

What Kuhn's description nicely demonstrated was his initiation into the historian's state of mind, gaining respect and empathy for the integrity of past science, learning how to enter into the "shoes" of past thinkers, realizing that past knowledge was ingeniously and carefully crafted. It was not an argument for inevitable holism. In fact, when describing how the idea of a unique central earth was "interwoven" (rather than later "interlocked") with fundamental concepts "within the fabric of Aristotelian thought" (1957, 84) before *Structure*, Kuhn was more careful: "Earth motion does not necessitate either the existence of a vacuum or the infinity of the universe. But it is no accident that both these views won acceptance shortly after the victory of the Copernican theory" (1957, 90).

Kuhn's argument in *The Copernican Revolution* has all the irresistibility of a rational reconstruction. Yet, as my previous discussion indicates, paradigms are not impenetrable. Neither are they necessarily incoherent if modified piecemeal. The following example will suffice: According to Bohr, the acausality and contextuality (dependence on the experimental setup) of measurement results are intrinsically linked. Bohr's numerous thought experiments supposedly showed the strict logical connection between the two. In Bohm's version of quantum theory, the contextuality of measurement is retained yet the inevitability of acausality is challenged—Bohm's theory can be deterministic. There is nothing incoherent in Bohm's framework—at least no more so than in the Copenhagen one. Bohm's version of quantum theory can, if one wishes, be constructed as a modification of the Bohrian paradigm, rather than as a radical alternative to it. Where then does the notion of "inevitable" holism come from?

When discussing Heisenberg's and Hanson's philosophy, I indicated the strong ideological drive behind the claim of holism for "closed theories" and "paradigms"—it was designed to discourage critical questioning of the orthodox view. I have also argued that Kuhn perhaps illegitimately translated his respect for the integrity of past knowledge into the notion of a binding holism. Yet the assertion of the holism of paradigms might have another source—the holism implied by the Duhem-Quine thesis.

As Robert Westman (1994) has noted, the overall image of science in Kuhn's *Copernican Revolution* (1957) resembles some of Duhem's theses, and Kuhn has acknowledged the "formative" role Quine's "Two Dogmas" played in the development of his own philosophy of science (1993, 313). The Duhem-Quine thesis asserts the intrinsic holism of theoretical frameworks: an experiment does not contradict a single theoretical statement (which consequently can be saved from refutation) but undermines the whole theoretical framework at once.

Yet this holism does not imply logical connections between different aspects of the theory. The fact that a theory is holistic in the sense that one rather than another aspect can be disproved does not mean that all aspects stand or fall together. Quite the contrary—the fact that we can choose at will which one of the elements is suspect without undermining the others means that theoretical frameworks do not have the logical cohesion implied by the notion of a holistic paradigm. Linda Wessels's recent analysis of the implications of Bell's theorem can be taken as a nice illustration of this point. Wessels enumerated a list of the usual philosophical assumptions about systems in interaction and asked which *one* of these assumptions must be given up in view of Bell-type experiments (1989, 83, 88). It is enough to give up only *one* of these deeply entrenched ideas, rather than all of them at once. Which of the assumptions one in fact discards depends on the local, theoretical, and sociopolitical circumstances.

It is this lack of a binding holism that creates room for fruitful dialogues between scientists with different approaches, and for the selective appropriation of ideas from opponents. From a dialogical perspective, the lack of paradigmatic holism is basic to the freedom (not arbitrariness!) of scientific theorizing.

Paradigms and Creativity

Kuhn's notion of dogmatic commitment to a paradigm was proposed not only as an adequate description of scientific activity; it has a prominent normative dimension as well. Paradigmatic science is "science as it ought to be done" (Masterman 1970, 60).[24] Only singleminded, concentrated, consensual effort can lead to progress. Without the rigid commitment of its practitioners, science can easily slip into wasteful anarchism—of the sort Feyerabend sometimes, tongue in cheek, advocated. Kuhn never changed his position: it is the conservatism of science, its rigid commitment to paradigms, that produces, in his words, "the greatest and most original bursts of creativity of all human endeavors" (quoted in Horgan 1991, 15).

Dogmatic adherence to paradigms ensures "rapid scientific advance" (Kuhn 1977, 232) because it alone allows the "loci of trouble" (234) to be restricted to those adamant points that refuse to surrender elegantly to authoritarian paradigmatic methods. When such trouble spots, or "anomalies," accumulate, then a turning point—the realization of the need for a new paradigm—comes. Scientific revolution is inaugurated.

An accumulation of troubles is not the only and perhaps not the best

---

24. Kuhn endorsed Masterman's dictum (Kuhn 1970, postscript).

way to grow, either in science or life. Flexibility and open-mindedness might prevent the accumulation of troubles altogether. The situation in my field—the history of science—can serve as an instructive example.

Much of the historiography of the past thirty years has been informed by the Kuhnian notions of revolution and paradigm. Historians, using Kuhn's concepts, increasingly got into trouble—first the notion of Gestalt switches and the concept of conversions had to be abandoned, even by Kuhn.[25] Yet the notions of revolution and paradigm were retained as adequate general descriptions of scientific change. Adherence to these ideas has led historians to perform an astonishing variety of ad hoc moves: We have a "quiet" revolution in chemistry. We have revolutions of Russian type and revolutions of Franco-British type (Gillies 1992a, 5): the Copernican revolution was "Russian," while the Einsteinian was "Franco-British."[26] In mathematics, revolutions are only of the Franco-British type (Gillies 1992a, 6). We have revolutions that last no less than 300 years (the Newtonian), and others that last only (!) 150 years (the probabilistic revolution; Krüger, Gingerenzer, and Morgan 1987). We also have somewhat sterile scholarly debates: should we call certain developments revolutionary even if they do not fit the Kuhnian framework (Gillies 1992b)?

Do we have to accumulate anomalies to such a degree before we can begin to question our concepts? Would we not be better off if we had, from the beginning, been open to the adequacy of other types of narratives? The idea that dogma is needed to prevent the wasting of effort is built on an erroneous presupposition—that the notion of a paradigm is adequate. If different options are constantly open to negotiation, examination, selective appropriation (as I have argued is often the case), there is no duplication of effort or dispersion of creative energy. Dispose of the notion of rigid paradigms, and the notion of "waste" disappears as well. In fact, opposition is vital for advance—not only as competition or incentive, but as a reference point around which one's own ideas are elaborated, modified, and strengthened.

Of course, a certain amount of perseverance is necessary for the success of any project. Yet perseverance does not imply dogmatic

25. As Westman (1994) has pointed out, in *Copernican Revolution* most "conversion" stories promised by Kuhn remained untold.

26. In the Russian type of revolution the old concepts and patterns of understanding are "irrevocably discarded." In 1917 the tsar was executed, and the monarchy eliminated. In the British and French revolutions of the seventeenth and eighteenth centuries, respectively, the monarchies were overthrown and the kings executed, but the monarchies were eventually restored. In the Franco-British type of revolution, the "old" persists but experiences "considerable loss of importance." According to Gillies, the Copernican and chemical revolutions were of the Russian type, while the Einsteinian was of the Franco-British type (1992a, 1, 5).

commitment. Similarly, mastery of one's tools is necessary in any field—but does it mean that the use of scientific tools (mathematical, experimental) is limited to puzzle solving? And whereas the importance of puzzle solving should not be underestimated, should it be glorified? Scientific activity is not merely puzzle solving, it is also problem solving. Yet without alternatives, problems sometimes cannot even be defined. Without Bohm's challenge, we would not have Bell's results (Bell 1966; 1987, 160).

The creativity and longevity of what Kuhn called "normal science" is due, I suggest, not to its dogmatism but rather to its lack of it. The resilience of normal science is possible because it fits the Kuhnian notion of normal science poorly. In contrast, scholars have found that Kuhn's notion of rigid paradigms fits political groups and religious communities especially well (see the papers in Gutting 1980).

Kuhn's notion of the scientist as a puzzle solver seems especially apt as a description of the experience of graduate and postgraduate students in science—such as Kuhn himself was before he turned to the history of science.[27] The experience of a mature creative scientist (not a scientist in a "mature" science) is perhaps better characterized by David Finkelstein: "Physics presently functions with *many of its practitioners* simultaneously seeking and doubting the existence of what they seek" (1987, 291, my italics).

This simultaneous seeking and doubting, when it occurs within a web of stimulating, unrestricted dialogue, reaps a rich scientific harvest. The dialogical approach, in contrast to the Kuhnian structure of incommensurable paradigms and paradigmatic commitments, perceives communication between friends and foes alike as the precondition of all scientific creativity. Unlike the concept of a paradigm, which by its nature excludes the "other," the dialogical approach celebrates the existence of other minds as indispensable for scientific advance and for intellectual self-identity. In science, as in all of our endeavors, giving up openness, tolerance, individuality, and freedom gains us nothing.

---

27. "But both *my own experience* in scientific research and my reading of the history of science lead me to wonder whether flexibility and open-mindedness have not been too exclusively emphasized as the characteristics for basic research" (Kuhn 1977, 226, my italics).

⤳

CHAPTER 15

Dialogical Philosophy and Historiography:
A Tentative Outline

Dissonance,
if you are interested,
leads to discovery.

William Carlos Williams 1936, 176

To be means to communicate.

Mikhail Bakhtin 1979, 312

The uncertainty of science, its open-endedness, and the lively disagreement among its practitioners are sources of never-ending vitality. Science is a remarkable human achievement because of its ability to turn both uncertainty and disagreement (supposedly disadvantages) into formidable strength. This ability to turn weakness into strength is due to the basic responsiveness of scientific efforts—a responsiveness that is institutionalized into an immense and amazingly fruitful system of scientific communication. This responsiveness, combined with ever-growing experimental ingenuity and mathematical sophistication, underlies the unparalleled achievement of science.

The dialogical analysis I articulated in part 1 of this book agrees remarkably well with the view of science as seeking and doubting simultaneously, as doing one's best today while being aware that it all can be overthrown tomorrow. When the freedom to doubt is suppressed or circumscribed, the conditions for the emergence of dogmatic ideologies prevail in science as in political life (part 2).

Observing that science is practiced in a communicative network, with cultural, political, and institutional dimensions, is to state a platitude of current studies of science.[1] Nor is there any dearth of excellent historical studies of scientific dialogues and controversies. However, the use of

---

1. For the most recent example of the wide-ranging contextual study of science, see Galison (1997).

rigid terms for the description of science, even if reduced to a local "here" and "now," does not permit an adequate expression of fruitful scientific freedom and constructive doubt. Such terms as "conceptual schemes" and "frameworks," when combined with the realization of theoretical freedom (the underdetermination of theory by data, the theory-ladenness of observation), lead to the notion of proliferation of frameworks, to relativism and arbitrariness (Feyerabend 1975).

In contrast, the dialogical approach acknowledges the freedom to theorize yet does not presuppose "conceptual frameworks" and thus does not lead to their plurality. When science is practiced in communicative interaction, there is no serious reason to expect the existence of independent, let alone incommensurable, theoretical alternatives (chapter 14).

While the idea that science is practiced in a complex communicative network is a commonplace, the basic notion of the dialogical approach developed in this book—the multidirectionality and addressivity of a single elementary thought—is far from being acknowledged, let alone assumed, in existing accounts of the practice of science.

In Praise of Disagreement

I propose to supplement current approaches in the history and philosophy of science with the dynamic idiom of the "flux of addressivity." I propose to treat the addressive response as the primary epistemological and social unit for the analysis of science. Thus the notion of a scientific thought presupposes the existence of an interlocutor to whom the thought is addressed or by whose statements the thought is triggered. Such "addressivity" is complex and multichanneled; scientists respond to many colleagues and at the same time address other, potential and actual interlocutors. This open-ended, multidirectional, constantly changing web is the source of the conceptual tensions and ambiguity that fuel creative scientific effort.[2] I adopt the term "dialogism" for this dialogical approach to the formation of scientific knowledge (later in this chapter I will comment on the source of this term). Dialogism is a process epistemology, not a monological structural theory of the growth of knowledge. The notion of addressivity is not a static concept; it is both directional and temporal. It introduces the locality, contextuality, and historicity of scientific practice in a fundamental way.

2. What I call the multidirectional web of addressivity is in fact a time-varying network, where by multidirectionality I mean the connectivity of the "I"-node to other nodes representing different interlocutors. I minimize the use of the term "network" because it has some connotations in the field of studies of science that I wish to avoid.

Dialogism as an epistemology of science is built on the notion of the "other." In the dialogical approach, the "social" is not necessarily shared or consensual, rather the social is that process or concept that by its nature assumes the existence of the "other." Dialogism does not deny the sociological aspect of science, nor does it merely supplement the basic cognitive substratum with "social factors." The approach of reducing the "cognitive" to the "social" is also alien to dialogism. In the dialogical approach, the social and the cognitive are fused in the notion of an addressed thought and in the notion of a communicative creative act.

Neither positivist nor post-Kuhnian conceptions of science have any epistemological function for the "other." Thus there is no place in the current historiography of science for scientific individuality. The notions of agreement and consensus, on either the global or the local scale, underlie most existing scholarship in studies of science and society. While in pre-Kuhnian historiography agreement among scientists was thought inevitable because most scientists followed the "rational method," post-Kuhnian scholarship examines the social mechanisms leading to consensus. And while excellent historical studies of controversies and disagreements abound, they assume that the state of disagreement is a temporary one—the disagreement is inconsequential after the "experiments end" (Galison 1987), once "stormy waters settle" (Collins 1992, 4), or when the controversy is "closed" (Latour 1987).

In the dialogical approach, the existence of the "other" establishes the preeminence of the concept of disagreement. This is not to say that agreement does not exist in science, or is not valuable in the growth of knowledge. But the explanatory power of the notion of agreement is limited. Excessive attention to consensus has become a barrier to the construction of a dynamic theory of scientific practice. Too much reliance on notions of consensus has resulted in a great loss to contemporary historiography—the almost total extinction of studies of individual scientific creativity and of analyses of detailed theoretical labor. Too much emphasis on agreement, combined with the philosophical thesis that theory is underdetermined by experimental data, has resulted in the excesses of the sociology of knowledge and the flattening of the cognitive to the social by social constructivists.[3]

3. The thesis of the underdetermination of theory by data is based on the realization that a finite set of experimental results does not determine a unique scientific theory. Many theories compatible with the same data are possible. Philosophers who refuse to draw relativistic conclusions from this analysis employ the notion of epistemic "virtues" (simplicity, consistency, generality) to single out a preferred theory. The founders of the sociology of scientific knowledge rightly point out that such epistemic virtues fail as explanatory categories for the description of scientific practice (Bloor 1981). They argue instead that the choice of scientific theories is determined by social categories such as in-

The notion of uniform agreement, or consensus, is rooted in prominent sources in the philosophy, sociology, anthropology, and history
of science. In addition to Kuhn (1970), these sources include such important names as Douglas (1975, 1986), Fleck (1979), Habermas (1972),
James (1907), Peirce (1931–35), Rorty (1979, 1991), and Wittgenstein
(1953). Those thinkers who aimed to replace the bankrupt notion of
truth as correspondence with reality employed instead the notion of
intersubjective agreement (Rorty 1991). They perceived such a consensual, community-based vision of science as having clear humanistic
advantages. In Rorty's (1991) work, science provides models of "unforced agreement" and "human solidarity." Yet the notion of paradigmatic science is incompatible with the virtues of tolerance and open-
mindedness. By its very definition, a paradigm excludes diversity of
opinion and impugns the search for viable alternatives. I have argued
that the notion of a paradigm legitimates the orthodox and silences the
opposition (chapter 14). It is only a totalitarian society that speaks with
one voice.

Nor is the notion of agreement between minds less problematic than
that of agreement between "idea" and "reality." Do we have any reliable epistemological tools to establish the existence of such agreement?
The testimony of people who say they "agree" is, of course, insufficient.
Bohr's delightful expression "We agree more than you think" exemplifies this point.[4] The notion of intersubjective agreement seems to me as
deeply flawed as the correspondence theory of truth.

The role of a scientific community in the Kuhnian and post-Kuhnian
historiography of science is to create and to implement agreement. Yet
in this case, the scientific collective is epistemologically superfluous.[5]
Moreover, something odd is going on in the scientific community all
the time. The more historians explore actual scientific practice, the more
they discover that controversy is the order of the day. If the aim is agreement, if the only way to progress is agreement, why do scientists spend
so much time in long, sometimes bitter confrontations? Should not the
aims and the means of scientific practice be in somewhat better harmony? Rather than explaining away the disagreements as anomalies,
we need a theory of knowledge in which disagreement finds its proper,
prominent, and permanent epistemological place.

Closely connected with the notion of consensus is the notion of

---

terest. They also claim that they do not intend to eliminate the importance of cognitive
aspects. The problem is not with their intentions but with the dearth of conceptual tools
for integrating the social and the cognitive.

4. Bohr's younger disciples in Copenhagen chose this expression as the epigraph for
the *Journal of Jocular Physics,* in which they expressed their disagreements with Bohr in a
subtle, good-hearted fashion (Beller 1999).

5. I elaborate on this point later in this chapter, when discussing the "lesser" scientists.

"belief." Scientific solidarity occurs when all participants arrive at the same belief—either by the uncontaminated use of their reasoning faculties ("unforced agreement") or by "brainwashing," propaganda, or "mob psychology."[6] The notions of "belief" and "commitment" are no less problematic than the notion of agreement. Scientists appear to be too opportunistic, they often betray their "beliefs" and break their "commitments," deviating from codes of scientific methodology and from collective norms. The notion of belief underlies the older, "internal" history of scientific ideas, and its mirror image—the strong program in the sociology of knowledge. In the strong program, "beliefs," which are socially, rather than cognitively, determined, are no less fundamental than in "internal," purely intellectual, history.

It is widely assumed in the current historiography of science that the only viable alternative to the solitary knower is the scientific collective, held together by shared belief and mutual trust. Another assumption is that such "trust" is mandatory for the formation of "shared belief," and for the existence of the "social and cognitive order" (Shapin 1994). Yet another basic notion in the historiography of science—both the positivist and the sociological—is the notion of "commitment." Those scientists who have strong "beliefs" are "committed" to their statements. For how is one to be trusted if one does not believe in (if one is not committed to) what one is doing?[7]

The dynamic sociology of knowledge should not rely on static notions of "fixed belief" and "commitment." We have to disassociate the idea of devotion to one's work from that of belief, perseverance from commitment, cooperation from consensus, trust from conformism. In a multidirectional dialogical response one can strike a creative balance between trust and doubt, acceptance and skepticism, agreement and disagreement. To reinstate the centrality of the notion of disagreement, we have to strip away its negative connotations. Intellectual disagreement need not be, and often it is not, quarrelsome or aggressive. While conflict among global "paradigms," "conceptual schemes," or "systems of thought" might imply threatening competition for survival, and the notion of pluralism of frameworks might suggest "wasted effort" (Kuhn's opinion, see chapter 14), the notion of local disagreement has none of these dark overtones. Local disagreement is fruitful and constructive, rather than threatening and wasteful.

6. In the social constructivist approach, the notion of "truth" is identical with the notion of "shared," or "fixed" belief: "For historians, cultural anthropologists, and sociologists of knowledge, the treatment of truth as accepted belief counts as [a] maxim of method, and rightly so" (Shapin 1994, 4).

7. "How could coordinated activity of any kind be possible if people could not rely upon other's undertaking? There would be no point to keeping engagements . . . with people who could not be expected to honor their commitment" (Shapin 1994, 8).

The notion of agreement permeates some of the most exciting work in the history of science. Shapin and Schaffer have provided a pioneering account of the origins of modern science as a social process aimed at ending dangerous disputes (Shapin and Schaffer 1985; Shapin 1994). They describe how Boyle created distinct "technologies" for generating collectively agreed upon "matters of fact." Yet what Boyle simultaneously aimed to establish were the rules for disagreeing in a gentlemanlike fashion: "how to speak of things with freedom," and yet in the manner "of persons with civility." Finding things to agree upon ("matters of fact") was possibly a necessary condition for the conducting of scientific inquiry, yet not a sufficient one. The sufficient condition, I suggest, was an institutional way to handle disagreement, thus turning a vice into a virtue, a predicament into an asset. Henry Oldenburg, the secretary of the Royal Society, who transformed his network of correspondence into a scientific periodical, *Philosophical Transactions*, realized that in order to get ahead one had "to clash good men's heads together." What became distinctive of modern science was not so much the scientific "method," or scientific "norms," but the strongly institutionalized communication system (scientific periodicals, scientific gatherings) to handle disagreement—a system that one cannot afford to ignore if one is to survive as a professional in a scientific community. This communication system increased the pace of scientific discovery exponentially, not merely because discoveries were no longer lost, but because it set the mechanism of communicative interaction into full motion.[8]

According to Rorty, scientific institutions can serve as exemplary models for the rest of the culture because they "give concreteness and detail to the idea of 'unforced agreement'" (1991, 39). Yet it is less unforced agreement than unforced response—be it agreement or disagreement—that is a crucial part of the scientific way of life. Those thinkers who identified consensus as a necessary condition for social order spoke of "our colonization of each other's mind" as the "price we pay for thought" (Douglas 1975, xx). Yet there is no need for such brutal "colonization"; it is uncoerced intellectual engagement and the right of free response that are the very preconditions of scientific thought. Disagreement is a necessary part of such a response. Full agreement stops conversation.

---

8. Studies of the scientific revolution (in the sixteenth to seventeenth centuries) center on experimentation, mathematization, and the political underpinnings of the "new science." These studies do not in general deal with the establishment of scientific communication systems, and with their impact on the emerging scientific effort. Such a discussion is absent from two recent comprehensive studies of the scientific revolution (Shapin 1994; Cohen 1994).

The Philosophical and Historiographical Advantages of Dialogism

Dialogism describes all novelty, both dramatic and relatively routine, by the same mechanism of communicability. There are no irrational "Gestalt switches," or jumps, in such a description. In this sense, dialogism is an unfolding description of the growth of knowledge, yet not a linear one. Dialogism retains the idea of the progress of knowledge without reducing it to one-dimensional accumulation. It also retains the fruitful notion of the continuity (the conservative aspect) of science. The notion of addressive response is both social and cognitive. It is in principle local and open ended. The notion of "best informed temporary presupposition" is more suitable to the description of dynamic scientific activity than the sociological notion of "fixed belief."

In the dialogical approach, the historical, philosophical, and sociological merge into a unified viewpoint, rather than being independent perspectives divorced from each other. The multidirectionality of dialogical responsiveness accounts for the spectrum of possibilities, from the internal history of "pure" ideas to the strong versions of social constructivism, obtainable as one-dimensional projections. The social determinism in the writings of sociologists of knowledge is not an adequate description because it is confined to a single dimension. The description of acausality in quantum mechanics as an expression of the zeitgeist (Forman 1971) therefore provides only a limited perspective. The notion of zeitgeist is itself a monological notion. The emergence of Heisenberg's verdict on the essentially acausal nature of the quantum formalism was a complicated story, characterized by vacillations and about-faces, not a one-dimensional adaptation (chapters 2, 3, 4, 5, 8, and 9).

The dialogical approach incorporates the main message of Kuhnian and post-Kuhnian historiography that an adequate philosophy of science must take into account the lessons of detailed sociohistorical studies. Kepler's statement (an epigraph to this book) can serve as an inspiration: "The roads by which men arrive at their insights . . . seem to me almost as worthy of wonder as these matters in themselves" (quoted in Koestler 1960, 59). Yet to understand how people arrive at their knowledge is not simply fascinating—it is directly relevant to the epistemological status of a knowledge claim. A successful scientific result is one that is strongly embedded in the practice of the day. This embedding is achieved by responding simultaneously to as many other scientific statements (read in papers, heard in conversations, and so forth) as possible—drawing on many possible supports (confirmations), taking into account many possible objections (refutations). Both Carnap's idea of confirmation and Popper's idea of refutation are, again, one-dimensional idealizations, and both can be present at the same time in a dialogical response.

Karl Popper's theory of the growth of knowledge, in contrast to the Kuhnian theory, incorporates the uncertainty and open-endedness of scientific theorizing admirably. Yet Popper's identification of "testability" with "refutability" is too narrow, as many critics did not fail to notice: scientists do other things besides trying to falsify their own theoretical constructs. Moreover, from Popper's perspective, scientific theorizing is fundamentally irrational, for there is no organic connection between the method (refutation) and the goal (truth; Newton-Smith 1981, 44–70).

Dialogism does not suffer from this weakness. From the dialogical perspective, the method (dialogical response) and the goal (strongest possible support for scientific result, taking into account as many potential refutations and confirmations as possible) are closely linked. The dialogical approach I am suggesting is not that of Lakatos (1970) or Elkana (1977), in which dialogues take place between global theoretical frameworks, such as "research programs" or "metaphysical systems." According to dialogism, such global constructs conceal rather than reveal the basic communicative nature of scientific creativity.

In the dialogical approach there is an organic, intimate connection between the philosophical and sociological accounts of scientific activity. The form of a scientific article that includes references to other papers appears to be an artificial convention in most theories of the growth of knowledge. In contrast, from the dialogical point of view, citations mirror the basic mechanism of creativity. Citations find a prominent place in Bruno Latour's (1987) account of scientific activity. Latour's analysis of citation as a means by which ideas survive is a penetrating description of the mechanism of the dissemination and consolidation of knowledge. Yet Latour's analysis is more about the politics of domination than about scientific cognition. His approach has little to offer if one wants to understand an individual creative act. The conflation of the political and the cognitive might sometimes take place—but it need not and often does not. Dialogues need not be, even if they sometimes are, militant struggles for ascendancy. Basic dialogical responsiveness is not about the elimination of an opponent. Such responsiveness can be both critical and sympathetic.

When I say that citations in a scientific paper mirror the basic mechanism of scientific creativity, I do not mean to imply that the list of references tells the story in a straightforward way. One can, for reasons of self-aggrandizement, conceal one's sources—a common cause of priority battles. One may be unable to include all sources because some are not yet published (as Dirac was unable to list Heisenberg's fluctuation paper in Dirac 1927, though he mentioned his oral discussions with Heisenberg). Or one can cite papers for reasons of etiquette alone (as did those scientists who cited matrix mechanical papers while using the wave theoretical approach). Citations might also be, as Latour has demonstrated, about power alliances. Yet citations are predominantly about the emergence of ideas, about the birth of novelty.

The art historian Ernst Gombrich (1963) has quoted one of his colleagues as saying that "all pictures owe more to other pictures than they do to nature." By analogy one could say that scientific papers owe perhaps more to other scientific papers than to nature. This is not to discount experiments in the development of science but to emphasize that the evaluation and incorporation of experimental results takes place in an overall communicative flux of ideas. In such a communicative web, scientists have the freedom to take an apparent incompatibility between a theory and observational facts not as a refutation of the theory but rather as a reason to reinterpret the meaning of the observational statements (theory-ladenness of observation). Such was Heisenberg's reinterpretation of the electron's path as a sequence of discrete points subject to uncertainty relations rather than a continuous trajectory (chapter 4). This freedom of interpretation does not imply arbitrariness, quite the opposite: the theory-ladenness of observation provides the leeway that allows a new result to relate to a larger number of loci in the dialogical web, thus entrenching it more firmly in the practice of the day.

The scientific community in the dialogical approach is not a gray army of indistinguishable puzzle solvers. Dialogism reevaluates the role of the "lesser" scientists in the development of scientific knowledge. In both evolutionary and revolutionary accounts lesser scientists are, epistemologically, dispensable. In internal accounts of the history of science, the evolution of ideas—"the great chain of being"—occurs exclusively in the minds of great scientific heroes. In Kuhnian accounts, lesser scientists obediently solve the puzzles posed by the great scientists, founders of the new paradigms. Expand the life span of the great scientists long enough, and they would accomplish everything themselves. The scientific community is, epistemologically, superfluous. In the dialogical approach the lesser scientists are as indispensable as the great ones. In chapter 4, I demonstrated that some of the most important insights pertaining to the formulation of the uncertainty principle by Heisenberg belong to scientists eliminated from the history of quantum mechanics—Sentfleben and Campbell.

From a dialogical perspective, the scientific community is composed of living, feeling individuals, rather than cognitive robots endowed only with reasoning faculties. From the perspective of philosophical accounts, which formulate the criteria for the growth of knowledge in pure epistemic terms ("degree of confirmations"—Carnap; "verisimilitude"—Popper; "positive heuristic"—Lakatos), the intrusion of emotions into scientific activity is a regrettable, if not incomprehensible, aberration. While emotions have a prominent place in the sociologically informed history of science, their existence remains epistemologically unsatisfactory because they are derived from sociological notions. Robert Merton (1957) connects emotional battles for priority with scientific norms (originality) and scientific rewards (fame). In the strong program, emotions are derived from the existence of "interests."

In dialogism, affect and cognition are closely linked. There is a natural, organic connection between emotion and cognition in the very act of the local dialogical response. If our intellectual activity is in principle responsive to other human beings, then cognition and emotion are often fused. The ideas we encounter are not "disembodied"; they are colored by the voice of the other. These voices are marginal or authoritative, remote or immediate, threatening or sympathetic. This is the reason that even the most abstract scientific notions can arouse fierce emotions (chapters 2 and 4). Henri Poincaré observed that most creative mathematical ideas are "those which, directly or indirectly, affect most profoundly our emotional sensibility" (quoted in Vernon 1970, 77–78). This emotional sensibility can be connected with personal ambition, as revealed in the confrontations between Heisenberg, Schrödinger, and Bohr (chapters 2, 4, and 6): emotions in scientific activity can be connected with great aesthetic satisfaction, as in Dirac's case (Kragh 1982; Darrigol 1992a), with meaningful existential issues, as for Bohr (chapter 12), or with religious and mystical issues, as for Pauli (Laurikainen 1988). While most recent historiography of science is hostile to individual psychology (concentrating instead on shared ideas, values, and norms), the dialogical approach centers on the creativity of individual scientists. Such individual psychology should be based not on "purported eternal verities of the mind," which historians, with good reason, reject (Daston 1995, 4); rather it should study the individual's complex set of responses. Dialogues, be they silent or vocal, are often fueled and colored by individual emotions.

Connected with the emotion-cognition issue is the philosophical distinction between the "context of discovery" and the "context of justification." The process of discovery can be contaminated with emotions, while the process of justification supposedly consists of objective, rational evaluation (Reichenbach 1959; Popper 1959). This distinction is increasingly challenged by historians, philosophers, and sociologists of science (Nersessian 1995; Nickles 1995; Knorr-Cetina 1981). From the dialogical point of view there is no essential distinction between the process of discovery and that of justification. If we do not formulate our thoughts in a way that is addressed to the concerns of others, we cannot expect them to respond. Heisenberg's uncertainty paper was permeated with his responses to Schrödinger's contention that the matrix (Heisenberg's) version of quantum mechanics was so abstract and nonintuitive that it would impede further progress in physics (chapters 3, 4, and 5). The need to meet Schrödinger's challenge, and to persuade the scientific community of the great value of the matrix approach (and so ensure its acceptance and dissemination), was both a motivation and a presupposition of Heisenberg's intellectual efforts. Thus discovery and justification in the dialogical approach are in fact the same kind of processes, permeated with responsiveness and addressivity.

This is not to say that there is no distinction between the process of discovery and the resulting achievement. The elegant formulation of matrix theory as an algebraic structure is different (a category difference) from the contrived path that led to its discovery. Yet this structure was justified, not by some general method comparing theory and experiment, but rather by the same communicative process of using matrix theory in further research. While the history of its birth is largely eliminated from the reported scientific result, a trained eye can sometimes discern the traces of the process of discovery even in abstract mathematical formulas. Thus the elements of the position matrix are a reminder of the reinterpretation procedure that led Heisenberg to the matrix formulation of quantum mechanics (chapter 2).[9]

Theory as Practice: Between Tools and Metaphors

Recently much work in the history of science has focused on the idea of scientific knowledge as practice rather than as static atemporal representation. The new approach explores the multiplicity and heterogeneity of scientific doing, describing the different aspects of scientific activity—the conceptual, the social, and the material—as fragmented and local. A prominent philosophical advocate of this approach is Arthur Fine (1986, and especially the new postscript in 1996b). In addition to Fine's work, this trend owes much to Ian Hacking's influential *Representing and Intervening* (1983). Hacking's book has inspired increased attention to the notion of experiment, and to a thorough reevaluation of the complex role and nature of scientific experimentation. Hacking's

9. Knowing the heuristics of discovery can also illuminate the physical understanding of a theory. This is the reason Pauli declined Born's offer to collaborate on the elaboration of Heisenberg's results into the matrix formulation: "You are only going to spoil Heisenberg's physical ideas by your futile mathematics" (quoted in van der Waerden 1967, 37).

work has been followed by such important books as Shapin and Schaffer (1985), Franklin (1986), Galison (1987, 1997), Buchwald (1994, 1995), and Pickering (1995).

The dialogical approach is well suited to emphasize the dynamic aspect of scientific theorizing, and to minimize the static elements of the representational idiom in the description of the growth of knowledge. Theory, no less than experimentation, is practice: scientists engage in theorizing, writers of textbooks and philosophers describe "theories." This is not to say that scientists do not use representations and visual imagery—graphs, photographs, computer images, diagrams, and visual models are indispensable scientific tools. Yet the meaning of such representations, like the meaning of observational terms, is embedded in the dialogical context. The idea of the de Broglie wave packet (superposed waves with different wavelengths) as a representation of a microscopic particle was employed extensively in quantum theory and its interpretation. Yet Schrödinger, Heisenberg, and Bohr used the idea of a wave packet each in his own way and each for his own ends. Each ascribed to the concept a different theoretical function and a different degree of reality, and each employed the wave packet imagery to criticize the use of this notion by the opponent (chapters 4, 6, and 11).

The notion of theorizing naturally introduces a temporal element into historical description;[10] in the dialogical approach theorizing inherently takes time. In chapter 2, I described the emergence of Born's probabilistic interpretation as a process "in flux," during which creation, deliberation, and reevaluation (all of which take time) occurred. In chapter 4, I described how Heisenberg's uncertainty principle emerged from a coalescence of insights, arrived at in different dialogues. Heisenberg's preference for acausality was shaped as a response to Schrödinger's competitive, causal version of quantum mechanics, to Campbell's discussion of time and chance in quantum theory, to Sentfleben's deliberations about causality, and to Jordan's doubts about the essentially statistical character of quantum theory. Causality issues also surfaced in Heisenberg's dialogues with Pauli and Dirac. Parts of these dialogues occurred concurrently, while others occurred at different times. We see here not the unfolding of a single timeless argument, but the gradual coalescence and confrontation of insights that, over time, reinforced, illuminated, and qualified each other. The notion of practice invokes the notion of tools (mathematical, experimental); and while dialogism, as developed here, does not provide adequate epistemological underpinnings for the instrumental aspect of science, it does not contradict

---

10. The real-time understanding of practice is, as several authors argue, a major historiographical challenge today (Pickering 1995; Stump 1996).

this aspect in any way. The dynamic, performative nature of scientific theorizing is common both to the instrumental and dialogical aspects of science.

When there was agreement in the community of quantum physicists, it was more often about the efficiency of tools than about metaphysical "paradigmatic" issues. Einstein did not doubt the unprecedented utility of quantum mechanical tools, only their philosophical, or rather ideological, interpretation by the orthodox. Einstein challenged the Copenhagen philosophy by skillfully using the mathematical tools of the quantum formalism (chapter 7). Einstein especially disliked Bohr's metaphor of complementarity, which he considered arbitrary and removed from scientific theorizing (chapter 12). Dyson similarly places the revolutionary aspect of science in the construction of new tools rather than in philosophical ideas or metaphors. One can build many different models, metaphors, and conceptual worlds with the same tools—such models and metaphors can be dispensable, while the tools are not.[11] The contrast between the solidity of tools and the fluidity of metaphorical structures represents a fascinating tension in the history of theoretical physics. The tools are used in everyday practice, while revolutionary philosophical ideas and metaphors are employed primarily for dissemination and consumption (chapters 8, 9, 11, 12, and 13).

Kuhn regarded a revolutionary change as metaphorical redescription: "The juxtaposed items are exhibited to a previously uninitiated by someone who can already recognize their similarity, and who urges that audience to learn to do the same" (1987, 20). Bohr's complementarity was just such a metaphorical redescription. Yet Bohr did not find the existing similarities—he imaginatively created them (chapter 12). Not surprisingly some of the most important textbooks on quantum mechanics omit the complementarity principle (Dirac 1930; Tomonaga 1962; Feynman, Leighton, and Sands 1969). The fact that quantum mechanics is such a potent theory and yet has no uniformly accepted interpretation demonstrates that a coherent metaphysics is not necessary to the advancement of theoretical science.[12] That philosophical agreement is dispensable in scientific theorizing is clear from Schweber's (1994) masterful study of the history of quantum electrodynamics. The

11. Though the application of tools is neither self-evident nor self-explanatory. Outside a dialogical web, formulas are mute. While the impossibility of the simultaneous determination of position and momentum was implied in the basic commutation formula $pq - qp = (h/2\pi i)\mathbf{1}$ from the beginning, it was only mathematically derived later by Heisenberg, during his struggles with the interpretation of quantum mechanics.

12. This conclusion is akin to Duhem's attitude, and to that of Duhem's most prominent follower today, van Fraassen.

outstanding theoretical physicists who founded quantum electrodynamics—Dyson, Feynman, Schwinger, and Tomonaga—had less motivation to build easily digestible, visualizable models and explanations
for wider consumption than their prominent predecessors, who erected
the nonrelativistic quantum mechanics.

To say that metaphors play a limited role in theoretical physics does
not imply that the search for metaphors is superfluous or illegitimate.
Metaphors can be powerful heuristics. In addition, the need to find
some broader meaning in scientific results, fluid as they are, the need to
reconcile one's intellectual activity with one's spiritual life is one many
scientists feel. It is this need that accounts for the wide appeal of Bohr's
complementarity (chapter 12).[13] Yet complementarity, I have argued,
is not merely an inspiring metaphor; it is also a potent ideological
weapon. Bohr and his followers imposed rigid strictures on the imagination of physicists, claiming that complementarity is not simply one
possible way to interpret quantum physics but the only possible way
(chapter 9).

Truth and Beauty

Seemingly irreconcilable tensions characterize the pronouncements
physicists make about scientific theorizing. On the one hand, they talk
about nonarbitrariness, well-groundedness, and the near inescapability
of scientific results. On the other hand, they describe the feeling of freedom, imagination, beauty, and even "poetry" that underlies creative
scientific effort. Those philosophers who sought theories of individual
rationality and a binding scientific method did not leave epistemological room for free individual creativity. Sociologists and social historians
of science dispensed with the idea of an overarching scientific method,
yet their accounts often imply too much cognitive arbitrariness.

Dialogism can incorporate both the freedom and the nonarbitrariness
of scientific work. The editors of *The Flight from Science and Reason*, summarizing the paper by Hershbach (1996), compare science to "a tapestry
of variable design, constantly under construction, with seemingly un-
related threads combining in unexpected and beautiful ways" (Gross,
Levitt, and Lewis 1996, 9). In dialogism, the threads are the creative
scientist's dialogical responses, directed or triggered by the multiplicity
of existing results reported by his or her peers. And while a working
scientist must be as informed as possible (in contrast to creative artists,

13. I distinguish between imaginative and vague extensions of scientific results by metaphors into wider domains (such as Bohr's complementarity principle) and well-defined
structural analogies used as heuristics (Bohr's correspondence principle).

say), still at each moment, in a contingent way, he or she is exposed only to a part of the existing world of results. Moreover, the scientist's responses—the threads—are not predetermined. Still more freedom comes into play when the scientist weaves these responses into a theory. Yet this imaginative, skillful weaving is not arbitrary—the creative scientist attempts to use as many threads as he or she has at hand each moment and aims to construct as tight, and as strong, and as structured a tapestry as possible. In this way the achieved result can be both persuasive to others and beautiful. The process of creation (weaving) and the result itself (tapestry) are both imaginative and well grounded.

Thus, in a dialogical process, scientific theorizing is both free and nonarbitrary. It is this nonarbitrariness that underlies a scientist's feeling that his or her theories are "touching" reality itself, that the best results are "true" and "objective." Dialogism respects the scientist's feeling that his best work is as well grounded ("objective") as humanly possible. Thus, while rejecting the philosophical notion of theories as "mirrors" of reality (the correspondence theory of truth), dialogism is compatible with scientists' intuitions of a high degree of truth and objectivity of their constructs. By concentrating on creative individuality, dialogism does not consider science subjective.

Not all aspects of scientific theorizing are equally well grounded— some are more arbitrary than others. While Heisenberg's uncertainty formula and Schrödinger's wave equation are indispensable for the working quantum physicist, he or she can create without the Copenhagen interpretation. Thus the Copenhagen interpretation is dispensable—many physicists engaged in successful scientific theorizing are opposed to it, while many others subscribing to it nevertheless never use it in their research. It is reasonable to consider such dispensable parts of science more "subjective" and arbitrary than the indispensable parts. It is illuminating to analyze these aspects of science in terms of the rhetorical strategies of persuasion (part 2).

While closely dealing with the details of scientific ideas, dialogism can be considered a social epistemology of science. It is social because, in its basic notions of addressivity and responsiveness, it presupposes the existence of other minds (other people) as a necessary condition of scientific process. While concentrating on the content of ideas, the notion of an overarching scientific method that guides an individual scientist is foreign to it. In this sense, dialogism is removed from the inspirations of logical positivists, followers of Popper, and those philosophers of science today who would resuscitate, in modified form, the idea of a self-sufficient individual rationality. The notion of a Divine Architect, or a Divine Creator, into whose wisdom and intentions a single scientific genius can penetrate (Newton, Einstein) is foreign to the dialogical approach (though dialogism recognizes such imagery as powerful, motivational, and heuristic).

Dialogism as social epistemology differs fundamentally from social constructivism. Dialogism does not hold that social determinants, in the sense of the usual sociological variables, are the primary determinants of scientific ideas. Unlike social constructivism, dialogism does not conflate beliefs and facts—dialogism shies away from the notion of belief altogether. Dialogism rejects the social constructivist notion of the consensus theory of truth: dialogism rejects the notion of consensus as mandatory and the notion of truth as explanatory.

Dialogism by its nature pays close attention to the specifics of the local situation. For that reason, dialogism does not reject the contributions achieved in other approaches, including social constructivism. Whether sociological or political factors are predominant (eugenics, Lysenko affair), mildly relevant, or virtually nonexistent (Dirac's theorizing in quantum theory) is to be determined by the specifics of the local theorizing. Dialogism accepts (and, in fact, requires) the important realization of social scholars of science that science is practiced in a rich variety of social, political, cultural, national, and institutional settings.

Arthur Fine has recently formulated an agenda for a new, adequate philosophy of science:

1. Bracket truth as explanatory concept.

2. Recognize the openness of science at every level, especially the pervasive activities of choice and judgment.

3. Concentrate on local practices without any presupposition as to how they fit together globally, or even as to whether they do fit together.

4. Remember that science is a human activity, so that its understanding involves frameworks and modalities for social action.

5. Finally, on the basis of all the above, try to understand the phenomena of opinion formation and dissolution in science in all its particularity. (1996a, 249)

We can conclude that dialogism meets Fine's challenge.[14]

The fruitfulness of the notion of "disunity" in science is powerfully illustrated in Peter Galison's (1997) recent book. Galison describes

---

14. With regard to item 1: bracketing the notion of truth does not mean that scientific theories are arbitrary and do not adequately reflect nature; rather it means that the general abstract notion of "truth" has proved bankrupt for the description of science. With regard to item 4: Dialogism does not, at this stage, offer more general social modalities, yet it is basically "friendly" to such an undertaking. The notion of basic addressivity and responsiveness, whether explicit or not, probably underlies any theory of social action.

twentieth-century physics as composed of different theoretical, experimental, and instrumental subcultures, each having a "life of their own." This collection of distinct subcultures is held together by "trading along their borders"—such trading is nothing else but local communicability, which does not require overarching agreements and thus avoids the problems caused by the use of homogeneous global frameworks (such as relativism).

Galison, following Peirce, offers a "cable" metaphor for the description of scientific process and scientific community—the distinct, partially independent fibers (subcultures) are laminated and intertwined, and the whole structure gains strength by their interaction and mutual support. The cable metaphor proves most fruitful for Galison's description of twentieth-century physics.

While there is much similarity between Galison's approach and the dialogical point of view I am articulating (an emphasis on local interactivity, a rejection of global "blocks" of knowledge, the articulation of strength following from multiplicity rather than homogeneity), there are also crucial differences between the two. From the dialogical point of view, a single strand of the cable (a subculture) is still too homogeneous; a more apt metaphor is that of (quantum) entanglement—the ideas of different scientists are so "entangled" that it is often difficult to talk about the separate individuality of such ideas. Galison introduces the cable metaphor in order to explain the cohesion of science: if one decomposes global conceptual schemes into local fragments, what holds scientific practice together? From the dialogical point of view, such a problem does not arise: the cohesive nature of the scientific enterprise, be it fragmented as it may, is assured by the dialogical connectivity of a single creative act.

The dialogical approach developed in this book bears some similarity to the ideas of Mikhail Bakhtin, the Russian literary theorist and philosopher, whose writings I encountered after most of this book was written.[15] The Bakhtinian approach is based on the concept of dialogue and the indispensability of the notion of the "other" in human existence and thinking. Every thought, according to Bakhtin, is conditioned by the actual and potential thought of the other. We are alive to the extent that we are responsive (and responsible), as well as addressive to the other.

In his classic *Problems of Dostoyevsky's Poetics*, published in 1929, Bakhtin (1994) singled out Dostoyevsky as the discoverer of a new literary

---

15. While Bakhtin has a large following in literary theory, history, and other human and social sciences, his ideas have had hardly any impact in the history and philosophy of science.

genre—the polyphonic novel. Dostoyevsky's novels and short stories display the open-endedness and irreducible plurality of human existence. Dostoyevsky's heroes are not clearly defined characters—their truth, never a completed one, emerges in confrontation with, and anticipation of, the other's point of view. Every idea, those of the characters and those of the author, is a rejoinder in unfinished and unfinalizable dialogues carried out simultaneously. The author does not have an objectifying privileged gaze: she or he can unfold the open-ended life of characters by letting them speak, yet cannot define or explain them.

Bakhtin contrasts such polyphonic thinking with the monologism of natural science. According to Bakhtin, in science there is no genuine place for the other: the other in science is the one who errs. In contrast, in art and spiritual life, two voices are the minimum for thought and existence. The Bakhtinian scholar Holquist (1990) coined the term "dialogism" to describe Bakhtin's approach. I adopted this term for a dialogical epistemology and historiography of science, despite Bakhtin's opinion that the dialogical approach does not apply to the natural sciences. Contra Bakhtin, I argue that the "other" is fundamental in science no less, and arguably even more, than in other human endeavors.

Dialogism, as developed in this book, differs substantially from Bakhtin's point of view. While it is a powerful tool for the analysis of scientific texts (chapters 5, 6, and 7), the dialogical approach to science goes beyond textual analysis. For Bakhtin, "the author" was a potent construct for the analysis of poetics and not a creative historical figure. Dialogism in science attempts to accurately reconstruct the process of discovery by creative scientists.

Nevertheless, the dialogical approach, with its awareness of the addressivity of our thinking, whether in science or in art, allows us to establish a meaningful bridge between the sciences and the humanities. There is notable similarity between my analysis of Heisenberg's uncertainty paper (chapter 5) and Bakhtin's analysis of Dostoyevsky's novels. As a novel by Dostoyevsky is shot through with rejoinders and anticipatory responses to the other, so discourse in a scientific paper carries an awareness of an interlocutor, through the author's anticipation of how the other might approach the problem. As in *The Brothers Karamazov* each of Ivan's speeches is a dialogue not only with Alyosha and himself but with other points of view that are addressed simultaneously, so Heisenberg's paper is permeated with addressivity to the "conservative Lords," to Pauli, Dirac, Bohr, Jordan, and others. As a novel by Dostoyevsky is a plurality that does not achieve resolution, so a scientific paper is an open text whose meaning will undergo change in future dialogues.

This similarity should not be taken too far. It does not mean that there is no difference between scientific and literary texts. The difference is in fact profound. While both scientists and artists respond to the works of others and address their audiences, working scientists cannot afford to be mute (if they are to survive as professionals) in the dialogical flux of ideas in their field. In particular and perhaps most important, theorists must respond to (and incorporate) the experimental findings reported by their colleagues.

The creative artist is not bound by what might be called a "dialogical imperative." If a working scientist withdraws from intense engagement with the recent work of others, his or her own work may be not merely superseded but even declared irrelevant. Thus, in addition to experimentation and mathematization, arts and sciences are distinguished by the difference in the extent and necessity of dialogical communicability.

Dialogism as developed here does not (and possibly cannot) adequately cover all aspects of scientific practice—dexterity and ingenuity in using scientific tools, whether experimental or mathematical, and instrument making may be outside its application. Nor does the dialogical approach permit a complete reconstruction of all facets of creativity in scientific theorizing—intellectual feats of imagination, risk taking, and serendipity remain an enigma.

Nevertheless, the awareness of dialogical communicability allows a deeper and more detailed understanding of the (rational) workings of a scientific mind. Dialogism is not a general "theory of the growth of knowledge"; rather it is an approach, an attitude, a realization that can inform and I hope prove valuable in the study of science.

As I have mentioned, many scientists are aware of the communicative nature of their own endeavors (chapters 1 and 5). Such awareness pervades Schrödinger's reflections on his scientific activity in his Nobel lecture: "In my scientific work (and also in my life), I have never followed one main line, one program defining a direction for a long time. . . . My work is never entirely independent, since if I am to have an interest in a question, others must also have one. My word is seldom the first, but often the second, and may be inspired by a desire to contradict or to correct, but the consequent extension may turn out to be more important than the correction" (quoted in Darrigol 1992b, 238).

Schrödinger's words eloquently capture the importance of disagreement, addressivity, and open-endedness in scientific theorizing. Or, in Bakhtin's phrase, "there is neither the first word nor the last word" (1979, 373). . . .

REFERENCES

〜

Abbreviations

AHQP   Archive for the History of Quantum Physics, American
         Philosophical Society, Philadelphia, assembled and edited
         by T. S. Kuhn, J. Heilbron, P. Forman, and Lini Allen.
         *Sources for History of Quantum Physics: An Inventory and
         Report,* Philadelphia: American Philosophical Society

APHK   Bohr, *Atomic Physics and Human Knowledge* (1958)

ATDN   Bohr, *Atomic Theory and the Description of Nature* (1934)

BCW   *Niels Bohr Collected Works,* followed by volume number.

Essays   Bohr, *Essays 1958–1962 on Atomic Physics and Human Knowl-
         edge* (1963)

PC   Hermann et al., *Wolfgang Pauli, Scientific Correspondence with
         Bohr, Einstein, Heisenberg* (1979)

Aharonov, Y., and D. Bohm. 1961. "Time in the Quantum Theory and
    the Uncertainty Relation for Time and Energy." *Physical Review* 122:
    1649–58. Reprinted in Wheeler and Zurek 1983.
Aris, R., H. T. Davis, and R. Stuewer, eds. 1983. *Springs of Scientific Cre-
    ativity.* Minneapolis: University of Minnesota Press.
Aronowitz, S. 1988. *Science as Power: Discourse and Ideology in Modern
    Society.* Minneapolis: University of Minnesota Press.
Asquith, P. D., and R. N. Giere, eds. 1981. *Proceedings of the 1980 Biennial
    Meeting of the Philosophy of Science Association.* Vol. 2. East Lansing,
    Mich.: Philosophy of Science Association.
Badash, L., T. O. Hirshfelder, and M. P. Broida, eds. 1980. *Reminiscences
    of Los Alamos, 1943–45.* Dordrecht: Reidel.

Bakhtin, M. 1979. *Estetica Slovesnovo Tvorchestva.* Moscow: Isskusstvo.

———. 1994. *Problems of Dostoyevsky's Poetics.* Edited and translated by C. Emerson. Minneapolis: University of Minnesota Press. [First published in 1929.]

Ballentine, L. 1970. "The Statistical Interpretation of Quantum Mechanics." *Reviews of Modern Physics* 42:358–81.

Barnes, M., and D. Edge, eds. 1982. *Science in Context: Readings in the Sociology of Science.* Cambridge, Mass.: MIT Press.

Basso, K. H., and H. A. Selby, eds. 1976. *Meaning in Anthropology.* Albuquerque: University of New Mexico Press.

Batens, D., and J. P. van Bendegem, eds. 1988. *Theory and Experiment: Recent Insights and New Perspectives on Their Relation.* Dordrecht: Reidel.

Bell, J. S. 1964. "On the Einstein-Podolsky-Rosen Paradox." *Physics* 1: 195–200.

———. 1966. "On the Problem of Hidden Variables in Quantum Mechanics." *Reviews of Modern Physics* 38:447–52.

———. 1987. *Speakable and Unspeakable in Quantum Mechanics.* Cambridge: Cambridge University Press.

Beller, M. 1983. "Matrix Theory before Schrödinger: Philosophy, Problems, Consequences." *Isis* 74:469–91.

———. 1985. "Pascual Jordan's Influence on the Discovery of Heisenberg's Indeterminacy Principle." *Archive for the History of Exact Sciences* 33 (4): 337–49.

———. 1988. "Experimental Accuracy, Operationalism, and the Limits of Knowledge, 1925 to 1935." *Science in Context* 2:147–62.

———. 1990. "Born's Probabilistic Interpretation: A Case Study of 'Concepts in Flux.'" *Studies in the History and Philosophy of Science* 21: 563–88.

———. 1992a. "The Birth of Bohr's Complementarity: The Context and the Dialogues." *Studies in History and Philosophy of Science* 23:147–80.

———. 1992b. "Schrödinger's Dialogue with Göttingen-Copenhagen Physicists: 'Quantum Jumps' and 'Realism.'" In Bitbol and Darrigol 1992, 277–306.

———. 1993. "Einstein and Bohr's Rhetoric of Complementarity." *Science in Context* 6:241–55.

———. 1996a. "Bohm and 'Inevitability' of Acausality." In Cushing, Fine, and Goldstein 1996, 211–30.

———. 1996b. "The Conceptual and the Anecdotal History of Quantum Mechanics." *Foundations of Physics* 26:545–57.

———. 1997a. "Against the Stream: Schrödinger's Interpretation of Quantum Physics." *Studies in History and Philosophy of Science* 28B (3): 421–32.

————. 1997b. "Criticism and Revolutions." *Science in Context* 10 (1): 13–37.

————. 1998. "The Sokal Hoax: At Whom Are We Laughing?" *Physics Today*, September, 29–34.

————. 1999. "Jocular Commemorations: The Copenhagen Spirit." *Osiris*, forthcoming.

————. Forthcoming. "Kant's Impact on Einstein Thought." In Howard and Stachel, forthcoming.

Beller, M., R. Cohen, and J. Renn, eds. 1993. *Einstein in Context.* Cambridge: Cambridge University Press.

Beller, M., and A. Fine. 1994. "Bohr's Response to EPR." In Faye and Folse 1994, 1–31.

Beltrametti, E., and B. C. van Fraassen, eds. 1981. *Current Issues in Quantum Logic.* New York: Plenum.

Ben-Menahem, Y. 1990. "Equivalent Descriptions." *British Journal for the Philosophy of Science* 41:261–79.

————. 1992. "Struggling with Realism: Schrödinger's Case." In Bitbol and Darrigol 1992, 25–40.

Bernhard, C. G., E. Crawford, and Per Sörbom, eds. 1982. *Science, Technology and Society in the Time of Alfred Nobel.* Oxford: Pergamon.

Beyler, R. H. 1996. "Targeting the Organism: The Scientific and Cultural Context of Pascual Jordan's Biology, 1932–47." *ISIS* 87: 248–73.

Biagioli, M. 1990. "Galileo's System of Patronage." *History of Science* 28:1–62.

Bitbol, M., ed. 1995. *The Interpretation of Quantum Mechanics.* Woodbridge, Conn.: Ox Bow.

————. 1996. *Schrödinger's Philosophy of Quantum Mechanics.* Dordrecht: Kluwer.

Bitbol, M., and O. Darrigol, eds. 1992. *Erwin Schrödinger: Philosophy and the Birth of Quantum Mechanics.* Paris: Editions Frontières.

Black, M. 1962. *Models and Metaphors.* Ithaca, N.Y.: Cornell University Press.

Blaedel, N. 1988. *Harmony and Unity: The Life of Niels Bohr.* Madison: Wisconsin Science Tech Publishers.

Bloch, F. 1976. "Heisenberg and the Early Days of Quantum Mechanics." *Physics Today* 29 (12): 23–27.

Bloor, D. 1981. "The Strengths of the Strong Programme." *Philosophy of the Social Sciences* 11:199–213.

Blum, W., H.-P. Dürr, and H. Rechenberg, eds. 1984. *Werner Heisenberg: Collected Works.* Munich: Piper.

Bohm, D. 1951. *Quantum Theory.* New York: Prentice-Hall. Reprint, New York: Dover, 1989.

———. 1952. "A Suggested Interpretation of the Quantum Theory in Terms of 'Hidden Variables,' I and II." *Physical Review* 85:166–93.

———. 1957. *Causality and Chance in Modern Physics.* London: Routledge.

———. 1964. "The Structure of Scientific Revolutions. By T. Kuhn." *Philosophical Quarterly* 14:377–79.

———. 1987. "Hidden Variables and the Implicate Order." In Hiley and Peat 1987, 33–46.

———. 1996. *On Dialogue.* Edited by Lee Nichol. London: Routledge.

Bohm, D., and B. J. Hiley. 1993. *The Undivided Universe: An Ontological Interpretation of Quantum Mechanics.* London: Routledge.

Bohm, D., and F. D. Peat. 1987. *Science, Order and Creativity.* New York: Bantam.

Bohr, N. 1913. "On the Constitution of Atoms and Molecules." *Philosophical Magazine* 26:1–25, 476–502, 857–75.

———. 1918. "On the Quantum Theory of Line Spectra": Part I, "On the General Theory," Part II, "On the Hydrogen Spectrum." *Det Kongelige Danske Videnskabernes Selskab, Matematisk-Fysiske Meddelser* 4 (1): 1–36, 36–100.

———. 1924. "On the Application of the Quantum Theory to Atomic Structure, Part I." *Proceedings of the Cambridge Philosophical Society,* suppl., 1–42.

———. 1925. "Über die Wirkung von Atomen Bei Stössen." *Zeitschrift für Physik* 34:142–57. Translated in *BCW,* 5:194–206.

———. 1926. "Atomic Theory and Mechanics." *Nature* 116:845–52.

———. 1927a. "Fundamental Problems of the Quantum Theory." Unpublished manuscript, 13 September. Bohr Manuscript Collection, AHQP. In *BCW,* 6:75–80.

———. 1927b. "Philosophical Foundations of the Quantum Theory." Unpublished manuscript. Bohr Manuscript Collection, AHQP. In *BCW,* 6:69–71.

———. 1927c. "The Quantum Postulate and the Recent Development of Atomic Theory." In *Atti del Congresso Internazionale dei Fisici, 11–20 Settembre 1927,* 565–88. Bologna: Nicola Zanichelli. Reprinted in *BCW,* 6:113–36. Page references are to *BCW.*

———. 1927d. "The Quantum Postulate and the Recent Development of Atomic Theory." Unpublished manuscript, 12–13 October 1927. Bohr Manuscript Collection, AHQP. In *BCW,* 6:89–98. [Reprinted in Sanders 1987, 75–89.]

———. 1928. "The Quantum Postulate and the Recent Development of Atomic Theory." *Nature* 121 (suppl.): 580–90. Reprinted in Wheeler and Zurek 1983, 87–126. [Facsimile in *BCW,* 6:148–58. Revised version in *ATDN*, 52–91. German version in *Naturwissenschaften* 16 (1928): 245–57.] Page references are to Wheeler and Zurek.

———. 1929a. "The Atomic Theory and the Fundamental Principles Underlying the Description of Nature." Lecture for the Scandinavian Meeting of Natural Scientists. Translated in *ATDN*, 102–19. [Originally published in Danish in *Fysik Tidsskrift*, 1929. Facsimile of translation in *BCW*, 6:236–53.]

———. 1929b. "The Quantum of Action and the Description of Nature" (in German). *Naturwissenschaften* 17:483–86. Translated in *ATDN*, 92–119. [Facsimile in *BCW*, 6:208–17.] Page references are to *ATDN*.

———. 1930. "Philosophical Aspects of Atomic Theory." Address to the Royal Society of Edinburgh, 26 May. Unpublished manuscript. Bohr Manuscript Collection, AHQP. In Sanders 1987, 129–46. [*BCW* has only facsimile of an abstract from *Nature* 125:958.]

———. 1931a. "Harald Høffding." Speech at the meeting of the Society of the Sciences, 11 December. Bohr Manuscript Collection, AHQP. In Sanders 1987, 173–80. [Originally published in Danish in *Oversigt over det Kongelige Danske Videnskabernes Selskabs Virksomhed*, 1931.]

———. 1931b. "Space-Time Description and Conservation Principles." Notes for unpublished paper, 16 November 1930–20 February 1931. Bohr Manuscript Collection, AHQP. In Sanders 1987, 147–60.

———. 1932. "H. Høffding's View on Physics and Psychology." Notes from lecture at the tenth International Psychology Congress, August. Bohr Manuscript Collection, AHQP. In Sanders 1987, 195–203.

———. 1933. "Light and Life." Address at the opening meeting of the International Congress on Light Therapy, Copenhagen, August. *Nature* 131:421–23, 457–59. Revised version in *APHK*, 3–12. [Danish version in *Naturens Verden* 17:49–54. German version in *Naturwissenschaften* 21:245–50.] Page references are to *APHK*.

———. 1934. *Atomic Theory and the Description of Nature*. Cambridge: Cambridge University Press. [Reprinted as *The Philosophical Writings of Niels Bohr*, Vol. 1, Woodbridge, Conn.: Ox Bow, 1961, 1987.]

———. 1935a. "Can Quantum-Mechanical Description of Physical Reality Be Considered Complete?" *Physical Review* 48:696–702.

———. 1935b. "Space and Time in Nuclear Physics." Notes from lecture at the Society for the Advancement of Physical Knowledge, Copenhagen, 21 March. Bohr Manuscript Collection, AHQP. In Sanders 1987, 205–20.

———. 1937a. "Analysis and Synthesis in Science." *International Encyclopedia of Unified Science* 1 (1): 28.

———. 1937b. "Biology and Atomic Physics." Address delivered at the Physical and Biological Congress in Memory of Luigi Galvani, Bologna, October. In *APHK,* 13–22. [Originally published in *Proceedings of the Galvani Congress,* Bologna].

———. 1937c. "Causality and Complementarity." *Philosophy of Science* 4: 289–98. [Originally published in German in *Erkenntnis* 6:293–303. English version reprinted in Sanders 1987, 241–52.] Page references are to Sanders.

———. 1937d. Hitchcock lectures at the University of California, Berkeley, 8–24 March. Unpublished manuscript. Raymond Birge Collection, American Institute of Physics. In Sanders 1987, 253–355.

———. 1938a. "The Epistemological Problems of Atomic Theory." 2 May 1938. Bohr Manuscript Collection, AHQP. In Sanders 1987, 377–80.

———. 1938b. "Natural Philosophy and Human Cultures." Address delivered at the International Congress of Anthropological and Ethnological Sciences, Kronborg Castle, Elsinore, August. In *APHK,* 23–31. [Originally published in *Comptes rendues du congrès international de science, anthropologie et ethnologie,* Copenhagen. English version in *Nature* 143 (1939): 268–72.]

———. 1939. "The Causality Problem in Atomic Physics." In *New Theories in Physics,* 11–45. Paris: International Institute of Intellectual Cooperation. Reprinted in Sanders 1987, 383–422. Page references are to Sanders.

———. 1948. "On the Notions of Causality and Complementarity." *Dialectica* 2:312–19. Reprinted in Sanders 1987, 445–55. Page references are to Sanders.

———. 1949. "Discussion with Einstein on Epistemological Problems in Atomic Physics." In Schilpp 1970, 201–41. Reprinted in Wheeler and Zurek 1983, 9–49.

———. 1950. Summary of the Gifford lectures, 27 December 1949–25 January 1950. Bohr Manuscript Collection, AHQP. In Sanders 1987, 505–17.

———. 1955a. "Atoms and Human Knowledge." In *APHK,* 83–93. [Also in *Daedalus* 87 (1958): 164–74.]

———. 1955b. "Unity of Knowledge." In *APHK,* 67–82. [Originally published in Leary 1955.]

———. 1956. "Mathematics and Natural Philosophy." *Scientific Monthly* 82 (1). Reprinted in Sanders 1987, 551–61. Page references are to Sanders.

———. 1957a. Drafts for a postscript to "Discussion with Einstein on Epistemological Problems in Atomic Physics [Bohr 1949]," August. Bohr Manuscript Collection, AHQP. In Sanders 1987, 668–76.

———. 1957b. "The Philosophical Lessons of Atomic Physics." Karl Taylor Compton lectures, Massachusetts Institute of Technology, 5–26 November. Niels Bohr Holdings, Niels Bohr Library, American Institute of Physics. In Sanders 1987, 593–648.

————. 1957c. "Position and Terminology in Atomic Physics." Notes, 11 March 1957. Bohr Manuscript Collection, AHQP. In Sanders 1987, 665–67.

————. 1958a. *Atomic Physics and Human Knowledge.* New York: Wiley. [Reprinted as *The Philosophical Writings of Niels Bohr,* Vol. 2, *Essays 1932–1957 on Atomic Physics and Human Knowledge,* Woodbridge, Conn.: Ox Bow, 1987.]

————. 1958b. "Atoms and Human Knowledge." Address at convocation, Roosevelt University, 4 February. Bohr Manuscript Collection, AHQP. In Sanders 1987, 707–18.

————. 1958c. "Quantum Physics and Philosophy: Causality and Complementarity." In *Essays,* 1–7. [Originally published in Klibansky 1958. Reprinted in Sanders 1987, 679–87.]

————. 1958d. "Rudjer Boskovic." Talk at the Rudjer Boskovic Celebrations, Zagreb. Bohr Manuscript Collection, AHQP. In Sanders 1987, 719–34.

————. 1958e. "The Unity of Knowledge." John Franklin Carlson lecture, Iowa State University, January. Bohr Manuscript Collection, AHQP. In Sanders 1987, 689–705.

————. 1960a. "The Connection between the Sciences." Address to the International Congress of Pharmaceutical Sciences, Copenhagen, 7 August. Translated in *Essays,* 17–22. [Reprinted in Sanders 1987, 757–66.]

————. 1960b. "Quantum Physics and Biology." In *Symposia of the Society for Experimental Biology.* Vol. 14, *Models and Analogues in Biology.* Reprinted in Sanders 1987, 781–90.

————. 1960c. "The Unity of Human Knowledge." Address delivered at a congress arranged by La Fondation Européenne de la Culture, Copenhagen, October. In *Essays,* 8–16. [Reprinted in Sanders 1987, 767–79.]

————. 1961. "The Genesis of Quantum Mechanics." In *Essays,* 74–78. [Originally published in German as "Die Entstehung der Quantenmechanik," in Bopp 1961.]

————. 1962. "Light and Life Revisited." Address delivered at the inauguration of the Institute for Genetics, Cologne, June. Unfinished manuscript in *Essays,* 23–29.

————. 1963. *Essays 1958–1962 on Atomic Physics and Human Knowledge.* New York: Wiley. [Reprinted as *The Philosophical Writings of Niels Bohr,* Vol. 3, Woodbridge, Conn.: Ox Bow, 1987.]

————. 1972, 1984, 1985. *Niels Bohr: Collected Works.* Vols. 1, 5, and 6. Amsterdam: North Holland.

Bohr, N., H. Kramers, and J. Slater. 1924. "The Quantum Theory of Radiation." *Philosophical Magazine* 47:785–822. Reprinted in van der Waerden 1967, 159–76. Page references are to van der Waerden.

Bohr, N., and L. Rosenfeld. 1933. "Zur Frage der Messbarkeit der Electromagnetischen Feldgrossen." *Det Konglelige Danske Videnskabernes Selskab, Matematisk-Fysiske Meddelser* 12. [English translation in Cohen and Stachel 1979. Reprinted in Wheeler and Zurek 1983, 479–534.]

Bopp, F., ed. 1961. *Werner Heisenberg und die Physik unserer Zeit.* Braunschweig: Friedrich Vieweg und Sohn.

Born, M. 1924. "Über Quantenmechanik." *Zeitschrift für Physik* 26:379–96. Translated in van der Waerden 1967, 181–98. Page references are to van der Waerden.

———. 1926a. *Problems of Atomic Dynamics.* Cambridge, Mass.: MIT Press.

———. 1926b. "Quantenmechanik der Stossvorgänge." *Zeitschrift für Physik* 38:803–27. Translated as "Quantum Mechanics of Collision Processes," in Ludwig 1968, 206–25. Page references are to Ludwig.

———. 1926c. "Zur Quantenmechanik der Stossvorgänge." *Zeitschrift für Physik* 37:863–67. Translated in Wheeler and Zurek 1983, 52–55. Page references are to Wheeler and Zurek.

———. 1927a. "Das Adiabatenprinzip in der Quantenmechanik." *Zeitschrift für Physik* 40:167–92.

———. 1927b. "Physical Aspects of Quantum Mechanics." *Nature* 119: 354–57. Reprinted in Born 1956, 6–13. Page references are to Born 1956.

———. 1928. "On the Meaning of Physical Theories." *Nachrichten der Gesellschaft der Wissenschaften zu Göttingen.* Translated in Born 1956, 17–36. Page references are to Born 1956.

———. 1936. "Some Philosophical Aspects of Modern Physics." *Proceedings of the Royal Society of Edinburgh* 57, pt. 1: 1–18. Reprinted in Born 1956, 37–54. Page references are to Born 1956.

———. 1950. "Physics and Metaphysics." *Memoirs and Proceedings of the Manchester Literary and Philosophical Society* 91. Reprinted in Born 1956, 93–108. Page references are to Born 1956.

———. 1951. "Postscript to the *Restless Universe.*" Reprinted in Born 1956, 225–32. Page references are to Born 1956.

———. 1953a. "The Conceptual Situation in Physics and the Prospects of Its Future Development." Guthrie lecture, March. *Proceedings of the Physical Society* A 66:501–13. Reprinted in Born 1956, 123–39. Page references are to Born 1956.

———. 1953b. "The Interpretation of Quantum Mechanics." *British Journal for the Philosophy of Science* 4. Reprinted in Born 1956, 140–50. Page references are to Born 1956.

———. 1953c. "Physical Reality." *Philosophical Quarterly,* 139–149. Reprinted in Born 1956, 151–164. Page references are to Born 1956.

———. 1955a. "Is Classical Mechanics in Fact Deterministic?" *Physicalische Blätter* 11 (9): 49–54. [Reprinted in Born 1956.]

———. 1955b. "Statistical Interpretation of Quantum Mechanics." Nobel Prize lecture. *Science* 122 (3172): 675–79. [Reprinted in Born 1956.]

———. 1956. *Physics in My Generation.* London: Pergamon.

———. 1961. "Erwin Schrödinger." *Physics* B1 17:85–87.

———. 1962. *Physics and Politics.* Edinburgh: Oliver and Boyd.

———. 1963. *Ausgewählte Abhandlungen.* Göttingen: Vanderhoek and Ruprecht.

———. 1964. *Natural Philosophy of Cause and Change.* New York: Dover.

———. 1969. *Atomic Physics.* London: Blackie.

———, ed. 1971. *The Born-Einstein Letters,* New York: Walker.

Born, M., and W. Biem. 1968. "Dualism in Quantum Theory." *Physics Today* 21:51–56.

Born, M., W. Heisenberg, and P. Jordan. 1926. "Zur Quantenmechanik II." *Zeitschrift für Physik* 35:557–615. Translated as "On Quantum Mechanics II," in van der Waerden 1967, 321–85. Page references are to van der Waerden.

Born, M., and P. Jordan. 1925a. "Zur Quantenmechanik." *Zeitschrift für Physik* 34:858–88. Translated in van der Waerden 1967, 277–306. Page references are to van der Waerden.

———. 1925b. "Zur Quantentheorie aperiodischer Vorgänge." *Zeitschrift für Physik* 33:479–505.

———. 1930. *Elementare Quantenmechanik.* Berlin: Springer.

Born, M., and N. Wiener. 1926. "Formulierung der Quantengesetze der Periodisher und Nichtperiodischer Vorgange." *Zeitschrift für Physik* 36:177–87.

Bothe, W., and H. Geiger. 1925a. "Experimentelles zur Theorie von Bohr, Kramers und Slater." *Naturwissenschaften* 13:440–41.

———. 1925b. "Über das Wesen des Comptoneffekts: Ein experimenteller Beitrag zur Theorie der Strahlung." *Zeitschrift für Physik* 32:639–63.

Bridgman, P. W. 1927. *The Logic of Modern Physics.* London: Macmillan.

———. 1930. "Permanent Elements in the Flux of Present-Day Physics." *Science* 71:21.

Bub, J. 1992. "Quantum Mechanics without the Projection Postulate." *Foundations of Physics* 22:737–54.

———. 1997. *Interpreting the Quantum World.* Cambridge: Cambridge University Press.

Buchwald, J. Z. 1994. *The Creation of Scientific Effects: Heinrich Hertz and Electric Waves.* Chicago: University of Chicago Press.

———, ed. 1995. *Scientific Practice.* Chicago: University of Chicago Press.

Bukharin, N. I., ed. 1971. *Science at the Cross Roads,* new ed. London: Cass.

Butterfield, H. 1931. *The Whig Interpretation of History.* London: Bell.

Campbell, N. R. 1921. "Atomic Structure." *Nature* 107 (2684):170.

———. 1926. "Time and Chance." *London, Edinburgh and Dublin Philosophical Magazine and Journal of Science,* 7th ser., 1:1106–17.

———. 1927. "Philosophical Foundations of Quantum Theory." *Nature* 119:779.

Cartwright, N. 1983. *How the Laws of Physics Lie.* Oxford: Clarendon.

———. 1990. "Philosophical Problems of Quantum Theory: The Response of the American Physicists." In Krüger, Gingerenzer, and Morgan 1987, 417–37.

Casimir, H. B. G. 1967. "Recollections from the Years 1929–1931." In Rozental 1967, 109–14.

———. 1983. *Haphazard Reality: Half a Century of Science.* New York: Harper and Row.

Cassidy, D. C. 1992. *Uncertainty: The Life and Science of Werner Heisenberg.* New York: Freeman.

Chevalley, C. 1988. "Physical Reality and Closed Theories in Werner Heisenberg's Early Papers." In Batens and van Bendegem 1988, 159–76.

———. 1994. "Niels Bohr's Words and the Atlantis of Kantianism." In Faye and Folse 1994, 33–53.

Cohen, H. F. 1994. *The Scientific Revolution: A Historiographical Inquiry.* Chicago: University of Chicago Press.

Cohen, R. S., and M. Wartofsky, eds. 1968. *Proceedings of the Boston Colloquium for the Philosophy of Science 1966/1968.* Boston Studies in the Philosophy of Science, vol. 5. Dordrecht: Reidel.

Collingwood, R. G. 1939. *An Autobiography.* Oxford: Oxford University Press.

Collins, H. 1992. *Changing Order: Replication and Induction in Scientific Practice.* Chicago: University of Chicago Press.

Colodny, R. G., ed. 1972. *Paradoxes and Paradigms: The Philosophical Challenge of the Quantum Domain.* University of Pittsburgh Series in the Philosophy of Science, vol. 5. Pittsburgh: University of Pittsburgh Press.

———, ed. 1986. *From Quarks to Quasars.* University of Pittsburgh Series in the Philosophy of Science, vol. 7. Pittsburgh: University of Pittsburgh Press.

Compton, A. H. 1923. "A Quantum Theory of the Scattering of X-rays by Light Elements." *Physical Review* 21:483–502.

———. 1927. In *Nobel Lectures in Physics 1965,* 174–90.

Crombie, A. C., ed. 1963. *Scientific Change.* New York: Basic.

Cushing, J. T. 1990. *Theory Construction and Selection in Modern Physics: The S Matrix.* Cambridge: Cambridge University Press.

———. 1994a. "A Bohmian Response to Bohr's Complementarity." In Faye and Folse 1994, 57–75.

————. 1994b. *Quantum Mechanics, Historical Contingency and the Copenhagen Hegemony.* Chicago: University of Chicago Press.

————. 1996. "The Causal Quantum Theory Program." In Cushing, Fine, and Goldstein 1996, 1–19.

Cushing, J. T., A. Fine, and S. Goldstein, eds. 1996. *Bohmian Mechanics and Quantum Theory: An Appraisal.* Dordrecht: Kluwer.

Cushing, J. T., and E. McMullin, eds. 1989. *Philosophical Consequences of Quantum Theory: Reflections on Bell's Theorem.* Notre Dame, Ind.: University of Notre Dame Press.

Darrigol, O. 1986. The Origin of Quantized Matter Waves." *Historical Studies in the Physical Sciences* 16, pt. 2: 197–253.

————. 1992a. *From c-Numbers to q-Numbers: The Classical Analogy in the History of Quantum Theory.* Berkeley and Los Angeles: University of California Press.

————. 1992b. "Schrödinger's Statistical Physics and Some Related Themes." In Bitbol and Darrigol 1992, 237–76.

Daston, L. 1995. "The Moral Economy of Science." *Osiris* 10:3–24.

Daumer, M. 1996. "Scattering Theory from a Bohmian Perspective." In Cushing, Fine, and Goldstein 1996, 87–97.

Davidson, D. 1984. *Inquiries into Truth and Interpretation.* Oxford: Clarendon.

de Broglie, L. 1923. "Ondes et quanta." *Comptes Rendus* 177:507–10.

De Witt, B., and N. Graham, eds. 1973. *The Many Worlds Interpretation of Quantum Mechanics.* Princeton, N.J.: Princeton University Press.

Dieks, D. 1994. "Modal Interpretation of Quantum Mechanics, Measurements, and Macroscopic Behaviour." *Physical Review* A 49:2290–300.

Diner, S., D. Fargue, G. Lochak, and F. Selleri, eds. 1984. *The Wave Particle Dualism.* Dordrecht: Reidel.

Dirac, P. A. M. 1925. "Fundamental Equations of Quantum Mechanics." *Proceedings of the Royal Society of London* A 109:642–53.

————. 1926. "Quantum Mechanics and a Preliminary Investigation of the Hydrogen Atom." *Proceedings of the Royal Society of London* A 110:561–79.

————. 1927. "The Physical Interpretation of Quantum Dynamics." *Proceedings of the Royal Society of London* A 113:621–41.

————. 1929. "Quantum Mechanics of Many Electron Systems." *Proceedings of the Royal Society of London* A 123:714–33.

————. 1930. *The Principles of Quantum Mechanics.* Oxford: Clarendon.

————. 1963. "The Evolution of the Physicist's Conception of Nature." *Scientific American* 208 (5): 45–53.

————. 1977. "Recollections of an Exciting Era." In Weiner 1977, 109–46.

Douglas, M. 1975. *Implicit Meanings: Essays in Anthropology.* London: Routledge and Kegan Paul.

————. 1986. *How Institutions Think.* Syracuse, N.Y.: Syracuse University Press.

Dresden, M. 1987. *H. A. Kramers: Between Tradition and Revolution.* Berlin: Springer.

Duane, W. 1923. "The Transfer in Quanta of Radiation Momentum to Matter." *Proceedings of the National Academy of Sciences* 9:158–64.

Duhem, P. 1991. *The Aim and the Structure of Physical Theory.* Princeton, N.J.: Princeton University Press. [First published in 1907.]

Dürr, O., S. Goldstein, and N. Zanghi. 1992a. "Quantum Equilibrium and the Origin of Absolute Uncertainty." *Journal of Statistical Physics* 67:843–907.

————. 1992b. "Quantum Mechanics, Randomness, and Deterministic Reality." *Physics Letters* A 172:6–12.

————. 1996. "Bohmian Mechanics as the Foundation of Quantum Mechanics." In Cushing, Fine, and Goldstein 1996, 21–44.

Earman, J. 1986. *A Primer on Determinism.* Dordrecht: Reidel.

————. 1993. "Carnap, Kuhn and the Philosophy of Scientific Methodology." In Horwich 1993, 9–36.

Easlea, B. 1981. *Science and Sexual Oppression.* London: Weidenfeld and Nicolson.

Einstein, A. 1905. "Über die Erzeugung und Verwandlung des Lichtes betreffenden heuristischen Gesichtspunkt." *Annalen der Physik* 17: 132–48.

————. 1917. "Quantentheorie der Strahlung." *Physicalische Zeitschrift* 18:121–28. Translated as "On the Quantum Theory of Radiation," in van der Waerden 1967, 63–77. Page references are to van der Waerden.

————. 1924. "Quantentheorie des einatomigen idealen Gases." *Sitzungsberichte der Preußischen Academie der Wissenschaften zu Berlin,* 261–267.

————. 1949a. "Autobiographical Notes." Reprinted in Schilpp 1970. 3–94. Page references are to Schilpp.

————. 1949b. "Reply to Criticisms." Reprinted in Schilpp 1970, 665–88. Page references are to Schilpp.

Einstein, A., and S. Freud. 1933. *Why War?* Paris: International Institute of Intellectual Co-operation, League of Nations.

Einstein, A., B. Podolsky, and N. Rosen. 1935. "Can Quantum-Mechanical Description of Physical Reality Be Considered Complete?" *Physical Review* 47:777–80.

Elkana, Y. 1970. "Helmholtz's 'Kraft': An Illustration of Concepts in Flux." *Historical Studies in the Physical Sciences* 2:263–98.

————. 1977. "The Historical Roots of Modern Physics." In Weiner 1977, 197–265.

Elsasser, W. 1925. "Bemerkungen zur Quantenmechanik freier Electronen." *Naturwissenschaften* 13:711.

Erikson, E. 1982. "Psychoanalytic Reflections on Einstein's Centenary." In Holton and Elkana 1982.

Everett, H. 1957. "'Relative State' Formulation of Quantum Mechanics." *Reviews of Modern Physics* 29:454–62. [Reprinted in Wheeler and Zurek 1983, 315–23.]

Faye, J. 1991. *Niels Bohr: His Heritage and Legacy. An Antirealism View of Quantum Mechanics.* Dordrecht: Kluwer.

Faye, J., and H. Folse, eds. 1994. *Niels Bohr and Contemporary Philosophy.* Dordrecht: Kluwer.

Feuer, L. S. 1963. *The Scientific Intellectual: The Psychological and Sociological Origins of Modern Science.* New York: Basic.

Feyerabend, P. K. 1960. "Professor Bohm's Philosophy of Nature." *British Journal for the Philosophy of Science* 10:321–38. [Reprinted in Feyerabend 1981b, 219–45.]

———. 1970. "Consolation for the Specialist." In Lakatos and Musgrave 1970, 197–230.

———. 1975. *Against Method.* London: New Left.

———. 1981a. "Niels Bohr's World View." In Feyerabend 1981b, 247–93.

———. 1981b. *Philosophical Papers.* Vols. 1 and 2. Cambridge: Cambridge University Press.

———. 1993. *Against Method,* rev. ed. London: Verso.

———. 1995. *Killing Time.* Chicago: University of Chicago Press.

Feynman, R. P. 1976. "Los Alamos from Below: Reminiscences of 1943–1945." *Engineering and Science* 39:11–30.

———. 1980. "Los Alamos from Below." In Badash, Hirshfelder, and Broida 1980, 105–32.

———. 1985. *QED: The Strange Theory of Light and Matter.* Princeton, N.J.: Princeton University Press.

———. 1998. *The Meaning of It All: Thoughts of a Citizen Scientist.* Reading, Mass.: Addison Wesley.

Feynman, R. P., R. B. Leighton, and M. Sands. 1969. *The Feynman Lectures on Physics.* London: Addison Wesley.

Fine, A. 1986. *The Shaky Game: Einstein, Realism and the Quantum Theory.* Chicago: University of Chicago Press.

———. 1989. "Do Correlations Need to Be Explained?" In Cushing and McMullin 1989, 175–95.

———. 1993. "Einstein's Interpretations of the Quantum Theory." *Science in Context* 6 (1): 257–74.

———. 1996a. "Science Made Up: Constructivist Sociology of Scientific Knowledge." In Galison and Stump 1996.

————. 1996b. *The Shaky Game: Einstein, Realism and the Quantum Theory,* 2d ed. Chicago: University of Chicago Press.

Fine, A., M. Forbes, and L. Wessels, eds. 1991. *PSA 1990: Proceedings of the 1990 Biennial Meeting of the Philosophy of Science Association.* Vol. 2. East Lansing, Mich.: Philosophy of Science Association.

Finkelstein, D. 1965. "The Logic of Quantum Physics." *Transactions of the New York Academy of Sciences* 25:621–37.

————. 1987. "All Is Flux." In Hiley and Peat 1987, 289–94.

Fischhoff, B. 1975. "Hindsight ≠ Foresight: The Effect of Outcome of Knowledge on Judgment under Uncertainty." *Journal of Experimental Psychology: Human Perception and Performance* 1:288–99.

Fleck, L. 1979. *Genesis and Development of a Scientific Fact.* Chicago: University of Chicago Press. [First published in 1935.]

Fock, V. 1958. *Uspechi fziceskich nauk* 66:599–602. Translation by Fock, Bohr Manuscript Collection, AHQP.

Folse, H. 1985. *The Philosophy of Niels Bohr.* Amsterdam: North Holland.

————. 1989. "Bohr on Bell." In Cushing and McMullin 1989, 254–71.

————. 1994. "Bohr's Framework of Complementarity and the Realism Debate." In Faye and Folse 1994, 119–39.

Forman, P. 1971. "Weimar Culture, Causality, and Quantum Theory, 1918–1927: Adaptation by German Physicists and Mathematicians to a Hostile Intellectual Environment." *Historical Studies in the Physical Sciences* 3:1–117.

————. 1979. "The Reception of Acausal Quantum Mechanics in Germany and Britain." In Mauskopf 1979, 11–50.

Franck, J., and G. Hertz. 1913. "Messung der Ionisierungsspannung in verschiedenen Gasen." *Vers. Deutsche Physikalische Gesellschaft* 15:34–44.

Franklin, A. 1986. *The Neglect of Experiment.* Cambridge: Cambridge University Press.

French, A., and P. J. Kennedy, eds. 1985. *Niels Bohr: A Centenary Volume.* Cambridge, Mass.: Harvard University Press.

Friedman, M., and H. Putnam. 1978. "Quantum Logic, Conditional Probability, and Interference." *Dialectica* 32:305–15.

Galison, P. 1987. *How Experiments End.* Chicago: University of Chicago Press.

————. 1997. *Image and Logic.* Chicago: University of Chicago Press.

Galison, P., and D. J. Stump, eds. 1996. *The Disunity of Science: Boundaries, Contexts, and Power.* Stanford, Calif.: Stanford University Press.

Gay, P. 1968. *Weimar Culture.* New York: Harper and Row.

Geertz, C. 1976. "From the Native Point of View." In Basso and Selby 1976, 221–36.

Gentner, D., and M. Jeziorsky. 1993. "The Shift from Metaphor to Analogy in Western Science." In Ortony 1993, 447–80.

Ghirardi, G. C., A. Rimini, and T. Weber. 1986. "Unified Dynamics for Microscopic and Macroscopic Systems." *Physical Review* D 34: 470–91.

Ghose, P., and D. Home. 1992. "Wave-Particle Duality of Single-Photon States." *Foundations of Physics* 22 (12): 1435–47.

Ghose, P., D. Home, and G. S. Agarwal. 1991. "An Experiment to Throw More Light on Light." *Physics Letters* A 153:403–6.

———. 1992. "An Experiment to Throw More Light on Light: Implications." *Physics Letters* A 158:95–99.

Gibbins, P. 1987. *Particles and Paradoxes.* Cambridge: Cambridge University Press.

Gillies, D. 1992a. "Are There Revolutions in Mathematics?" In Gillies 1992b.

———, ed. 1992b. *Revolutions in Mathematics.* Oxford: Clarendon.

Gombrich, E. H. 1963. *Meditations on a Hobby Horse, and Other Essays on the Theory of Art.* London: Phaidon.

Gordon, W. 1927. "Der Comptoneffect nach der Schrödingerschen Theorie." *Zeitschrift für Physik* 40:117.

Gross, P. R., N. Levitt, and M. W. Lewis, eds. 1996. *The Flight from Science and Reason.* New York: New York Academy of Sciences.

Guillemin, V. 1968. *The Story of Quantum Mechanics.* New York: Scribner's.

Gutting, G. 1980. *Paradigms and Revolutions: Appraisals and Applications of Thomas Kuhn's Philosophy of Science.* Notre Dame, Ind.: University of Notre Dame Press.

Habermas, T. 1972. *Knowledge and Human Interests.* Boston: Beacon.

Hacking, I. 1983. *Intervening and Representing.* Cambridge: Cambridge University Press,

———. 1992a. "The Self-Vindication of the Laboratory Sciences." In Pickering 1992, 29–65.

———. 1992b. "Statistical Language, Statistical Truth and Statistical Reason: The Self-Authentification of a Style of Scientific Reasoning." In McMullin 1992.

Hanson, N. R. 1958. *Patterns of Discovery.* Cambridge: Cambridge University Press.

———. 1959. "Copenhagen Interpretation of Quantum Theory." *American Journal of Physics* 27:1–15. [Revised version of paper delivered at the meeting of the American Philosophical Association, 1958.]

Healey, R. 1989. *The Philosophy of Quantum Mechanics: An Interactive Interpretation.* Cambridge: Cambridge University Press.

Heilbron, J. L. 1982. "Fin-de-Siècle Physics." In Bernhard, Crawford, and Sörbom 1982.

———. 1987. "The Earliest Missionaries of the Copenhagen Spirit." In Ullman-Margalit 1987, 201–33.

Heilbron, J. L., and T. S. Kuhn, 1969. "The Genesis of the Bohr Atom." *Historical Studies in the Physical Sciences* 1:211–90.

Heisenberg, W. 1925. "Quantentheoretische Umdeutung kinematischer und mechanischer Beziehungen." *Zeitschrift für Physik* 33:879–93. Translated as "Quantum-theoretical Reinterpretation of Kinematical and Mechanical Relations," in van der Waerden 1967, 261–76. Page references are to van der Waerden.

―――. 1926a. "Mehrkörperproblem und Resonanz in der Quantenmechanik." *Zeitschrift für Physik* 38:411–26.

―――. 1926b. "Quantenmechanik." *Naturwissenschaften* 14:989–94.

―――. 1927a. "Schwankungerscheinungen und Quantenmechanik." *Zeitschrift für Physik* 40:501–6.

―――. 1927b. "Über den anschaulichen Inhalt der quantentheoretischen Kinematik und Mechanik." *Zeitschrift für Physik* 43:172–98. Translated in Wheeler and Zurek 1983, 62–84. Page references are to Wheeler and Zurek.

―――. 1928. "Erkenntnistheoretische Probleme in der Modernen Physik." In Blum, Dürr, and Rechenberg 1984, 22–28.

―――. 1930. *The Physical Principles of Quantum Theory.* Chicago: University of Chicago Press.

―――. 1931a. "Die Rolle der Unbestimmtheitsrelationen in der modernen Physik." *Monatshefte für Mathematik und Physik* 38:365–72.

―――. 1931b. "Kausalgesetz und Quantenmechanik." *Erkenntnis zugleich Analen der Philosophie* 2:172–82. [Reprinted in Blum, Dürr, and Rechenberg 1984, 29–39.]

―――. 1933. "The Development of Quantum Mechanics." In *Nobel Lectures in Physics* 1965, 290–301.

―――. 1934. "Wandlungen in den Grundlagen der exacten Naturwissenschaft in jüngster Zeit." *Naturwissenschaften* 40. Translated as "Recent Changes in the Foundations of Exact Sciences," in Heisenberg 1979, 11–26. Page references are to Heisenberg 1979.

―――. 1935. "Questions of Principle in Modern Physics." In Heisenberg 1979, 41–52.

―――. 1948. "Der Begriff 'Abgeschlossene Theorie' in der modernen Naturwissenschaft." *Dialectica* 2:331–36.

―――. 1952. *Philosophical Problems of Nuclear Science.* New York: Pantheon.

―――. 1958. *Physics and Philosophy.* New York: Harper and Row.

―――. 1967. "Quantum Theory and Its Interpretation." In Rozental 1967, 94–108.

―――. 1971. *Physics and Beyond.* New York: Harper and Row.

―――. 1973. "Development of Concepts in the History of Quantum Theory." In Mehra 1973, 264–75.

————. 1977. "Remarks about the Uncertainty Principle." In Price and Chissick 1977, 3–6.

————. 1979. *Philosophical Problems of Quantum Physics.* Woodbridge, Conn.: Ox Bow. [Reprint of *Philosophical Problems of Nuclear Science,* New York: Pantheon, 1952.]

Heisenberg, W., and P. Jordan. 1926. "Anwendung der Quantenmechanik auf das Problem der anomalen Zeemaneffecte." *Zeitschrift für Physik* 37:263–77.

Heisenberg, W., and W. Pauli. 1929. "Zur Quantendynamik der Wellenfelder." *Zeitschrift für Physik* 56:1–61.

Hendry, J. 1984. *The Creation of Quantum Mechanics and the Bohr-Pauli Dialogue.* Dordrecht: Reidel.

Hermann, A., K. v. Meyenn, and V. F. Weisskopf, eds. 1979. *Wolfgang Pauli: Scientific Correspondence with Bohr, Einstein, Heisenberg, A.O.* Vol. 1. Berlin: Springer.

Hermann, G. 1935. "Die naturphilosophischen Grundlagen der Quantenmechanik." *Abhandlungen der Friesschen Schule* 6:75–152.

Hershbach, D. R. 1996. "Imaginary Gardens with Real Toads." In Gross, Levitt, and Lewis 1966.

Hessen, B. 1931. "The Social and Economic Roots of Newton's 'Principia.'" In Bukharin 1971.

Hiley, B. T., and F. D. Peat, eds. 1987. *Quantum Implications: Essays in Honor of David Bohm.* New York: Routledge.

Holquist, M. 1990. *Dialogism: Bakhtin and His World.* London: Routledge.

Holton, G. 1970. "The Roots of Complementarity." *Daedalus,* fall, 1015–55. [Reprinted in Holton 1973.]

————. 1973. *Thematic Origins of Scientific Thought.* Cambridge, Mass.: Harvard University Press.

————. 1991. "Quanta, Relativity, and Rhetoric." In Pera and Shea 1991.

Holton, G., and Y. Elkana, eds. 1982. *Albert Einstein: Historical and Cultural Perspectives.* Princeton, N.J.: Princeton University Press.

Home, D., and M. A. B. Whitaker. 1992. "Ensemble Interpretation of Quantum Mechanics: A Modern Perspective." *Physics Report* 210:224–317.

Honner, J. 1987. *The Description of Nature: Niels Bohr and the Philosophy of Quantum Physics.* Oxford: Clarendon.

Hooker, C. A. 1972. "The Nature of Quantum Mechanical Reality: Einstein vs. Bohr." In Colodny 1972.

————. 1991. "Physical Intelligibility, Projection and Objectivity: The Divergent Ideals of Einstein and Bohr." *British Journal for the Philosophy of Science* 42:491–511.

————. 1994. "Bohr and the Crisis of Empirical Intelligibility: An Essay on the Depth of Bohr's Thought and Our Philosophical Ignorance." In Faye and Folse 1994, 155–74.

Horgan, J. 1991. "Profile: Reluctant Revolutionary." *Scientific American* 15:14–15.

Horwich, P. 1993. *World Changes: Thomas Kuhn and the Nature of Science.* Cambridge, Mass.: MIT Press.

Howard, D. 1990. "Einstein and Duhem." In *Pierre Duhem: Historian and Philosopher of Science,* edited by R. Ariev and P. Barker. *Synthese* 83: 363–84.

———. 1993. "Was Einstein Really a Realist?" *Perspectives on Science* 1:204–51.

———. 1994. "What Makes a Classical Concept Classical? Toward a Reconstruction of Niels Bohr's Philosophy of Physics." In Faye and Folse 1994, 201–30.

Howard, D., and J. Stachel, eds. Forthcoming. *The Young Einstein: Einstein Studies.*

Hughes, R. I. Y. 1989. "Bell's Theorem, Ideology and Structural Explanation." In Cushing and McMullin 1989, 195–208.

Ising, G. 1926. "A Natural Limit for the Sensibility of Galvanometers." *Philosophical Magazine* 1:827–34.

Jagger, A. M., and S. R. Bordo, eds. 1989. *Gender/Body/Knowledge: The Effect of Outcome of Knowledge on Judgement under Uncertainty.* New Brunswick, N.J.: Rutgers University Press.

James, W. 1907. *Pragmatism.* New York: Longmans, Green. [Reissued, Indianapolis: Hackett, 1981.]

Jammer, M. 1966. *The Conceptual Development of Quantum Mechanics.* New York: McGraw-Hill.

———. 1974. *The Philosophy of Quantum Mechanics.* New York: Wiley.

Jordan, P. 1927a. "Die Entwicklung der neuen Quantenmechanik." *Naturwissenschaften* 15:614–23, 636–49.

———. 1927b. "Philosophical Foundations of Quantum Theory." *Nature* 119:566–69.

———. 1927c. "Reply to Campbell." *Nature* 119:779.

———. 1927d. "Schrödinger, 'Abhandlungen zur Wellenmechanik.'" *Die Naturwissenschaften* 6.

———. 1927e. "Über quantenmechanische Darstellung von Quantensprungen." *Zeitschrift für Physik* 40:661–66.

———. 1927f. "Über eine neue Begründung der Quantenmechanik." *Zeitschrift für Physik* 40:809–38.

———. 1944. *Physics in the 20th Century.* New York.

———. 1972. *Der Naturwissenschaftler vor der religiösen Frage.* Hamburg: Stalling.

Jordan, P., and E. P. Wigner. 1928. "Über das Paulische Äquivalentverbot." *Zeitschrift für Physik* 47:631–51.

Kaiser, D. 1994. "Bringing the Human Actors Back on Stage: The Personal Context of the Einstein-Bohr Debate." *British Journal for the History of Science* 27:129–52.

Kalckar, J. 1967. "Niels Bohr and His Youngest Disciples." In Rozental 1967, 227–40.

Klein, M. 1970. "The First Phase of the Bohr-Einstein Dialogue." *Historical Studies in the Physical Sciences* 2:1–39.

———. 1977. "The Beginnings of Quantum Theory." In Weiner 1977, 1–39.

Klibansky, R., ed. 1958. *Philosophy in Mid-Century: A Survey.* Florence: La Nuova Italia Editrice.

Knorr-Cetina, K. 1981. *The Manufacture of Knowledge: An Essay on the Constructivist and Contextual Nature of Science.* Oxford: Pergamon.

Kochen, S. 1985. "A New Interpretation of Quantum Mechanics." In Lahti and Mittelstaedt 1985, 151–169.

Koestler, A. 1960. *The Watershed.* New York: Anchor.

Kojevnikov, A. 1987. "Electrodynamics in Matrix Mechanics: Discord in Interpretation of the Theory" (in Russian). Preprint no. 1. Moscow: Institute for History of Science and Technology.

———. 1997. "Niels Bohr and the Copenhagen Network in Physics, Part I." Manuscript.

Kojevnikov, A., and O. Novik. 1989. "Analysis of Information Ties Dynamics in Early Quantum Mechanics, 1925–1927. *Acta Historiae Rerum Naturalium necnon Technicarum* (Prague) 20:115–59.

Koyré, A. 1943. "Galileo and Plato." *Journal for the History of Ideas* 4:400–47. [Reprinted in Koyré 1968.]

———. 1957. *From the Closed World to the Infinite Universe.* Baltimore: Johns Hopkins University Press.

———. 1964. *Melanges Alexandre Koyré.* Vol. 1. Paris: Hermann.

———. 1968. *Metaphysics and Measurement.* Cambridge, Mass.: Harvard University Press.

Kragh, H., 1982. "Erwin Schrödinger and the Wave Equation: The Crucial Phase." *Centaurus* 26:154–97.

———. 1990. *Dirac: A Scientific Biography.* Cambridge: Cambridge University Press.

Kramers, H. A. 1924. "The Quantum Theory of Dispersion." *Nature* 114:310–11.

Kramers, H. A., and W. Heisenberg. 1925. "Über die Streuung von Strahlung durch Atome." *Zeitschrift für Physik* 31:681–707.

Krips, H. 1987. *The Metaphysics of Quantum Theory.* Oxford: Clarendon.

———. 1996. "Quantum Mechanics and the Post-Modern in One Country." *Cultural Studies* 10 (1): 78–114.

Krüger, L., G. Gingerenzer, and M. S. Morgan. 1987. *The Probabilistic Revolution.* Vols. 1 and 2. Cambridge, Mass.: MIT Press.

Kuhn, T. S. 1957. *The Copernican Revolution: Planetary Astronomy in the Development of Western Thought.* Cambridge, Mass.: Harvard University Press.

———. 1959. "The Essential Tension: Tradition and Innovation in Scientific Research." In Taylor 1959. [Reprinted in Kuhn 1977.]

———. 1963. "The Function of Dogma in Scientific Research." In Crombie 1963, 347–369, 386–395.

———. 1970. *The Structure of Scientific Revolutions,* 2d ed. Chicago: University of Chicago Press. First edition published in 1962.

———. 1971. "The Relations between History and the History of Science." *Daedalus* 100:271–304. Reprinted in Kuhn 1977, 127–61. Page references are to Kuhn 1977.

———. 1977. *The Essential Tension: Selected Studies in Scientific Tradition and Change.* Chicago: University of Chicago Press.

———. 1978. *Black-Body Theory and the Quantum Discontinuity, 1894–1912.* New York: Oxford University Press.

———. 1987. "What Are Scientific Revolutions?" In Krüger, Gingerenzer, and Morgan 1987, 7–22.

———. 1991. "The Road since *Structure.*" In Fine, Forbes, and Wessels 1991, 3–13.

———. 1993. "Afterwords." In Horwich 1993, 311–41.

Kuhn, T. S., J. L. Heilbron, P. Forman, and L. Allen, eds. 1967. *Sources for History of Quantum Physics: An Inventory and Report.* Philadelphia: American Philosophical Society.

Lahti, P., and P. Mittelstaedt, eds. 1985. *Symposium on the Foundations of Modern Physics.* Singapore: World Scientific.

———, eds. 1987. *Symposium on the Foundations of Modern Physics.* Singapore: World Scientific.

Lakatos, I. 1970. "Falsification and the Methodology of Scientific Research Programmes." In Lakatos and Musgrave 1970, 91–196.

Lakatos, I., and A. Musgrave, eds. 1970. *Criticism and the Growth of Knowledge.* Cambridge: Cambridge University Press.

Lakoff, G. 1993. "The Contemporary Theory of Metaphor." In Ortony 1993, 202–52.

Lanczos, C. 1926. "Über eine feldmässige Darstellung der neuen Quantenmechanik." *Zeitschrift für Physik* 35:812–30.

Landé, A. 1955. *Foundations of Quantum Theory.* New Haven, Conn.: Yale University Press.

———. 1965. "Quantum Fact and Fiction." *American Journal of Physics* 33:123–27.

———. 1967. *New Foundations of Quantum Mechanics.* Cambridge: Cambridge University Press.

———. 1968. "Replies." *Physics Today* 21:56.

Latour, B. 1987. *Science in Action: How to Follow Scientists and Engineers through Society.* Cambridge, Mass.: Harvard University Press.

Laurikainen, K. V. 1988. *Beyond the Atom.* Berlin: Springer.

Lazarus, R. S., A. D. Kanner, and S. Folkman. 1980. "Emotions: A Cognitive-Phenomenological Analysis." In Plutchik and Kellerman 1980, 189–243.

Leary, L. G., ed. 1955. *The Unity of Knowledge.* New York: Doubleday.

London, F. 1926. "Über die Jacobischen Transformationen der Quantenmechanik." *Zeitschrift für Physik* 37:915–25.

Ludwig, G. 1968. *Wave Mechanics.* Oxford: Pergamon.

MacKinnon, E. M. 1977. "Heisenberg, Models and the Rise of the Matrix Mechanics." *Historical Studies in the Physical Sciences* 8:137–88.

———. 1982. *Scientific Explanation and Atomic Physics.* Chicago: University of Chicago Press.

———. 1994. "Bohr and the Realism Debates." In Faye and Folse 1994, 279–302.

Maddox, J. 1988. "License to Slang Copenhagen." *Nature* 332:581.

Masterman, M. 1970. "The Nature of a Paradigm." In Lakatos and Musgrave 1970, 59–90.

Maudlin, T. 1997. "Kuhn Defanged: Incommensurability and Theory Choice." *Revue Philosophique de Louvain* 94:428–46.

Mauskopf, H., ed. 1979. *The Reception of Unconventional Science.* Boulder, Colo.: Westview.

Mauskopf, S. In press. "The Chemical Revolution." In *Reader's Guide to the History of Science,* edited by Arne Hessenbruch. London: Fitzroy Dearborn. Derived from a paper presented at the History of Chemistry Workshop, Edelstein Center, Hebrew University of Jerusalem, fall 1995.

McMullin, E. 1991. "Rhetoric and Theory of Choice in Science." In Pera and Shea 1991, 55–76.

———, ed. 1992. *The Social Dimension of Science.* Notre Dame, Ind.: University of Notre Dame Press.

Mehra, J., ed. 1973. *The Physicists' Conception of Nature.* Chicago: University of Chicago Press.

———. 1976. "The Birth of Quantum Mechanics." Talk given at the Cern Colloqium, Geneva.

Mehra, J., and H. Rechenberg. 1982. *The Historical Development of Quantum Theory.* Vols. 1–5. New York: Springer.

Merleau-Ponty, M. 1973. *Adventures of the Dialectic.* Evanston, Ill.: Northwestern University Press.

Merton, R. K. 1957. "Priorities in Scientific Discovery." *American Sociological Review* 22:635–59.

Merzbacher, E. 1983. "Energetic Ion-Atom Collisions: Early Beginnings and Recent Advances." Invited paper at the thirteenth international conference on the Physics of Electronic and Atomic Collisions, Berlin.

Miller, A. I. 1978. "Visualization Lost and Regained: The Genesis of the Quantum Theory in the Period 1913–1927." In Wechsler.

———. 1984. *Imagery in Scientific Thought.* Basel: Birkhäuser.

Moll, W. G. M., and H. C. Burger. 1925. "'The Thermo-Relay' and 'The Sensitivity of a Galvanometer and Its Amplification.'" *Philosophical Magazine* 6:624–31.

Moore, W. 1989. *Schrödinger, Life and Thought.* Cambridge: Cambridge University Press.

Muller, F. M. 1997. "The Equivalence Myth of Quantum Mechanics, Part II." *Studies in the History and Philosophy of Modern Physics* 28 (2): 219–47.

Murdoch, D. 1987. *Niels Bohr's Philosophy of Physics.* Cambridge: Cambridge University Press.

Nersessian, N. T. 1995. "Opening the Black Box: Cognitive Science and History of Science." *Osiris* 10:194–215.

Newton-Smith, W. H. 1981. *The Rationality of Science.* Boston: Routledge.

Nickles, T. 1995. "Philosophy of Science and History of Science." *Osiris* 10:139–64.

Nobel Lectures in Physics, 1922–1941. 1965. Vol. 2. Amsterdam: Elsevier.

Ortony, A., ed. 1993. *Metaphor and Thought,* 2d ed. Cambridge: Cambridge University Press.

Pais, A. 1967. "Reminiscences from the Post-War Years." In Rozental 1967, 215–27.

———. 1982. "Max Born's Statistical Interpretation of Quantum Mechanics." *Science* 218:1193–98.

———. 1986. *Inward Bound: Of Matter and Forces in the Physical World.* Oxford: Oxford University Press.

———. 1991. *Niels Bohr's Times, in Physics, Philosophy and Polity.* Oxford: Clarendon.

Pauli, W. 1926a. "Quantentheorie." In *Handbuch der Physik,* 23:1–278. Berlin: Springer. [Reprinted in Pauli 1964, 1:269–549.]

———. 1926b. "Über das Wasserstoffspectrum vom Standpunkt der neuen Quantenmechanik." *Zeitschrift für Physik* 36:336–63. [Translated as "On the Hydrogen Spectrum from the Stand Point of the New Quantum Mechanics," in van der Waerden 1967, 387–415.]

———. 1927. "Über Gasentartung und Paramagnetismus." *Zeitschrift für Physik* 41:81–102.

———. 1932. "Review of Born's and Jordan's *Elementare Quantenmechanik.*" *Naturwissenschaften* 18:602.

———. 1933. "Die allgemeinen Principien der Wellenmechanik." In *Handbuch der Physik,* 24 (1): 83–272. Berlin: Springer.

———. 1945. "Niels Bohr on His 60th Birthday." *Reviews of Modern Physics* 17:97–101.

———. 1950. "Die Philosophische Bedeutung der Idee der Komplementarität." In Pauli 1964, vol. 2.

———. 1954. "Wahrscheinlichkeit und Physik." *Dialectica* 8:112–24.

———. 1955a. "Exclusion Principle, Lorentz Group and Reflection of Space Time and Charge." In Pauli 1955b, 30–51.

———, ed. 1955b. *Niels Bohr and the Development of Physics.* New York: McGraw-Hill.

———. 1964. *Wolfgang Pauli: Collected Scientific Papers.* Vols. 1 and 2. Edited by R. Kronig and V. F. Weisskopf. New York: Wiley.

———. 1973. *Lectures on Wave Mechanics.* Cambridge, Mass.: MIT Press.

Peirce, C. S. 1931–35. *Collected Papers.* Edited by C. Hartshorne and P. Weiss. Cambridge, Mass.: Harvard University Press.

Pera, M., and W. Shea. 1991. *Persuading Science: The Art of Scientific Rhetoric.* Canton, Mass.: Science History.

Pickering, A. 1982. "Interests and Analogies." In Barnes and Edge 1982.

———, ed. 1992. *Science as Practice and Culture.* Chicago: University of Chicago Press.

———. 1995. *The Mangle of Practice: Time, Agency and Science.* Chicago: University of Chicago Press.

Pitowsky, I. 1994. "George Boole's Conditions of Possible Experience and the Quantum Puzzle." *British Journal for the Philosophy of Science* 45:95–125.

Plutchik, R., and H. Kellerman, eds. 1980. *Emotion Theory, Research and Experience.* New York: Academic.

Popper, K. 1959. *The Logic of Scientific Discovery.* New York: Basic. [First published in 1935.]

———. 1963a. *Conjectures and Refutations.* London: Routledge.

———. 1963b. "Three Views Concerning Human Knowledge." In Popper 1963a, 97–119.

Price, W. C., and S. Chissick, eds. 1977. *The Uncertainty Principle and Foundation of Quantum Mechanics.* New York: Wiley.

Price, W. C., S. Chissick, and T. Ravensdale, eds. 1973. *Wave Mechanics: The First Fifty Years.* New York: Wiley.

Przibram, K. 1967. *Letters on Wave Mechanics.* New York: Philosophical Library.

Putnam, H. 1968. "Is Logic Empirical?" In Cohen and Wartofsky 1968, 181–241.

Rashevsky, N. V. 1926. "Einige Bemerkungen zu Heisenbergschen Quantenmechanik." *Zeitschrift für Physik* 36:153–58.

Redhead, M. 1987. *Incompleteness, Non-locality and Realism: A Prolegomenon to the Philosophy of Quantum Mechanics.* Oxford: Clarendon.

Reichenbach, H. 1959. *Modern Philosophy of Science.* London: Routledge and Kegan Paul.

Renneberg, M., and M. Walker, eds. 1993. *Science, Technology, and National Socialism.* Cambridge: Cambridge University Press.

Rorty, R. 1979. *Philosophy and the Mirror of Nature.* Princeton, N.J.: Princeton University Press.

—. 1991. *Objectivity, Relativism and Truth. Philosophical Papers.* Vol. 1. Cambridge: Cambridge University Press.

Rosenfeld, L. 1956. "Review of Landé's *Foundations of Quantum Theory.*" *Nuclear Physics* 1:133–34.

—. 1961. "Foundations of Quantum Theory and Complementarity." *Nature* 190:384–88.

—. 1967. "Niels Bohr in the Thirties: Consolidation and Extension of the Conception of Complementarity." In Rozental 1967, 114–36.

—. 1971. "Men and Ideas in the History of Atomic Theory." *Archive for the History of Exact Sciences* 7:69–90.

Rozental, S., ed. 1967. *Niels Bohr: His Life and Work as Seen by His Friends and Colleagues.* New York: Wiley.

Ruark, A. E. 1935. "Is the Quantum-Mechanical Description of Physical Reality Complete?" *Physical Review* 48:466–67.

Rüger, A. 1987. "Atomism from Cosmology: Erwin Schrödinger's Work on Wave Mechanics and Space-Time Structure." *Historical Studies in the Physical Sciences* 18:377–401.

Sanders, J. T. 1987. *Niels Bohr: Essays and Papers.* Two volumes of Bohr's writing on epistemological matters, gathered from published papers and the unpublished manuscripts in the Bohr Manuscript Collection, AHQP. Translations from Danish by Else Mogensen. Deposited at the Niels Bohr Archive in Copenhagen, the Niels Bohr Library at the American Institute of Physics, the Bancroft Library at the University of California at Berkeley, the Baillien Library at the University of Melbourne, the Edelstein Center at the Hebrew University of Jerusalem, the Science Museum Library in London, the Institute of Technology Library at the University of Minnesota, the Accademia Nazionale della Scienze in Rome, the Widener Library at Harvard University, the Centre de documentation in Paris, the Hillman Library at the University of Pittsburgh, the Deutsches Museum Library in Munich, the American Philosophical Society Library in Philadelphia, the Biblioteca Universitaria at Madrid, and the Mugar Memorial Library at Boston University.

Schilpp, P. A., ed. 1970. *Albert Einstein: Philosopher-Scientist.* Library of Living Philosophers, vol. 7. Cambridge: Cambridge University Press. [First published, La Salle, Ill.: Open Court, 1949.]

Schmitt, C. 1983. *Aristotle and the Renaissance.* Cambridge, Mass.: Harvard University Press.

Schrödinger, E. 1924. "Bohr's neue Strahlungshypothese und der Energiesatz." *Naturwissenschaften* 12:720–24.

———. 1926a. "Quantisierung als Eigenwertproblem I." *Annalen der Physik* 79:361–76. Translated in Schrödinger 1927a. Page references to Schrödinger 1927a.

———. 1926b. "Quantisierung als Eigenwertproblem II." *Annalen der Physik* 79:489–527. Translated in Schrödinger 1927a. Page references to Schrödinger 1927a.

———. 1926c. "Quantisierung als Eigenwertproblem III." *Annalen der Physik* 80:437–90. Translated in Schrödinger 1927a. Page references to Schrödinger 1927a.

———. 1926d. "Quantisierung als Eigenwertproblem IV." *Annalen der Physik* 81:109–39. Translated in Schrödinger 1927a. Page references to Schrödinger 1927a.

———. 1926e. "Der stetige Übergang von der Micro- zur Macromechanik." *Naturwissenschaften* 14:664–66. Translated in Schrödinger 1927a. Page references to Schrödinger 1927a.

———. 1926f. "Über das Verhältnis der Heisenberg-Born-Jordanschen Quantenmechanik zu der meinen." *Annalen der Physik* 79:734–59. Translated in Schrödinger 1927a. Page references to Schrödinger 1927a.

———. 1926g. "Zur Einsteinschen Gastheorie." *Physikalische Zeitschrift* 27:95–101.

———. 1927a. *Collected Papers*. London: Blackie.

———. 1927b. "Energieaustausch nach der Wellenmechanik." *Annalen der Physik* 83:956–68. Translated in Schrödinger 1927a. Page references to Schrödinger 1927a.

———. 1935. "Die gegenwärtinge Situation in der Quantenmechanik." *Naturwissenschaften* 23:807–12, 812–28, 844–49. Translated in *Proceedings of the American Philosophical Society* 124:323–38; reprinted in Wheeler and Zurek: 152–67. Page references are to Wheeler and Zurek.

———. 1950. "What Is an Elementary Particle?" *Endeavour* 9 (35): 109–16.

———. 1952a. "Are There Quantum Jumps?" *British Journal for the Philosophy of Science* 3:109–23, 233–42.

———. 1952b. "July Colloquium 1952, AHQP." In Bitbol 1995, 19–38.

———. 1952c. *Science and Humanism*. Cambridge: Cambridge University Press.

———. 1952d. "Transformation and Interpretation in Quantum Mechanics." In Bitbol 1995, 39–95.

———. 1954. "Science, Philosophy and the Sensates." Manuscript draft for three William James lectures, AHQP. In Bitbol 1995, 123–49.

———. 1955. "The Philosophy of Experiment." *Nuovo Cimento* 1:5–15.

———. 1958. "Might Perhaps Energy Be a Merely Statistical Concept?" *Nuovo Cimento* 9:162–70.

Schweber, S. S. 1994. *QED and the Men Who Made It: Dyson, Feynman, Schwinger and Tomonaga*. Princeton, N.J.: Princeton University Press.

Scully, M. O., B. G. Englert, and H. Walther. 1991. "Quantum Optical Tests of Complementarity." *Nature* 351:111–16.

Seelig, C. 1954. *Albert Einstein: Eine dokumentarische Biographie*. Zurich: Europa.

Sentfleben, H. A. 1923. "Zur Grundlagen Quantentheorie." *Zeitschrift für Physik* 22:127–56.

Serwer, D. 1977. "Unmechanischer Zwang: Pauli, Heisenberg, and the Rejection of the Mechanical Atom, 1923–1925." *Historical Studies in the Physical Sciences* 8:189–256.

Shapin, S. 1994. *A Social History of Truth*. Chicago: University of Chicago Press.

Shapin, S., and S. Schaffer. 1985. *Leviathan and the Air-Pump*. Princeton, N.J.: Princeton University Press.

Shapiro, A. 1989. "The Evolving Structure of Newton's Theory of Color." *Isis* 71:197–210.

Shimony, A. 1985. "Review of Folse's *The Philosophy of Niels Bohr*." *Physics Today* 38:109.

———. 1989. "Search for a World View Which Can Accommodate Our Knowledge of Microphysics." In Cushing and McMullin 1989, 25–38.

Skinner, Q. 1969. "Meaning and Understanding in the History of Ideas." *History and Theory* 8:3–53.

Sklar, L. 1992. *Philosophy of Physics*. Oxford: Oxford University Press.

Slater, J. 1975a. *Solid State and Molecular Theory: A Scientific Biography*. New York: Wiley.

———. 1975b. "Wave Mechanics in the Classical Decade, 1923–1932." In Slater 1975a.

Sommerfeld, A. 1927. "Zum gegenwärtigen Stande der Atomphysik." *Physikalische Zeitschrift* 28:231–39.

Stachel, J. 1986. "Einstein and the Quantum: Fifty Years of Struggle." In Colodny 1986, 349–87.

———. 1993. "The Other Einstein." In Beller, Cohen, and Renn 1993, 275–91.

Storr, A. 1988. *Solitude: A Return to the Self*. New York: Free Press.

Stuewer, R. 1970. "Non-Einsteinian Interpretations of the Photoelectric Effect." In *Historical and Philosophical Perspectives of Science*. Minnesota Studies in Philosophy of Science, vol. 5. Minneapolis: University of Minnesota Press.

———. 1974. *The Compton Effect: Turning Point in Physics.* New York: Science History.

Stump, D. J. 1996. "New Directions in the Philosophy of Science Studies." In Galison and Stump 1996, 443–53.

Suppes, P., ed. 1980. *Studies in the Foundations of Quantum Mechanics.* East Lansing, Mich.: Philosophy of Science Association.

Taylor, C. W., ed. 1959. *The Third University of Utah Research Conference on the Identification of Scientific Talent.* Salt Lake City: University of Utah Press.

Teller, P. 1989. "Relativity, Relational Holism, and the Bell Inequalities." In Cushing and McMullin 1989, 208–24.

Tolansky, S. 1968. *Revolution in Optics.* Harmondsworth: Penguin.

Tomonaga, S. 1962. *Quantum Mechanics.* Translated by Koshiba. Amsterdam: North Holland.

Turner, M. 1987. *Death Is the Mother of Beauty: Mind, Metaphor, Criticism.* Chicago: University of Chicago Press.

Ullman-Margalit, E., ed. 1987. *Science in Reflection.* Boston Studies in the Philosophy of Science, vol. 110. The Israel Colloquium: Studies in History, Philosophy and Sociology of Science, vol. 3. Dordrecht: Kluwer.

Valentini, A. 1996. "Pilot-Wave Theory of Fields, Gravitation and Cosmology." In Cushing, Fine, and Goldstein 1996, 45–66.

———. 1998. *On the Pilot-Wave Theory of Classical, Quantum and Subquantum Physics.* Berlin: Springer.

van der Waerden, B. L., ed. 1967. *Sources of Quantum Mechanics.* New York: Dover.

van Fraassen, B. C. 1980. *The Scientific Image.* Oxford: Clarendon.

———. 1981. "A Modal Interpretation of Quantum Mechanics." In Beltrametti and van Fraassen 1981, 229–58.

———. 1989. "The Charybdis of Realism: Epistemological Implications of Bell's Inequality." In Cushing and McMullin 1989, 97–114.

———. 1991. *Quantum Mechanics: An Empiricist View.* Oxford: Clarendon.

van Vleck, J. 1973. "Central Fields in Two vis-à-vis Three Dimensions: An Historical Divertissement." In Price, Chissick, and Ravensdale 1973, 26–37.

Vernon, P. 1970. *Creativity.* Harmondsworth: Penguin.

Vickers, B., ed. 1984. *Occult and Scientific Mentalities in the Renaissance.* Cambridge: Cambridge University Press.

Vigier, J. P. 1982. "Non-Locality, Causality, and Aether in Quantum Mechanics." *Astronomische Nachrichten* 303:55–80.

Vygotsky, L. 1962. *Thought and Language.* Cambridge, Mass.: MIT Press.

Wechsler, J., ed. 1978. *On Aesthetics in Science.* Boston: Birkhäuser.

Weinberg, S. 1992. *Dreams of a Final Theory.* New York: Pantheon.

Weiner, C. 1977. *History of Twentieth Century Physics.* New York: Academic.

Weizsäcker, C. F. von. 1985. "Reminiscences from 1932." In French and Kennedy 1985, 183–90.

———. 1987. "Heisenberg's Philosophy." In Lahti and Mittelstaedt 1987, 277–93.

Wentzel, G. 1926a. "Die mehrfach periodischen Systeme in der Quantenmechanik." *Zeitschrift für Physik* 37:80–91.

———. 1926b. "Eine Verallgemeinerung der QuantenBedingungen für die Zwecke der Wellenmechanik." *Zeitschrift für Physik* 38:518–29.

Wertheimer, M., ed. 1959. *Productive Thinking,* New York: Harper.

Wessels, L. 1979. "Schrödinger's Route to Wave Mechanics." *Studies in History and Philosophy of Science* 10:311–40.

———. 1980. "The Intellectual Sources of Schrödinger's Interpretations." In Suppes 1980, 59–76.

———. 1981. "What Was Born's Statistical Interpretation?" In Asquith and Giere 1981, 187–200.

———. 1983. "Erwin Schrödinger and the Descriptive Tradition." In Aris, Davis, and Stuewer 1983, 254–78.

———. 1989. "The Way the World Isn't: What the Bell Theorems Force Us to Give Up." In Cushing and McMullin 1989, 80–96.

Westfall, R. S. 1985. "Science and Patronage: Galileo and the Telescope." *Isis* 76:11–30.

Westman, R. S. 1994. "Two Cultures or One? A Second Look at Kuhn's *The Copernican Revolution.*" *Isis* 85:79–115.

Wheaton, B. 1983. *The Tiger and the Shark: Empirical Roots of Wave-Particle Duality.* Cambridge: Cambridge University Press.

Wheeler, J. A., and W. H. Zurek, eds. 1983. *Quantum Theory and Measurement.* Princeton, N.J.: Princeton University Press.

Williams, W. C. 1936. *Paterson.* New York: New Directions.

Wilshire, D. 1989. "The Uses of Myth, Image, and the Female Body in Re-Visioning Knowledge." In Jagger and Bordo 1989.

Wise, N. J. 1987. "How Do Sums Count? On the Cultural Origins of Statistical Causality." In Krüger, Gingerenzer, and Morgan 1987.

———. 1993. "Pascual Jordan: Quantum Mechanics, Psychology, National Socialism." In Renneberg and Walker 1993, 224–54.

Wittgenstein, L. 1953. *Philosophical Investigations.* New York: Macmillan.

Zernike, F. 1926. "Die natürliche Beobachtungsgrenze der Stromstärke." *Zeitschrift für Physik* 40:628–38.

INDEX

〜